VOLUME FIFTY SEVEN

Advances in
CATALYSIS

ADVISORY BOARD

VOLUME FIFTY SEVEN

ADVANCES IN
CATALYSIS

Edited by

FRIEDERIKE C. JENTOFT
University of Oklahoma
Norman, Oklahoma, USA

AMSTERDAM • BOSTON • HEIDELBERG • LONDON
NEW YORK • OXFORD • PARIS • SAN DIEGO
SAN FRANCISCO • SINGAPORE • SYDNEY • TOKYO
Academic Press is an imprint of Elsevier

Academic Press is an imprint of Elsevier
32 Jamestown Road, London NW1 7BY, UK
525 B Street, Suite 1800, San Diego, CA 92101-4495, USA
225 Wyman Street, Waltham, MA 02451, USA
The Boulevard, Langford Lane, Kidlington, Oxford OX5 1GB, UK

First edition 2014

Notices
Knowledge and best practice in this field are constantly changing. As new research and
experience broaden our understanding, changes in research methods, professional practices,
or medical treatment may become necessary.

Practitioners and researchers must always rely on their own experience and knowledge in
evaluating and using any information, methods, compounds, or experiments described
herein. In using such information or methods they should be mindful of their own safety and
the safety of others, including parties for whom they have a professional responsibility.

To the fullest extent of the law, neither the Publisher nor the authors, contributors, or editors,
assume any liability for any injury and/or damage to persons or property as a matter of
products liability, negligence or otherwise, or from any use or operation of any methods,
products, instructions, or ideas contained in the material herein.

ISBN: 978-0-12-800127-1
ISSN: 0360-0564

British Library Cataloguing-in-Publication Data
A catalogue record for this book is available from the British Library

Library of Congress Cataloging-in-Publication Data
A catalog record for this book is available from the Library of Congress

For information on all Academic Press publications
visit our website at store.elsevier.com

Working together
to grow libraries in
developing countries

www.elsevier.com • www.bookaid.org

CONTENTS

Color Plate Section at the end of the Book

CONTRIBUTORS

Simon R. Bare
UOP LLC, a Honeywell Company, Des Plaines, Illinois, USA

Guido Busca
Dipartimento di Ingegneria Civile, Chimica e Industriale, Università di Genova, Genova, Italia

Konstantin Hadjiivanov
Institute of General and Inorganic Chemistry, Bulgarian Academy of Sciences, Sofia 1113, Bulgaria

Laszlo Nemeth
University of Nevada, Las Vegas, Nevada, USA

PREFACE

The first pages of this volume are dedicated to the obituary of Helmut Knözinger, an eminent surface spectroscopist who made significant contributions to the analysis of oxide surfaces and supported species in catalysts. He served as Editor of *Advances in Catalysis* from 1998 to 2011 and thereafter as an Advisory Board Member until his passing in early 2014.

In Chapter 1, Nemeth and Bare review the science and technology of framework metal-containing microporous and mesoporous materials. Many of these materials are characterized by the same framework types as zeolites, but the substitution of framework silicon by gallium, tin, or titanium leads to catalytic properties much different from those of the aluminum-substituted frameworks. The authors provide a comprehensive account, including extensive information from the patent literature, of methods of preparation of such materials; their structures; and their catalytic properties, with an emphasis on applications. Framework metal-containing materials are among the few developed in the preceding decades that have led to new catalytic processing technologies. The authors describe the key elements of these technologies and explain how the novel processes can be integrated into existing process schemes. The chapter is concluded with recommendations for future research on these materials.

Chapter 2, by Hadjiivanov, is focused on the characterization of surface OH groups by infrared spectroscopy. Under ambient conditions, OH groups are present on many surfaces; they serve as catalytically active sites in various processes and also as anchoring sites for supported species in many catalyst preparations. Because the stretching frequencies of OH groups are determined by two, often opposing, influences—the strength of the surrounding electrostatic field and the covalency of the bond to the surface site—these frequencies are of limited use for assessing the OH group reactivities. The author presents methods for capturing acid–base and other properties of OH groups; these include the "hydrogen-bond method," the "base protonation method" and its counterpart for basic sites, and transformations of OH groups in reactions with alcohols and other test reactants. The author also discusses OH group accessibility, which can be assessed with probe molecules of varying bulkiness, and H–D and $^{16}O-^{18}O$ exchange reactions, which allow unambiguous identification of OH groups and provide valuable insights into their participation in catalytic

reactions. Numerous tables with reported band positions and shifts characterizing the OH groups on oxides relevant to catalysis will help make this
chapter a valuable resource.

In Chapter 3, Busca summarizes the current state of knowledge of
aluminas, the various polymorphs of which constitute some of the most
commonly used catalyst components. The author starts with a discussion
of the bulk structures of transition aluminas, which are the intermediate
phases formed in the thermal transformation of aluminum oxyhydroxides
into the thermodynamically most stable modification, α-alumina. Crucial
are the definitions of the various phases, which are based on the methods
of preparation rather than on the structural properties. The understanding
of many alumina structures is incomplete, and progress, even with modern
analytical methods and theory, is hampered by the defective and disordered
nature of these materials. The stabilities of the various phases are governed
by both thermodynamics and kinetics, either of which can be affected by
impurities. The uncertainties in the surface structures are even greater than
those of the bulk structures. Numerous models of alumina surface structures
have been formulated over decades, but the true structures seem to become
even more elusive. Busca concludes his chapter with a list of research needs.

We report a change in the editorial team. After 18 years as Editor, Bruce
Gates is stepping aside and joining the Advisory Board. His vision of the
serial and discipline as an editor have shaped *Advances in Catalysis*; he constantly strived for critical, in-depth reviews and for clarity in language. His
insightfulness, enthusiasm, and team spirit made him an excellent
fellow editor.

FRIEDERIKE C. JENTOFT

HELMUT KNÖZINGER
1935–2014

Photograph by Thomas Knözinger.

Helmut Knözinger, a leader in surface spectroscopy and catalysis, passed away in January, 2014, in Munich.

Born on July 10, 1935, in Weilheim, Knözinger was at home for his whole life in his native Bavaria and its capital Munich. He studied physics at the Ludwig-Maximilians-University of Munich and began his graduate studies at the Institute of Physical Chemistry under the mentorship of Georg-Maria Schwab, receiving his doctoral degree in 1961. During this period, the Munich Institute was a Mecca for heterogeneous catalysis, and it maintained that status for decades, with Helmut Knözinger becoming a professor and playing a principal role as an influential researcher and leader. His vigorous activity continued for years after his formal retirement in 2000.

Before setting out on an academic career in science, Helmut Knözinger was an accomplished photographer, an enthusiastic outdoorsman, and an outstanding athlete, having placed first in a German national triple jump competition. As his research career developed, he focused on the characterization of catalyst surfaces, pioneering the application of elegantly complementary spectroscopies, highlighting infrared and Raman. He was drawn to metal oxides, as well as other classes of catalytic materials including zeolites, metal sulfides, and supported metal and metal–oxide clusters. He did path-breaking research in the use of probe molecules for spectroscopic investigations of catalyst surfaces, demonstrating the value of probes that (1) are small, to limit the steric hindrance of probe–surface interactions, and (2) interact as weakly as possible with the surface in order not to modify its properties. The wide applications of CO and N_2 for assessing surface Lewis and Brønsted acidity are largely based on Knözinger's work. He also developed other molecular probes, including C–H acids for the difficult challenge of understanding surface basicity, and he led the way in the use of small, weakly adsorbed probes for determination of the coordination states of cations through the formation of geminal metal complexes. His early research with Schwab (whose photographic portrait always hung on his office wall) involved investigations of catalytic reaction kinetics, and his work increasingly linked catalyst spectra with performance as a basis for elucidating catalytically active species. Examples of the catalysts Knözinger investigated include tungstated zirconia for low-temperature alkane isomerization, supported molybdena for hydroprocessing and related reactions, and supported vanadia and related catalysts for $DeNO_x$ chemistry and selective oxidation.

Among his hundreds of papers, an article on transition aluminas including model surface structures, published in 1978 with P. Ratnasamy (*Catal. Rev.—Sci. Eng.*, *17*, 31, 1978), has been cited more than 1500 times and remains a widely appreciated classic, not only a basis for investigating the properties of alumina-containing catalysts but also a guide for elucidating models of other oxide-containing catalysts. Numerous other Knözinger publications are also highly influential, perhaps none more than the multivolume *Handbook of Heterogeneous Catalysis*, a Wiley publication that has recently appeared in a second edition, which he coedited with G. Ertl, F. Schüth, and J. Weitkamp.

Helmut Knözinger energetically served the catalysis community, being a member, Vice Chair, and Chair of EUROCAT (1979–1991); working on the organization of the International Congress on Catalysis in Berlin in 1984; serving in the catalysis section of Dechema (Chair from 1996–1999); being the German representative to the Council of the European Federation of Catalysis Societies (1993–1999); serving as a member (1984–1996) and then Vice President (1995–1996) of the Council of the International Congress on Catalysis; and being President of the International Association of Catalysis Societies (1996–2000). He was an editor of *Advances in Catalysis* (1998–2011) and served on numerous editorial boards, including those of *Applied Catalysis B*, *Catalysis Letters*, *Catalysis Reviews*, *Catalysis Today*, *Journal of Catalysis*, and *Topics in Catalysis*.

Knözinger's work brought him many honors, including the Ciapetta Lectureship of the North American Catalysis Society, the Dechema Alwin Mittasch Prize, the Gay-Lussac Prize of the Alexander von Humboldt Foundation, and the Max Planck Research Award. He was elected a member of Academia Europaea in 2000.

Typical of Helmut Knözinger's science was his engagement with an extensive, worldwide group of coworkers—some of his collaborations carried on for decades. He welcomed many members of the catalysis community to Munich, which for years was a favorite sabbatical-leave destination. He also welcomed many students and postdoctoral researchers to complement his core group of advanced-degree students, many of whom now hold prominent positions. He mentored his team with gentle encouragement and support and unfailing good humor, building an ever-expanding network of students, colleagues, and alumni who often gathered at the Knözinger home, where Helmut and his wife Rosemarie famously entertained with the best Munich traditions of food, libations, and merriment. Other research group traditions were visits to the Munich beer gardens and summer weekend excursions to a remote hut in the Austrian Alps.

Helmut Knözinger's energy and enthusiasm were infectious. He rejoiced in his life, his family, and his extended family of coworkers. Everyone who worked with him appreciated his honesty, generosity, modesty, and kindness. He will be deeply missed by many friends and colleagues and a catalysis community that will long benefit from his thoughtful and innovative science.

Michel Che
Gerhard Ertl
Bruce C. Gates
Konstantin Hadjiivanov

CHAPTER ONE

Science and Technology of Framework Metal-Containing Zeotype Catalysts

Laszlo Nemeth*, Simon R. Bare[†]
*University of Nevada, Las Vegas, Nevada, USA
[†]UOP LLC, a Honeywell Company, Des Plaines, Illinois, USA

Contents

Advances in Catalysis, Volume 57
ISSN 0360-0564
http://dx.doi.org/10.1016/B978-0-12-800127-1.00001-1

1

Abstract

Since the discovery of titanium silicalite more than 30 years ago, framework metal-containing zeotype materials have become an important class of catalyst, finding application in several industrial processes. Incorporation of cations of titanium, tin, iron, and other elements into zeotype frameworks (and also into ordered mesoporous materials) has led to both scientific progress and engineering innovations in catalysis. As a result of these developments, framework metal-containing zeotype materials have been implemented in the preceding decade in new commercial, by-product-free green processes, which have improved sustainability in the chemical industry. Based on a comprehensive analysis of the recent literature including patents, this review is a summary of the current knowledge of the science and technology of framework metal-containing zeotype materials. The synthesis of these materials is summarized, followed by an account of state-of-the-art characterization results. The key catalytic chemistries, which can be classified into oxidation reactions such as olefin epoxidation, aromatic hydroxylation and ammoximation, and weak Lewis acid-catalyzed reactions, are discussed. Mechanisms proposed for these transformations are reviewed, together with the theoretical and modeling tools applied in this context. An overview of the technologies associated with the use of framework metal-containing zeotype materials demonstrates how these processes are linked with each other through key chemicals involved.

ABBREVIATIONS

AlPO aluminophosphate
BV Baeyer–Villiger
CCR continuous catalytic regeneration
CHP cumene hydroperoxide
CP-MAS cross polarization magic angle spinning
CTABr/OH cetyltrimethylammonium bromide/hydroxide
DFT density functional theory
EXAFS extended X-ray absorption fine structure
FTIR Fourier transform infrared
HDTMABr hexadecyltrimethylammonium bromide
HMDS hexamethyldisilazane
HMI hexamethyleneimine
HPPO hydrogen peroxide-based propylene oxide
mCPBA meta-chloroperbenzoic acid
MTA metric tons per annum

MTBE methyl tertiary-butyl ether
PO propylene oxide
SAPO silico-aluminophosphate
SBU secondary building unit
SDA structure-directing agent
TBAOH tetrabutylammonium hydroxide
TBHP *tert*-butyl hydroperoxide
TBOT tetrabutyl orthotitanate
TEA tetraethylammonium
TEOS tetraethyl orthosilicate or tetraethoxysilane
TEOT tetraethyl orthotitanate
TET tetraethyltitanate
TMAdaOH trimethyladamantammonium hydroxide
TPA tetrapropylammonium
TPAOH/Br tetrapropylammonium hydroxide/bromide
XRD X-ray diffraction

Note on zeolite nomenclature: Each unique zeolitic structure is assigned a three-letter code by the Structure Commission of the IZA; for example, FAU stands for the structure faujasite, which is composed of double six-membered ring, and single four-membered and six-membered ring secondary building units (SBUs). In addition to the IZA designation, a particular zeolite may also be given a name by the inventor who first synthesized the material; often the company or academic institution is incorporated in such names. For example, ZSM-5, which has the IZA framework designation MFI, was invented by Mobil and given the name Zeolite Socony Mobil-5 (ZSM-5) by its inventors. The IZA and more common names are used interchangeably throughout the review.

1. INTRODUCTION

1.1. Framework metal-containing zeotype materials: Classification

This review is focused on the science and technology of framework metal-containing zeotype materials, a critical and increasingly important class of catalyst. In the following paragraphs, first the structural characteristics of zeolites, which are close and well investigated relatives of zeotype materials, will be briefly summarized. Then the role of the framework metal for the structural and surface chemistry of zeolite and zeotype materials will be introduced.

Zeolites are crystalline aluminosilicates that are composed of an infinitely extending three-dimensional framework (which makes them formally tectosilicates) and charge-balancing cations. The framework consists of four-connected tetrahedral $[AlO_{4/2}]^-$ and $[SiO_{4/2}]$ subunits, that is, each subunit is linked in a fourfold coordination by sharing all of its oxygen atoms with other such tetrahedra. Each $[AlO_{4/2}]^-$ in the framework carries a net negative charge, which is balanced by an extra-framework cation (e.g., K^+, Na^+). Zeolites are typically synthesized through crystallization, under hydrothermal conditions, from strongly basic and thus hydroxide-rich aqueous reaction mixtures. The extra-framework cations, usually alkali, organoammonium or both, serve as structure-directing agents (SDAs) and the regular pore network forms around these cations. Over the course of the synthesis, the primary tetrahedral subunits are assembled into SBUs, which are simple polyhedra such as cubes, hexagonal prisms, or cubooctahedra. The final framework structure then consists of a repeating arrangement of these SBUs. Because there are many possibilities for the SBUs to assemble in three dimensions, a large number of crystallographically unique structures result; indeed over 130 different framework structures are now known and identified by three-letter codes given by the International Zeolite Association. Such structure codes are also used to describe zeotype frameworks, and the total number of known framework types amounts to 218 if the structures of aluminophosphates (AlPOs) and other compositions not classifying as zeolites are also counted.

Acidic properties of zeolites, of both Brønsted and Lewis type, are linked to the presence of aluminum or other cations with similar behavior. Acid sites can be created in a zeolite by exchanging the charge-balancing extra-framework cations and replacing them with protons (H^+). In the case of zeolites containing alkali cations, this postsynthesis modification is usually accomplished by ammonium ion-exchange followed by decomposition of the ammonium ion to a proton and the volatile by-product ammonia in a calcination step. Similarly, zeolites containing charge-balancing organoammonium cations may also be converted to the proton form via calcination. The proton balancing the anionic charge of the framework aluminum will constitute a Brönsted acid site, and there will be one such acid site per aluminum in the zeolite structure. It is also possible for aluminosilicate zeolites to exhibit Lewis acidity, resulting, for example, from the electron pair acceptor capability of a trigonally coordinated aluminum ion.

The framework metal-containing zeotype materials that are the focus of this review do not contain aluminum and thus do not classify as zeolites. Zeotype materials are characterized by properties that make them alike to zeolites and properties that make them markedly different. Silicates with Ga^{3+} or Fe^{3+} incorporated in the framework have some structural and chemical properties that are similar to those of aluminosilicates. Incorporation of Ti^{4+}, Sn^{4+}, and V^{5+} into the silicate framework generates distinctly different properties. The latter three metals often exhibit octahedral coordination in their metal oxides; tetrahedral coordination is particularly rare for tin and titanium in their oxides. Incorporation of V^{5+} in a tetrahedral silicate network as $[VO_{4/2}]^+$ would lead to a positively charged framework. The occurrence of such positively charged (three-dimensional) frameworks is rare. In contrast, two-dimensional varieties are stable, for example, layered double hydroxides. This knowledge suggests that direct substitution of V^{5+} into the framework may not be a viable strategy. Synthesis procedures must be developed with awareness of these propensities, and in contrast to the synthesis of aluminosilicates, alkali metals must be avoided, especially in the synthesis of Ti- and Sn-substituted silicates. The presence of alkali metal cations in the reaction mixtures promotes the formation of highly charged frameworks containing octahedral $[Ti^{4+}O_{6/2}]^{2-}$ and $[Sn^{4+}O_{6/2}]^{2-}$ species, a more natural coordination environment for these metals. The resulting frameworks, with a mixture of octahedral and tetrahedral coordination, do not correspond to zeolite or zeotype frameworks; and while such networks of mixed coordination may be porous, they are generally not as porous as most zeolite frameworks. To prevent the formation of networks with both octahedral and tetrahedral units, and incorporate the metals as, for example, neutral tetrahedral $[Ti^{4+}O_{4/2}]^0$ moieties, the syntheses of the desired metal silicates require a special protocol: As SDAs, exclusively organoammonium cations are used, and the conditions are chosen such that the SDA may promote the formation of a microporous silicate. Because of the larger ionic radii of the substituting metals in comparison to Si^{4+}, the amount of metal that is incorporated into the framework during synthesis is typically limited, and a highly siliceous material is obtained with a metal content of only a few percent. In the ideal case of a defect-free framework, these materials will possess no, or only very weak, Brönsted acidity, since the framework is essentially neutral and does not contain sites like the Al–O(H)–Si moiety of zeolites. Tin, titanium, and vanadium cations have a tendency towards octahedral coordination; in the tetrahedral state, they have empty orbitals available that can accept electrons and consequently, the metal-substituted frameworks exhibit Lewis acidity. The strength of the Lewis acidity is governed by the nature of the metal ion in the framework structure. Another key distinguishing property of framework metal-containing zeotype materials, which derives from their highly siliceous nature, is the high hydrophobicity relative to their aluminosilicate counterparts. Finally, metals capable of changing oxidation state, such as iron and vanadium, have the added capability of engaging in redox chemistry, which is not seen for aluminosilicates.

1.2. Significance of framework metal-containing zeotype materials

Zeolites, as a result of their unique physical and chemical properties, have had a major impact on the world having directly or indirectly affected everyone alive today. Judging from the number of publications and patents on this topic, it is evident that both the scientific and industrial interest in zeolites continues unabated. The invention of synthetic zeolites (including those with the framework types FAU, MFI, and others), and their catalytic applications revitalized the petroleum refining and petrochemical industry in the 1960's, and the development of framework metal-containing zeotype materials has the potential to have a comparable impact in this decade. The incorporation of metals such as titanium, iron, gallium, or tin into zeotype structures was a major development in molecular sieve chemistry, and these new catalysts, which are distinguished by their high product selectivities, have contributed to the scientific understanding of zeolite chemistry and provided new opportunities to revamp some key industrial processes. The following examples demonstrate the success of framework metal-containing zeotype materials:

- The invention of titanium silicalite, TS-1, resulted in a number of new industrial selective partial oxidation processes including those for the hydroxylation of phenol, the epoxidation of propylene, and the ammoxidation of cyclohexanone.
- The invention of framework tin-containing zeotype materials created new opportunities for reactions requiring weak Lewis acid sites. An example is Sn-beta, which has exhibited extraordinary selectivity in partial oxidation reactions like the Baeyer–Villiger (BV) oxidation of ketones to lactones, and also has been shown to be highly selective in the isomerization of glucose to fructose.
- A high selectivity of iron-containing ZSM-5 has been demonstrated for the direct oxidation of benzene to phenol with N_2O as the oxidant (which can be obtained from the waste streams of other processes).

- Gallosilicates have found application for light alkane aromatization in the UOP-BP Cylar™ process.

A grand challenge facing catalytic technology in the twenty-first century is the development of processes that provide 100% selectivity to the desired products. Success will result in cleaner manufacturing without significant production of by-products, with the attendant benefit of minimizing wastes and potential pollutants. Such catalyst-mediated processes would be efficient, cost-effective, and environmentally sound. Presently, catalysis science and technology are far from reaching this goal, but framework metal-containing zeotype materials offer the potential to minimize the formation of by-products and to produce key chemicals with high selectivity. The resulting processes can be considered as examples of sustainable "green chemistry," as will become evident from the commercial examples discussed in this review.

1.3. Scope and organization

This review is restricted to zeotype materials with cations of titanium, tin, vanadium, iron, or gallium incorporated into their frameworks and to related mesoporous materials such as MCM-41 with the same cations incorporated into their wall structures. The structure types MFI, BEA, and MCM-41 were selected because of their significance for catalytic applications. A discussion of the following types of materials is excluded, as their properties are reported elsewhere: impregnated and ion-exchanged zeolites, zeolites containing metal complexes with organic ligands (including the so-called "ship-in-a-bottle" catalysts), and metal-containing AlPOs and silico-aluminophosphates (SAPOs). Although the characterization of such materials and in particular the question of whether the metals are actually incorporated in the framework are important points, they are beyond the scope of this review (and we recognize that given improvement of characterization methods some of the structural attributions stated here may not stand the test of future scrutiny). Catalytic reactions carried out with acidic zeolites have been well reviewed *(1, 1b)*, and comprehensive summaries of oxidation reactions catalyzed by framework metal-containing zeolites *(2)* and TS-1 *(3)* are also available.

This chapter focuses on the application of framework metal-containing zeotype materials as catalysts and is a report of the state of the associated science, presenting a summary of current and future challenges. The chapter is organized as follows: First, the general synthesis procedures of framework metal-containing zeotype materials are described, followed by specific information regarding zeotype materials incorporating various metals. Next, the transformations catalyzed by framework metal-containing zeotype materials are summarized, and mechanistic details are provided when these are available. The technological applications of these materials, presented in Section 4, are the major focus of this chapter. A summary assessment and a list of challenges and opportunities conclude the chapter.

1.4. Methodology of literature analysis

To obtain the raw data for this review, data bases were created to analyze the literature of metal-containing zeolites, and then STN AnaVist and Endnote X5 were used to collect scientific papers, patents, and public announcements. STN Anavist and STN Express are a fully integrated software package designed to easily and efficiently search Chemical Abstracts and other data bases (including patents) online for essential science and technology through STN. STN AnaVist offers a variety of ways to analyze the search results and also visualize patterns and trends in the research. In the context of this review, STN AnaVist was used to efficiently analyze and interpret patent and non-patent literature retrieved from the Chemical Abstracts database using STN Express. An Endnote database was formulated using the results from Scifinder, Science Direct, and Delphion databases. The search results were converted to a uniform Endnote format. The following search strategy and keywords were used to search in databases: Metals (titanium, tin, vanadium, iron, gallium, germanium, etc.) and zeolite structures (silicalite, MFI, beta, MTW, MWW, MCM-41, etc.) together with selected chemistry and process names.

The Anavist database collected 6289 hits, and the biweekly updates typically added a couple of dozen new publications. On the basis of the clustering concept, the following numbers of hits were counted: titanium ($\sim$2000), iron ($\sim$1500), tin ($\sim$350), vanadium ($\sim$450), and gallium ($\sim$500). The distribution was $\sim$4800 non-patents and $\sim$1300 patents. The primary organizations that were identified include the following (listed in order of number of publications): ENI/Snamprogetti, the Technical University of Valencia, ARCO/Lyondell, Dow, BASF, Sumitomo, China Petroleum, East China University, and Dalian Institute. Key researchers included Peng Wu, Bruno Notari, Giuseppe Bellussi, Takashi Tatsumi, Avelino Corma, Paul Ratnasamy, Rajiv Kumar, Min Lin, and Junpei Tsuji.

2. SYNTHESIS AND STRUCTURAL PROPERTIES

2.1. General synthesis procedures

The so-called primary synthesis is the most commonly applied method used to synthesize framework metal-containing zeotype materials. The general scheme of the synthesis is similar to that of aluminosilicate (i.e., zeolite) synthesis, with respect to principal ingredients and procedure. As illustrated in Figure 1.1, the materials are crystallized under hydrothermal conditions, from a mixture of solutions of a silicon source, a metal source, a SDA, and a mineralizing agent. Only when appropriate conditions are applied the desired zeotype structure with metal incorporated in the framework position(s) will form. The selection of SDA, temperature, time, and pH are particularly critical to obtain high crystallinity and uniform materials. Part of the challenge (and still almost an art) in the synthesis of these materials is the development of a robust and reproducible procedures.

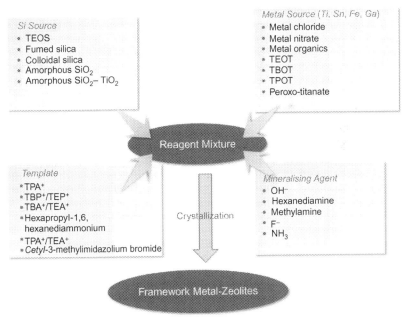

Figure 1.1 The principal synthesis routes used to prepare metal framework-containing zeotype materials. The four major components are a silicon source, a metal source, a template, and a mineralizing agent. These give the overall reagent mixture that then undergoes crystallization under hydrothermal conditions to produce the desired zeotype material.

An alternative, less common, approach is the so-called secondary synthesis, whereby the metals are introduced into an already existing zeolite framework. This approach was introduced by Skeels and Flanigen *(4)* and was originally used to obtain defect-free zeolites with a silicon-enriched framework, which is achieved through reacting zeolites with fluorosilicates. In a modified procedure, zeolites are reacted under mild conditions with aqueous metal fluoride salt solutions (e.g., titanium fluoride or ammonium hexafluorotitanate). The secondary synthesis process has been successfully applied to incorporate titanium, iron, tin, and chromium (using aqueous fluoride salts) into the framework of a number of zeolites.

The method used predominantly in the literature is primary synthesis, and there is ongoing debate in the literature regarding the effectiveness of secondary synthesis. It is not possible to make a statement about the preferred method for large-scale synthesis, as the commercial manufacture of these materials is typically a closely guarded trade secret.

Tables 1.1–1.5 are summaries of the key synthesis papers together with the detailed synthesis conditions for framework metal-containing (Ti, V, Sn, Ga, Fe) zeotype materials.

2.2. Titanium-containing zeotype and ordered mesoporous materials

2.2.1 TS-1

2.2.1.1 History

More than 30 years ago, Taramasso *et al.* of Snamprogetti SPA filed a milestone patent, US Patent 4,410,501 *(5)*. This patent marked the beginning of a new era of environmentally friendly oxidation chemistry with hydrogen peroxide as the oxidant. The researchers described the primary synthesis of titanium-containing alumina-free ZSM-5, now usually referred to as TS-1. The significance of this invention was recognized by the International Zeolite Association in 1992 when G. Bellussi, M. Clerici, V. Fattore, B. Notari, C. Perego, F. Buonomo, A. Esposito, F. Maspero, C. Neri, and U. Romano (all of Eniricerche S.p.A., Milan, Italy) were honored with the Donald Breck Award "for advancing our knowledge of the structures and properties of titanium MFI zeolites (TS-1) and for demonstrating both the potential and applications of these novel catalysts for partial oxidation reactions." In the 30 years since the invention more than a thousand papers and patents have been published in this area, as determined by use of the keyword "TS-1" in the above-described search methodology. Several extensive review papers *(3b,d, 86)* have appeared that summarize the current knowledge of TS-1 and of other zeolites with titanium incorporated into the framework.

Table 1.1 Synthesis of titanium-containing zeotype and ordered mesoporous materials

Material	Structure	Metal source	Silicon source	Template	Specific information	Crystallization conditions (temperature, time)	Reference (first author, year)
TS-1	MFI	TEOT (tetraethyl orthotitanate)	TEOS	TPAOH	Solution in H_2O	175 °C, 10 days	Taramasso, 1983 (5)
TS-1	MFI	$Ti(OBu)_4$	TEOS	TPAOH	Solution in H_2O	170 °C, 3–5 days	Thangaraj, 1991 (6)
TS-1	MFI	TiO_2	SiO_2	TPAOH	Microwave/impregnated SiO_2–TiO_2 xerogel	170 °C, 5 h	Padovan, 1989 (7)
TS-1	MFI	TEOT	TEOS	TPAOH	Solution in H_2O	175 °C, 12 days	Van der Pol, 1992 (8)
TS-1	MFI	$TiCl_4$	Fumed silica	TPABr	Ammonium fluoride solution in water is added	175 °C, 5 days	Dwyer, 1992 (9)
TS-1	MFI	TBOT (tetrabutyl orthotitanate)	TEOS	Hexapropyl-1,6-hexanediammonium	Using di-TPA cation to stabilize the MFI structure	170 °C, 3 days	Tuel, 1994 (10)
TS-1	MFI	TBOT	Fumed silica	Hexamethyleneimine (HMI)	Under rotation conditions	170 °C, 1–6 days	Zhang, 2009 (11)
TS-1	MFI	TiF_4	TEOS	TPAOH	Under rotation conditions	170 °C, 2 days	Jorda, 1997 (12)
PQ-TS-1	MFI	TBOT	TEOS	TPAOH	Cost-effective commercial manufacture	160 °C, 2 days	Senderov, 2003 (13)

Continued

Table 1.1 Synthesis of titanium-containing zeotype and ordered mesoporous materials—cont'd

Material	Structure	Metal source	Silicon source	Template	Specific information	Crystallization conditions (temperature, time)	Reference (first author, year)
TS-1 Core shell	MFI	TBOT	TEOS	TPAOH	Modified with polydiallyldimethylammonium chloride	170 °C, 2 days	Xu, 2011 (14)
TS-1@meso silica Core shell	MFI	TEOT	TEOS	Triblock copolymer surfactant P123 (EO$_{20}$PO$_{70}$EO$_{20}$, Mw = 5800)	TS-1 is added to the solution mixture and the resulting suspension is vigorously stirred and ultrasonicated	70–100 °C, 24 h	Peng, 2012 (15)
Hierarchical TS-1	MFI	TBOT	TEOS	TPAOH/Xerogel	Incipient wetness impregnation of SiO$_2$–TiO$_2$ xerogels	170 °C, 8 h with microwave heating	Sanz, 2011 (16)
Hierarchical mesoporous TS-1	MFI	Tetrabutyl titanate	TEOS	TPAOH	Steam-assisted crystallization method (SAC)	Dried under stirring at 80 °C, then crushed into powder and placed in autoclave for crystallization	Zhou, 2010 (17)
Hierarchical TS-1	MFI	TEOT	TEOS	TPAOH	Precrystallized in a reflux system under stirring at 90 °C for 24 h	170 °C, 8 h under microwave heating	Serrano, 2010 (16,18)

Hierarchical TS-1	MFI	TBOT	Silica sol	Carbon–TPABr	Carbon material from sucrose carbonization	170 °C, 7 days	Wang, 2011 (19)
Micro-meso-materials from TS-1 seeds	MFI	TBOT	TEOS	TPAOH/CTABr	CTABr addition after aging for 24 h at 21 °C, then CTABr removed from solid with ethanol and acetic acid	Dried at 40 °C, 24 h	Reichinger, 2007 (20)
Hollow TS-1	MFI	TBOT	TEOS	TPAOH	With postsynthesis treatment in a solution containing TPAOH	170 °C, 3 days	Wang, 2007 (21)
Mesoporous TS-1	MFI	Ti(OPr)$_4$	TEOS	TPAOH	TPAOH is mixed with an appropriate amount of carbon particles as template	Microwave irradiation (1200 W) with stepwise temperature increase from 80 to 165 °C	Ok, 2011 (22)
TS-1 membrane	MFI	TEOT	TEOS	TPAOH	The TS-1 membrane is grown on preseeded porous stainless steel support using hydrothermal synthesis method.		Au, 2001 (23)
TS-1 ceramic membrane	MFI	TBOT	TEOS	TPAOH	Aging under stirring for 3 h at room temperature	Secondary growth is set at 180 to 190 °C, with duration of 2–2.5 h	Sebastian, 2010 (24)

Continued

Table 1.1 Synthesis of titanium-containing zeotype and ordered mesoporous materials—cont'd

Material	Structure	Metal source	Silicon source	Template	Specific information	Crystallization conditions (temperature, time)	Reference (first author, year)
Titania-supported TS-1	MFI	TBOT	TEOS	TPAOH	High purity titania-support is first impregnated with TBOT dissolved in isopropyl alcohol	Mixed Ti,Si gel crystallized at 175 °C for 3 days under stirring in presence of impregnated titania	Nemeth, 1994 (25)
Extruded TS-1	MFI	TEOT	TEOS	TPAOH	Extrusion process employs inorganic and organic binders, TS-1 powder, and water	170 °C, 24 h or under microwave irradiation at 170 °C, 2 h	Serrano, 2009 (26)
TS-1 alumina binder and polystyrene dispersion	MFI	Tetraisopropyl orthotitanate	TEOS	TPAOH	Calcined TS-1 is mixed with aluminum oxide, formic acid, polystyrene, and water	175 °C, 92 h	Muller, 2002 (27)
TS-1/SiO$_2$ egg-shell	MFI	TBOT	TEOS	TPABr and TPAOH	Solution aged overnight at room temperature.	Synthesis solution and silica beads placed in Teflon-lined autoclave at 175 °C, 2 days under rotation	Wang, 2011 (28)

Material	Structure	Ti source	Si source	Template/base	Notes	Conditions	Reference
TS-2	MEL	TBOT	TEOS	TBAOH	Hydrothermal under static conditions	170 °C, 2 days	Reddy, 1992 (29)
TS-2	MEL	TBOT	TEOS	TBAOH	Microwave heating	250 W, 0.5–15 h	Serrano, 2004 (30)
TS-2	MEL	TBOT	TEOS	TBAOH	10% Titanium incorporation	170 °C, 8 days	Reddy, 1990 (31)
Double substitution of silicalites	MFI	Tetra ethyl titanate	Tetra ethyl silicate	TPAOH	The catalyst contains both Ti and a trivalent element (Al, Ga, Fe) in the framework	175 °C, 24 h	Bellussi, 1991 (32)
Ti–Sn–S1	MFI	Ti(O–iPr)$_4$	TEOS	TPAOH	Ti and Sn incorporations	150 °C, 72 h	Nemeth, 1998 (33)
Ti–ZSM-12	MTW	TBOT	TEOS	(Et$_2$MeN+C$_3$H$_6$)$_2$	Prepared under stirring	180 °C, 6–10 days	Tuel, 1995 (34)
Ti-beta	BEA	TEOT	Fumed silica	TEAOH	Carried out in the absence of alkali cations	135 °C, 1–10 days	Camblor, 1993 (35)
ETS-10/ETS-4	ETS	Crystalline titanium oxide, e.g., P25 (Degussa)	Colloidal silica	No organic additive	Under hydrothermal conditions	200 °C, Hydrothermal pressure, 7–10 days	Davis, 1996 (36) Kuznicki, 2002 (37)
ETS-14	Penkvilksite like	TiCl$_3$ or titanium oxychloride	Fumed silica	No organic additive	Hydrothermal decomposition of ETS-10	200 °C, 7 days	Kuznicki, 1998 (38)

Continued

Table 1.1 Synthesis of titanium-containing zeotype and ordered mesoporous materials—cont'd

Material	Structure	Metal source	Silicon source	Template	Specific information	Crystallization conditions (temperature, time)	Reference (first author, year)
Ti-ITQ-2	MWW	Titanocene dichloride	Layered silica precursor from fumed silica	HMI, TMAdaOH, hexadecyltrimethylammonium bromide	Made by first synthesizing MCM-22(P)	135 °C, 11 days	Corma, 1999 (39)
Ti-ITQ-7	ISV	TEOT	TEOS	1,3,3-trimethyl-6-azonium tri-cyclo[3.2.1.4^{6,6}] dodecane	The crystalline silica polymorph of lowest density	150 °C, 12 days	Díaz-Cabañas, 2000 (40)
Ti-MWW	MWW	TBOT	Fumed silica	Piperidine or HMI	Acid treatment of lamellar precursor	At 130 °C then 150 °C each for 1 day and further at 170 °C for 5 days	Wu, 2001 (41)
Ti-MCM-41	Hexagonal lattice	Ti(OEt)$_4$	Fumed silica	CTABr	Ultralarge pore Ti-zeolite	140 °C, 28 h	Corma, 1994 (42)
Silated Ti-MCM-41	Hexagonal lattice	Ti(OEt)$_4$	Fumed silica	CTABr	Silylated with hexamethyldisilazane (HMDS)	100 °C, 2 days	Corma, 1998 (43)
Ti-MCM-48	Three-dimensional ordered cubic structure	Ti(OEt)4	Amorphous silica	CTABr/OH	Silylated material	150 °C, 6 h	Pena, 2001 (44)

	Structure	Ti source	Si source	Template	Description	Conditions	Reference
Ti-UTD-1	DON	TEOT	Fumed silica	Cp$^*_2$CoOR	High silica, large pore material	175 °C, 6 days	Balkus, 1995 (45)
Ti-UTD-8	Novel phase structure	TEOT	Fumed silica	Cp$^*_2$CoOH	Large pore molecular sieve, with one-dimensional channels	175 °C, 6 days	Balkus, 1995 (45b)
Ti-SBA-15	Sponge-like pore structure	TBOT	TEOS		Postsynthesis method (Ti added to SBA-15)	100 °C, 2 days	Wu, 2002 (46)
Ti-TUD-1	Sponge-like pore structure	TBOT	TEOS	TEA	TEA, Ti n-butoxide, TEAOH and aging and heating dry gel in autoclave	Room temperature, 1–4 days	Shan, 2001 (47)
Ti-JDF-L	Tetragonal layered structure	TBOT	Fumed silica	TBABr	Layered titanosilicate catalyst	180 °C, 10 days	Thomas, 2007 (48)

Table 1.2 Synthesis of vanadium-containing zeotype and ordered mesoporous materials

Material	Structure	Metal source	Silicon source	Template	Specific information	Crystallization conditions (temperature, time)	Reference (first author, year)
VS-1	MFI	$VOSO_4 \cdot 5H_2O$	Fumed silica	TPAOH	Thermally stable with a low aluminum content	180 °C, 3 days	Rigutto, 1991 (49)
VS-2	MEL	$VOSO_4$ (V^{4+})	TEOS	TBAOH	Synthesis carried out in inert atmosphere	170 °C, 2 days	Sen, 1996 (50)
VS-2	MEL	$VOSO_4 \cdot 5H_2O$	TEOS	TBAOH,	Hydrothermal synthesis	170 °C, 2 days	Rao, 1993 (51), Ramaswamy (52)
V-HSM	Hexagonal molecular sieve	$VOSO_4 \cdot 5H_2O$	TEOS	Dodecylamine	Crystalline mesoporous molecular sieves	Room temperature, 18 h	Reddy, 1996 (53)
V-MTW	MTW	$VOSO_4 \cdot 3H_2O$	Fumed silica	1,6-Hexamethylene-bis (benzyldimethylammonium hydroxide)	Aluminosilicate, with pure silica polymorph	160 °C, 6–8 days	Bhaumik, 1995 (54)
V-NCL-1	MFI	$VOSO_4 \cdot H_2O$	Fumed silica	Hexamethylene bis (triethylammonium bromide)	Hydrothermal synthesis	170 °C, 15 days	Reddy, 1993 (55), Kumar, 1996 (56)

Material	Structure	V source	Si source	Template	Synthesis method	Conditions	Reference
V-MCM-41	Hexagonal lattice	$VOSO_4$	Fumed silica	Dodecyltrimethyl ammonium bromide	Hydrothermal synthesis	100 °C, 6–7 days in Teflon-lined autoclaves tumbled at 25 rpm	Sayari, 1995 (57)
V-MCM-41	Hexagonal lattice	Vanadyl oxalate	Sodium silicate	HDTMABr	Sol–gel method synthesis	120 °C, 96 h	Zhang, 2001 (58)
V-MCM-48	Three-dimensional ordered cubic structure	$VOSO_4$ or $VOCl_3$	Fumed silica	CTAOH/Br	One–pot synthesis	150 °C, 6 h	Pena, 2001 (59)
V-ZSM-48	MRE	$VOSO_4 \bullet 5H_2O$	Fumed silica	1,8-Diaminooctane	Hydrothermal synthesis	180 °C, 5 days	Tuel, 1994 (60)
V-MCM-41	Hexagonal lattice	$VOSO_4 \bullet 5H_2O$	Fumed silica	Alkyltrimethylammonium chloride or bromide	$(C_{16}H_{33})(C_nH_{2n+1})(CH_3)_2NBr$, with $n = 1$–22 to control pore size	100 °C, 7 days	Sayari, 1995 (57)
V-beta	BEA	V_2O_5	Amorphous silica	TEAOH	Direct hydrothermal synthesis	140 °C, 20 days	Chien, 1997 (61)
V-beta	BEA	$VOSO_4 \bullet H_2O$	Fumed silica	TEAOH	Zeolite beta seeds were added	140 °C, 12 days	Corma, 2003 (62)

Table 1.3 Synthesis of tin-containing zeotype and ordered mesoporous materials

Material	Structure	Metal source	Silicon source	Template	Specific information	Crystallization conditions (temperature, time)	Reference (first author, year)
Sn-S1	MFI	$SnCl_4 \cdot 5H_2O$	TEOS	TPAOH	Under static conditions	170 °C, 2 days	Mal, 1997 (63)
Sn–MFI sec. synthesis	MFI	SnO, SnF_4, $SnCl_2 \cdot 2H_2O$	SiO_2/Al_2O_3 material	SiO_2/Al_2O_3 material	Extracting aluminum and substituting by tin	75 °C, in water	Skeels, 1989, 1995 (4,64)
Sn-silicalite	MFI	$SnCl_4 \cdot 5H_2O$	TEOS and fumed silica	TPAOH	With pH adjustment during synthesis	170 °C, 5 days	Fejes, 1996 (65)
Sn-S2	MEL	$SnCl_4 \cdot 5H_2O$	TEOS	TBAOH	Al-free Sn-silicalites	160 °C, 2 days	Mal, 1995 (66)
Sn-ZSM-12	MTW	$SnCl_4 \cdot 5H_2O$	TEOS	1,6-hexamethylene bis(benzyl dimethyl ammonium hydroxide)	Large pore Sn-silicate	160 °C, 5 days	Mal, 1995 (67)
Sn-beta	BEA	$SnCl_4 \cdot 5H_2O$	TEOS	TEAOH	In fluoride medium with dealuminated zeolite beta seeds addition	100 °C	Corma, 2001 (68)
Sn-beta secondary synthesis	BEA	$SnCl_4 \cdot 4H_2O$	NH_4–beta zeolite	—	Postsynthesis from highly dealuminated beta zeolite	140 °C, 10–40 days	Li, 2011 (69)
Sn-MCM-41	Hexagonal lattice	$SnCl_4 \cdot H_2O$	Fumed silica	CTAOH/Br	Under static conditions	135 °C, 24 h	Corma, 2005 (70)

Table 1.4 Synthesis of gallium-containing zeotype materials

Material	Structure	Metal source	Silicon source	Template	Specific information	Crystallization conditions (temperature, time)	Reference (first author, year)
Ga–MFI	MFI	Ga-nitrate or Na-gallate	Na-trisilicate	TPABr	Hydrothermal synthesis	160–200 °C, 1 days	Choudhary, 1997 (71)
GaAPO-11 and GaAPSO-11	AEL	Gallium nitrate	Colloidal silica	Di-*n*-propyl amine	Synthesis with or without silicon as an additional accompanying element	200 °C, 13–40 h	Machado, 1997 (72)
Ga–MFI	MFI	Ga$_2$O$_3$	Fumed silica	TPABr	Hydrothermal method	160 °C, 8 h	Ma, 2001 (73)
Ga–MFI	MFI	Ga(NO$_3$)$_3$ and Ga$_2$O$_3$	From H-MFI	From H-MFI	H-MFI impregnated	75–100 °C	Dooley, 1996 (74)
Ga–MFI	MFI	Ga(NO$_3$)$_3$	Fumed silica	ZSM-5 seeds	ZSM-5/matrix material	160 °C, 26 h	Degnan, 1996 (75)
Ga–MFI	MFI	Ga(NO$_3$)$_3$	Fumed silica	TPABr	Direct synthesis and postsynthetic modification of boralites	175 °C, 8 days	Klik, 1997 (76)

Continued

Table 1.4 Synthesis of gallium-containing zeotype materials—cont'd

Material	Structure	Metal source	Silicon source	Template	Specific information	Crystallization conditions (temperature, time)	Reference (first author, year)
Ga/H-ZSM-5	MFI	Gallium nitrate	Na-trisilicate	TPABr	Impregnated by incipient wetness technique	Crushed to 52–72 mesh size particles and calcined under static air at 600 °C for 1 h	Choudhary, 1997 (5b)
Ga-beta	BEA	Gallium nitrate nonahydrate	Amorphous silica	TEAOH	Prepared in the absence of alkali cations	135 °C under rotation conditions	Camblor, 1992 (77)
Ga–NCL-1	MFI	Gallium sulfate	Fumed silica	Hexamethylene-bis-(triethylammonium bromide)	With H_2SO_4 addition	170 °C, 5–7 days	Sasidharan, 1998 (78)
Ga/Al-EMT	EMT	Aluminogallosilicate hydrogel	Silica sol	Crown ether (18-crown-6, Aldrich)	Hexagonal analogue of cubic Y faujasite	110 °C, 15 days	Fechete, 2005 (79)
Ga-MCM-22	MWW	$Ga(NO_3)_3 \cdot 9\,H_2O$	Sodium silicate	Hexamethyleneimine	Obtained by ion-exchange of H-MCM-22	160 °C, 3 days	Kumar, 1996 (80)

Table 1.5 Synthesis of iron-containing zeotype materials

Material	Structure	Metal source	Silicon source	Template	Specific information	Crystallization conditions (temperature, time)	Reference (first author, year)
Fe–MFI	MFI	$Fe(NO_3) \cdot 9H_2O$	Fumed silica	TPABr	Fe–MFI hydrothermal synthesis	170 °C, 3 days	Milanesio, 2003 (81)
Fe–MFI	MFI	Iron nitrate or iron phosphate	Fumed silica or silica gel	TPABr	Hydrothermal method under static conditions	170 °C, 5–7 days	Centi, 2006 (82)
Fe–MFI	MFI	$FeCl_3 \cdot 6H_2O$	Sodium silicate	TPAOH	With pH adjustment during synthesis	150 °C, 8 h	Fejes, 1996 (65)
Fe–MFI	MFI	$Fe(NO_3)_3 \cdot 39H_2O$	Fumed silica	TPABr	In fluoride-containing media	170 °C, 7 days	Testa, 2003 (83)
Fe–NCL-1	MFI	Ferric sulfate	Fumed silica	Hexamethylene-bis-(triethylammonium bromide)	Hydrothermal synthesis	170 °C, 5–7 days	Sasidharan, 1998 (78)
Fe–beta	BEA	Ferric sulfate	TEOS	TEAOH	Hydrothermal synthesis	120 °C, 12 days	Kumar, 1990 (84)
Fe–TUD-1	Sponge-like pore structure	Ferric chloride	TEOS	TEAOH	Hydrothermal synthesis	180 °C, 8 h.	Hamdy, 2005 (85)

2.2.1.2 TS-1 synthesis

Two main methods for synthesizing TS-1 by direct crystallization are described in the early literature *(5a,87)*. Both methods are variations of the general recipe for primary synthesis (as described in Section 2.1) and rely on the same template, namely, the tetra-*n*-propylammonium (TPA) ion. The methods differ mainly in the details regarding the preparation of the precursor mixture for the crystallization of the titanium-containing silicalite.

The first method uses the controlled concurrent hydrolysis of tetraethoxytitanium (IV) and tetraethoxysilane. This procedure has been labeled the "mixed alkoxide" method. Frequently, the acronyms TET (tetraethyltitanate) and TEOS (tetraethyl orthosilicate or tetraethoxysilane) are used for the respective reactants; these are derived from the alternative names tetraethyl titanate and tetraethylorthosilicate. In examples described in patents, the synthesis involves adding TET to TEOS and then combining the alkoxide mixture with an aqueous solution of a SDA, which is typically tetrapropylammonium hydroxide (TPAOH). The resulting precursor mixture is then heated to a temperature of 175 °C to initiate crystallization. Subsequent washing of the crystallized solid with water, drying, and air calcination produces framework titanium-containing silicalite.

In the second synthesis method, the precursor is obtained by mixing colloidal silica with a solution containing TPAOH, hydrolyzed titanium alkoxide, and hydrogen peroxide. This procedure has been labeled the "dissolved titanium" method.

The secondary synthesis procedure, described earlier, has also been applied to prepare titanium silicates *(4,88)*. This method of incorporation of framework metals into zeolites was part of Union Carbide's zeolite research.

The invention of TS-1 created significant scientific interest. The material's unique properties and the resulting catalytic performance are believed to be associated with the isolated titanium sites, which are highly active and selective in oxidation reactions with hydrogen peroxide as the oxidant. The scientists from the ENI group *(3e,86,89)* have since published more details about the material's synthesis and its physical and catalytic properties. However, as more than 600 papers were obtained when using "TS-1" as a search keyword, we restrict our summary to only a small fraction of them, which we regard as essential papers from the early years (i.e., 1983–1995) and others that appeared in the subsequent 10–15 years.

A "dry" synthesis procedure was developed by Padovan *et al. (90)* In this procedure, silica is impregnated with a solution of a titanium precursor in aqueous TPAOH. Pure and well-crystallized Ti silicates are obtained, except at the longest synthesis times.

Many papers address the preparation of TS-1 via the mixed alkoxide method. Notari *(91)*, Millini and Perego *(92)*, Bellussi *et al.* *(3a,93)*, and van der Pol *et al.* *(8)* optimized the sources of titanium and silicon and investigated the influence of various synthesis parameters. Noteworthy for the synthesis of TS-1 is the temperature of crystallization, which has been varied over a wide range. The effectiveness of titanium incorporation increases with increasing temperature, up to the point where titania (more precisely, the anatase modification of titania) starts to form, at approximately 200 °C. Crystallization at low temperatures favors the formation of colloidal TS-1 crystals with an average diameter in the range of 80–110 nm. At a fixed temperature, the titanium content of the crystalline product is always lower than that of the precursor mixture. This observation implies that the solution becomes enriched in titanium relative to the solid during synthesis and that longer crystallization times ultimately lead to the deposition of extra-framework titanium.

Bordiga *et al.* *(94)* published a review in which they asserted that the mode of titanium incorporation into the silicalite framework is the substitution of a silicon ion by a titanium ion, resulting in the formation of tetrahedral $[TiO_4]$ units. The maximum concentration of titanium that can be incorporated into the silicalite structure has been debated intensely, and the controversy is ongoing. For example, Thangaraj *et al.* *(6)* published a synthesis of TS-1 that resulted in a high titanium content, with an upper limit of the $Ti/(Ti+Si)$ atomic ratio of 0.10. In contrast, Millini *et al.* *(92b)* reported a much lower degree of titanium incorporation, with the $Ti/(Ti+Si)$ atomic ratio generally being in the range of 0.03–0.04. In 2006, Phonthammachai *et al.* *(95)* reported a Si:Ti atomic ratio of 5.0, corresponding to a $Ti/(Ti+Si)$ atomic ratio of 0.2, with only a small amount of non-framework titanium present.

UV–vis, laser Raman (under resonance conditions), and X-ray absorption spectroscopy have been the dominant techniques used both to determine the maximum level of titanium incorporation in the framework, and also to identify the structure of the species formed by adsorption of H_2O_2 on titanium centers within the TS-1 framework. Unfortunately, the majority of these techniques are not quantitative, and efforts have rarely been made to quantitatively cross-correlate the results from the different characterization methods. It is therefore difficult for the researcher to read the literature in this area and arrive at a conclusion. Research to resolve these inconsistencies would be beneficial to the field.

TPAOH is a relatively expensive templating agent, and so it is no surprise that several investigations have been carried out with the goal of replacing it.

Synthesis of TS-1 was accomplished with templating agents such as hexapropyl-1,6-hexanediammonium salts *(60)*, 1,6-hexanediamine, methylamine, and TPA bromide *(96)*. Moreover, Kooyman *et al.* *(97)* attempted to synthesize TS-1 in the presence of fluoride (using TiF_3 and the rutile modification of TiO_2 as precursors) but observed anatase impurities, with evidence arising from extended X-ray absorption fine structure (EXAFS) data that were consistent with octahedral—and thus extra-framework—titanium. Jorda *et al.* *(12)* used TiF_4 for the synthesis and conclusively demonstrated that it is possible to synthesize TS-1 by using the air- and water-stable TiF_4. These authors were able to incorporate up to two titanium atoms per unit cell, corresponding to a $Ti/(Ti+Si)$ ratio of 0.2, but their samples contained trace amounts of residual fluoride from the synthesis.

To elucidate the causes of the unique properties and reactivity of TS-1, Drago *et al.* *(98)* performed combined adsorption and calorimetry experiments. The data showed that TS-1 does not have strong acid sites. The authors concluded that the entire surface of TS-1 is hydrophobic and that the small number of hydrogen bonding sites creates a hydrophobic environment that enables TS-1 to adsorb hydrocarbons in the presence of water.

In the preceding decade, several new morphologies and other modifications of TS-1 have been developed. The synthesis and catalytic properties of TS-1 materials with mesoporous–microporous hierarchical structures *(16–18,99)*, obtained in the presence of amphiphilic organosilanes, were reported. Hierarchical porosity was also obtained by preparing titanosilicate beads from an anion-exchange resin; this material was designated TiSil-HPB-60 *(99a)*. Other procedures to prepare unique morphologies or composite forms of TS-1 have been reported. These include the following: (i) the synthesis of hollow TS-1 crystals *(21)* formed via a dissolution–recrystallization process, (ii) the synthesis of nanocrystalline TS-1 *(100)*, (iii) the fabrication of TS-1–based membrane reactors *(101)*, (iv) the synthesis of supported titanium silicalite *(25)* with titania as the support, (v) the extrusion of TS-1 *(26a,102)*, (vi) the manufacture of a spherical silica–zeolite composite *(103)*, and (vii) the development of an extruded catalyst for fixed-bed reactor applications using an alumina binder and a polystyrene dispersion *(27)*. Furthermore, Onimus *et al.* of Lyondell patented a spray-dried catalyst *(104)*, and a procedure for the successful regeneration of titanium-containing catalysts has also been claimed *(105)*. Other novel results include a zinc-treated TS-1 catalyst *(106)*, the nanocasting of TS-1 *(107)*, a template-free synthesis *(108)*, and hierarchical TS-1 *(109,110)*. Furthermore, a method for supporting noble metals on TS-1 *(111)* and the synthesis and catalytic properties of gold supported on TS-1 *(112,113)* have been disclosed.

2.2.1.3 Location of titanium in the framework of TS-1

As stated above, Ti^{4+} replaces Si^{4+} in the silicalite structure. However, the MFI framework topology in the orthorhombic form, which is the stable form under ambient conditions, has 12 crystallographically distinct tetrahedral cation sites, as shown in Figure 1.2.

Numerous investigations have led to reports that titanium preferentially occupies one or more sites in the silicalite framework, rather than being randomly distributed across all positions. The proposed preferred locations and the associated references are listed in Table 1.6.

Some of the analyses to determine the titanium locations exploited the differing signs of the silicon and titanium coherent scattering lengths in neutron powder diffraction to refine the site occupancy of titanium within the framework. Hijar *et al.* *(115)* found that the favored sites were T3, T7, T8, T10, and T12.

An investigation by Henry *et al.* *(116)* was based on the refinement of combined neutron and X-ray diffraction (XRD) data to examine the titanium locations in the MFI structure. The authors stated that the interpretation of the diffraction data was complicated by the presence of silicon

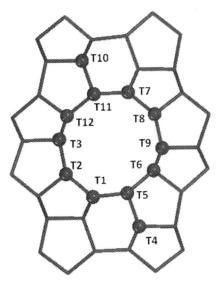

Figure 1.2 The 12 T-sites in a unit cell of silicalite (MFI).

Table 1.6 Preferred titanium location in TS-1

Preferred site	Method	Reference (first author, year)
None (random distribution)	None provided	ENI Flyer from TS-1, 2009 *(114b)*
T3, T7, T8, T10, and T12	Rietveld analysis of powder neutron diffraction data	Hijar, 2000 *(115b)*
T8 and T10, with a small amount in T3	Neutron and X-ray refinement	Henry, 2001 *(116)*
T6, T7, T11, and T10	Rietveld analysis of powder neutron diffraction data	Lamberti, 2001 *(117)*
T8 and T10 with lower concentrations in T4 and T11	Periodic density functional theory	Gale, 2006 *(118)*
T10, T4, T8, and T11 with T6 vacancy T9 and T5	*ab initio* density functional theory	Yuan, 2011 *(119)*

vacancies within the framework, which, like titanium, reduce the neutron scattering from the tetrahedral sites. Thus, the presence of vacancies leads to difficulties with the refinement of titanium site occupancies. Only by employing a variety of titanium isotopes that are characterized by different scattering lengths did it become feasible to correctly refine the site occupancies. The authors concluded that titanium is preferentially sited at T8 and T10 positions, with the possibility of a small amount in the T3 position, whereas the silicon vacancies are associated mainly with the T1 and T5 positions.

In addition to the experimental investigations, there have been numerous theoretical attempts to identify the locations of titanium in MFI. In early work, force field methods were applied *(120)*. Later, Gale *(118)* evaluated the thermodynamics of the substitution of titanium into the silicalite framework to form TS-1 by using periodic density functional theory (DFT). In contrast to the results obtained with force field and *ab initio* cluster calculations, the DFT calculations led to the inference that the favored tetrahedral sites are T8 and T10 when the degree of substitution is equal to one titanium ion per unit cell, in good agreement with the best information currently available from diffraction analysis. At lower titanium concentrations (i.e., less than one atom per unit cell), T4 and T11 sites may also be populated. The calculated lengths of the bonds characterizing titanium and neighboring oxygen atoms were found to be in good agreement with EXAFS data.

Comparison of Gale's work with experimental results suggests that the titanium distribution in TS-1 is substantially in agreement with the site preferences expected from thermodynamics, as had been proposed previously on the basis of theoretical investigations (118).

On the basis of ab initio DFT calculations, Yuan et al. (119) extended the theoretical understanding by considering the stabilities of various local arrangements, which were silicon vacancies, regular titanium sites in the framework denoted as [Ti(OSi)$_4$], and defective titanium sites denoted as [Ti(OSi)$_3$OH]. [Ti(OSi)$_4$] stands for regular sites in the MFI framework characterized by titanium coordinated to four silicon atoms via a bridging oxygen in tetrahedral coordination, and [Ti(OSi)$_3$OH] stands for a defective site in the framework characterized by only three Ti—O—Si bonds and a fourth bond to a hydroxyl group. The authors showed that the four energetically favorable sites for silicon vacancies are T6, T12, T4, and T8, whereas favorable sites for titanium centers of the type [Ti(OSi)$_4$] are T10, T4, T8, and T11. Thus, vacancies and titanium to some extent are preferentially associated with the same sites. Independent of whether titanium fills a silicon vacancy or substitutes for a fully coordinated silicon ion, the most preferred site for titanium is T10. This result indicates that the insertion mechanism does not affect the location of titanium in the MFI lattice. When defective titanium sites of the [Ti(OSi)$_3$OH] type are included in the calculations, the silicon vacancy at T6 with a titanium at its neighboring T9 site (designated as T6-def-T9-Ti pair) is the most energetically favorable location, followed by a T6-def-T5-Ti pair, with the energy gap between the two structures being small (1.64 kcal/mol).

As is evident from the various results discussed above, there is no general consensus regarding the location of titanium inside the MFI structure, notwithstanding more than a decade of research on this question. To characterize TS-1 and determine the titanium location, UV–vis, Raman, and Fourier transform infrared (FTIR) spectroscopy, EXAFS analysis, X-ray and neutron diffraction, and ab initio DFT calculations have all been used. Some of the analytical difficulties encountered are associated with properties inherent to titanium, and the situation is better when the heteroatom has a higher atomic number such as tin. In this case, characterization techniques that depend strongly on the atomic number such as EXAFS analysis can be used to precisely define the site in the framework that is occupied by the heteroatom (see Section 2.4).

Regarding TS-1, only conclusions that are supported by a combination of results from several techniques may be considered to have some reliability.

It is also possible that there is not a single correct answer because the titanium location may vary with synthesis conditions, loading, and other parameters.

2.2.2 Ti-MCM-41

Mesoporous materials with pore diameters in the range of 10–150 Å readily allow the access of bulky molecules and are potentially attractive catalysts for chemical synthesis. A significant advance in this area was made by Vincent *et al.* of Sylvania *(121)* who synthesized a mesoporous material by using a silica–alumina gel as silicon source and an alkyl-trimethylammonium cation with a long alkyl chain as the SDA. The resulting material was found to have a regular structure with a hexagonal array of pores. It was named MCM-41 *(122)*. Although the walls of these siliceous materials are amorphous, the long-range structures are highly ordered. Cubic (MCM-48) and lamellar (MCM-50) forms of the material have also been synthesized *(122a)*. Given the discovery of such a new structure type with its large pores, it was not surprising that research was conducted to try to introduce titanium into the structure. The first titanium-containing high-silica material of this class (Ti–MCM-41) was synthesized by Corma *et al. (42,123)* A tetrahedral Ti (IV) site was confirmed by a combination of EXAFS and UV–vis spectroscopy measurements *(124)*.

Following this initial success, new strategies were developed *(43)* to improve both the catalytic epoxidation activity and selectivity of Ti-MCM-41. It was found that anchoring trimethylsilyl groups to the silanol groups led to a remarkably increased activity. This result suggests that highly hydrophobic materials are desirable for the catalytic epoxidation. The materials were prepared as follows. The authors first made a sample from a gel having the following molar composition: 1 SiO_2:0.015 $Ti(OEt)_4$:0.26 CTABr:0.26 TMA-OH:24.3 H_2O. The obtained Ti-MCM-41 was silylated with HMDS and the product designated as Ti-MCM-41S. The MCM-41 structure was preserved after silylation, as demonstrated by the presence of three well-defined reflections in XRD patterns that are characteristic of the hexagonal array of the MCM-41 material. The surface area of the silylated sample was found to be roughly 1000 $m^2\ g^{-1}$, which indicated that the silylation resulted in only a minimal loss of surface area.

Samples with various degrees of silylation were obtained by changing the HDMS/Ti-MCM-41 ratio as shown in Table 1.7. The carbon content, determined by elemental analysis, was used to calculate the degree of surface coverage by trimethylsilyl groups. The space required for the trimethylsilyl groups was found to be 47.6 $Å^2$ molecule^{-1}.

Table 1.7 Degree of silylation of Ti-MCM-41 catalysts depending on the amount of hexamethyldisilazane (HMDS) applied

Sample	Ti content (Wt% TiO$_2$)	HMDS/sample (Wt/Wt)	Carbon content (Wt%)	Surface coverage (%)
Ti-MCM-41	1.9	–	–	0
Ti-MCM-41S1	1.9	0.026	4.2	33
Ti-MCM-41S2	1.8	0.034	7.9	63
Ti-MCM-41S3	2.1	0.123	10.3	82
Ti-MCM-41S4	2.0	0.260	11.9	95

Adapted from Ref. *(43)*, with permission of the Royal Society of Chemistry.

Yamamoto *et al. (125)* investigated the preparation of Ti-MCM-41 from a synthesis gel derived from Ti(OC$_4$H$_9$)$_4$ and various silicon sources. By using TEOS, which does not contain aluminum or sodium, as a silicon source, they were able to incorporate titanium into the framework of MCM-41 without the formation of anatase. Organically modified mesoporous titanium-substituted MCM-41 materials (generally termed Ti-MCM-41-R, with R = Ph, Me) have also been synthesized. The organic group is attached to the external surface of Ti-MCM-41.

2.2.3 Ti-SBA-15

Titanium has also been incorporated into ordered mesoporous silicas of the SBA-15 type *(126)*. The influence of synthesis parameters on the properties of titanium-substituted SBA-15 silicas prepared by a direct one-step synthesis (cocondensation) was systematically investigated *(127)* through characterizing the products by N$_2$ physisorption, XRD, diffuse reflectance UV–vis spectroscopy, and elemental analysis. The results showed that when a low titanium precursor concentration (i.e., less than 0.05 mol/l) was used in the initial synthesis gel, incorporation of titanium into the solid was not completed. Under these conditions, titanium ions were well dispersed in the silica framework and present mainly in tetrahedral coordination in the product. When the isolated titanium species present in the gel reached a critical concentration, an increase in titanium incorporation into the solid was observed as a consequence of the formation of anatase clusters on the material's surface. The titanium loading at which anatase formation was observed was found to be strongly influenced by the synthesis conditions. Moreover, when the formation of anatase takes place, the amorphous titanium species dispersed in the siliceous solid are resolubilized in the acidic solution and redeposited, thus ultimately increasing the fraction of bulk titania (anatase phase) in the mesoporous silica matrix.

Markowitz et al. (128) investigated the effects of adding the organosilane n-(trimethoxysilylpropyl)-ethylenediaminetriacetic acid during hydrothermal synthesis on the properties of titanium-containing SBA-15 mesoporous silica. X-ray photoelectron spectra showed that the Si/Ti atomic ratio decreased significantly with the addition of the organosilane compared with that characterizing the nonfunctionalized silicate, implying that the organosilicate contained more surface titanium. XRD and N_2 gas adsorption analysis revealed that the functionalized titanosilicate was an ordered mesoporous material with a narrow pore size distribution. Solid-state cross polarization magic angle spinning (CP-MAS) NMR experiments demonstrated that addition of organosilane during hydrothermal synthesis produced mesoporous silica with hexacoordinated Ti(IV). Application of the same synthesis procedure without the addition of the organosilane resulted in mesoporous silica with tetracoordinated Ti(IV).

2.2.4 Ti-UTD

Freyhardt et al. (129) synthesized a 14-membered ring zeolite using the new SDA bis(pentamethyl-cyclopentadienyl)cobalt and incorporated titanium into this zeolite, to form the structures Ti-UTD-1 (45a) and Ti-UTD-8 (45b). UTD-1 has monodimensional, large, elliptically shaped pores (with the dimensions of 7.5×10 Å) which are delineated by the 14-membered ring.

2.2.5 Other titanium-containing zeotype materials

Doubly substituted analogues of TS-1 have also been reported. Trong et al. (130) synthesized bifunctional molecular sieves with titanium and various trivalent ions, for example, Ti-MFI that also contained B^{3+}, Al^{3+}, or Ga^{3+}. Tin and vanadium have also been incorporated into the titanium silicalite structure (33,131) by a primary synthesis method. The incorporation of a second metal changes the redox properties of the materials as well as their morphology. Incorporation of tin into titanium silicalite improved the epoxidation selectivity of the catalyst compared with that of (monosubstituted) TS-1.

TS-2, a small-pored titanium silicalite with the MEL structure, was synthesized by Reddy et al. (29,132) Their synthesis method was a modification of that used for TS-1 and differed with respect to the SDAs; for example, tetrabutylammonium hydroxide (TBAOH) was employed. Other authors *(133)* also investigated the synthesis and optimized the incorporation of titanium into the framework of TS-2.

Kuznicki *et al.* *(37b, 134)* synthesized crystalline titanium molecular sieves called ETS-10 and ETS-4. These materials have open structures that contain both tetrahedral and octahedral primary building units.

Titanium-beta (Ti-BEA) is a large pore zeotype material with intersecting 12 MR pores. These large pores facilitate the diffusion of bulky molecules to the active titanium sites. The tetraethylammonium (TEA) ion is used as the SDA in the synthesis of Ti-BEA. A variety of synthesis methods have been reported *(135)*.

There have been reports and patents describing the synthesis of other titanium-containing zeolite structures such as Ti-MWW (ZSM-22) *(41b, 136)*, Ti-ITQ-2 *(39)*, Ti-ITQ-6 *(137)*, Ti-SSZ-33 *(138)*, and Ti-YNU-1 *(139)*.

2.2.6 Summary: Titanium-containing materials

In summary, titanium-containing molecular sieves have been the subject of significant research and commercialization, especially for environmentally friendly processes (as summarized below). The types of materials range from microporous to mesoporous and include a wide range of zeolitic structures. A large number of synthesis methods have been described, whereby the focus has been on achieving highly crystalline materials while using low-priced SDAs. The incorporation of titanium into various zeotype and mesoporous structures has been thoroughly investigated. The degree of control of the material properties is already impressive, and it is expected that further research will lead to even more precise control. Furthermore, we believe that in this area there is a significant amount of unpublished knowledge, including trade secrets developed during the commercialization of these materials.

Today it is well accepted that the degree of titanium incorporation is highly dependent on the synthesis conditions, and, for selective oxidation catalysis, framework incorporation is necessary to achieve both high activity and high selectivity. There appears to be a limit of titanium incorporation into the frameworks, which is in the percentage level (3 mol%, corresponding to a Ti/(Si + Ti) atomic ratio of 0.03). Higher titanium concentrations result in the presence of titanium in non-framework sites and/or the formation of anatase. A significant research effort has been invested into determining the specific locations of titanium in various zeotype structures;

no consensus has been reached on this issue, and it cannot be excluded that the titanium location depends on the synthesis procedure.

2.3. Vanadium-containing zeotype and ordered mesoporous materials

The incorporation of V^{5+} into a tetrahedral silicate network is a challenge because the insertion of $[V^{5+}O_{4/2}]^+$ results in a framework with a net positive charge. Such a material would be an anion exchanger, which are generally not stable in the form of microporous oxides. The stability of the materials is limited because of the charge situation that the exchange anion (i.e., an extra-framework anion) encounters inside the pore: the positive charge associated with the framework cation is screened by the surrounding negatively charged framework oxide ions. Hence, it is not surprising that several investigations have demonstrated that only a small fraction of the vanadium ions are actually incorporated into the framework during direct hydrothermal synthesis *(56,140)*. The vanadium species are typically loosely bound to the pore walls and easily washed out with an aqueous solution *(50)*. In vanadium-containing microporous materials, V^{5+} and/or V^{4+} can replace framework silicon but are often not present in ideal tetrahedral coordination to the framework; instead a tetrahedral coordination with partial coordination to the framework or square pyramidal coordination may be found *(141)*.

Bellussi and Rigutto *(86)* reviewed the synthesis of various vanadium-containing silicates and pointed out that, in addition to the gel composition and the crystallization conditions, the vanadium source plays an important role in determining the incorporation of vanadium ions. Rather than using a V^{5+} source, more success with V–incorporation has been attained when cationic V^{4+} is used in the form of VO^{2+}. In this manner, vanadium has been incorporated into silicalite and other framework types by hydrothermal synthesis, and the chemical properties of these materials have been reviewed. *(142)* These materials are easily oxidized to redox-active V^{5+}-silicates via calcination in air.

The syntheses of VS-1 *(49)*, VS-2 *(51)*, V-BEA *(61)*, V-MTW *(54)*, V-MCM-22 *(143)*, V-MCM-41 *(57,144)*, and V-UTD-1 *(145)* have been reported, and these materials have been characterized. The chemical nature and the reactivity of these framework vanadium-containing silicates depend on the valence and coordination environment of the vanadium species in the framework structure. Theory-based calculations *(142,146)* have indicated that vanadium(V) is stabilized in zeotype frameworks in two different tetrahedral coordinations: one is characterized by a highly stable vanadyl (V=O)

group that is tricoordinated to the silicate framework, the other is characterized by a $[V^{5+}O_{4/2}]^+$ unit balanced by a hydroxide ion hydrogen bonded to
one of the bridging V–O–Si oxygen ions and is thus an example of the above
mentioned anion exchanger. Tetrahedral vanadium(IV) sites, although
much less favorable, are also present, with three V–O–Si linkages to the silicate framework and the fourth coordination site occupied by an acidic
hydroxyl group. Water does not affect the coordination of vanadium and
remains in the center of the zeolitic cavity. These results are fully consistent
with experimental data and allow the identification of the molecular structures of vanadium sites in zeolite-like frameworks.

In summary, vanadium-containing silica-based molecular sieves have
been synthesized under conditions similar to those applied for preparation
of the analogous titanium-containing materials. However, the syntheses
are significantly more difficult with vanadium(V) because of the high solubility of vanadium oxoanions in the hydroxide-rich solutions used to synthesize silicate frameworks and the fact that V^{5+} incorporation leads to
positively charged frameworks, which are less stable. The latter problem
has been addressed by using cationic VO^{2+} for incorporation during synthesis; the V^{4+} can be subsequently oxidized to redox-active V^{5+} by calcination.
It has been shown that the maximum amount of vanadium that can be incorporated into the framework is approximately 2.5 wt%, corresponding to a
$V/(Si + V)$ atomic ratio of 0.028. At ratios exceeding this limit, dispersed
vanadium oxide particles are formed. While the vanadium-containing
molecular sieves are scientifically interesting materials, they have limited
promise as catalysts for use in aqueous phase because of the propensity of
the vanadium to leach out of the framework.

2.4. Tin-containing zeotype and ordered mesoporous materials

Although tin is an important element in both adsorbent applications and
catalysis science, procedures for the incorporation of Sn^{4+} into zeotype
structures at framework positions were not readily evident (147). The
postsynthesis modification of zeolites using SnF_4 or $SnCl_4$ was attempted,
but the crystalline structures of the zeolites were damaged (4, 148). However, Mal et al. (66a) incorporated Sn^{4+} into silicalite (MFI) and silicalite-2
(MEL) structures. Tin incorporation into the frameworks of MTW
(66a, 149), beta (68), and MCM-41 (150) has also been accomplished.

The discovery of Sn-beta, which has unique acidic properties,
allowed the demonstration of new selective catalytic transformations.

The new catalytic chemistries include Meerwein–Ponndorf–Verley, BV, and Oppenauer oxidation reactions, and more recently, glucose isomerization. These catalytic reactions stand out because of exceptional selectivity. Details are discussed in Section 3.5 of this review.

The synthesis of Sn–beta *(68,151)* with isolated single tin sites was accomplished in a fluoride medium: TEOS was hydrolyzed in a stirred aqueous solution of TEAOH. Then a solution of $SnCl_4 \cdot 5H_2O$ in water was added and the mixture stirred until the ethanol formed by hydrolysis of the TEOS had evaporated. HF was added to the resulting clear solution, and a thick paste formed. Then a suspension of nanocrystalline seeds (20 nm in diameter) of dealuminated zeolite beta in water was added. The crystallization was carried out in a Teflon-lined stainless-steel autoclave, which was heated to a temperature of 140 °C and continuously rotated for 20 days. After drying and calcining of the resultant solid, XRD showed that a highly crystalline material with BEA structure was obtained, which contained 1.6 wt% Sn according to chemical analysis. Further characterization by XRD and UV, IR, and ^{119}Sn MAS NMR spectroscopies verified the structure and the isomorphous substitution of silicon by tin.

Bare *et al. (152)* investigated the structure of Sn–beta by using Sn K–edge EXAFS spectroscopy to better understand the uniqueness of the tin sites and thereby to try to explain why the catalyst exhibits high selectivity (sometimes termed "enzyme-like") in the Meerwein–Ponndorf–Verley, BV, and Oppenauer oxidation reactions *(68)*. A schematic representation of the Sn–beta structure derived from EXAFS data is shown in Figure 1.3. It could be demonstrated that the tin atoms do not randomly insert into the framework of the beta structure. Instead, the only tin distribution that is fully consistent with the EXAFS data is one where pairs of tin atoms occupy opposite vertexes of the six-membered rings in the BEA structure. Figure 1.3 shows a pair of T5 sites (highlighted in red), with the required 5.1 Å separation, representing one possible tin pair within the beta structure.

Li *et al. (69)* reported the preparation of nanosized Sn–beta particles by a postsynthesis modification. Highly dealuminated beta zeolite underwent a gas–solid reaction with $SnCl_4$ vapor at elevated temperature. The properties of the resultant Sn–beta sample were characterized by various techniques. It was inferred that the tin species were inserted into the framework via reaction of $SnCl_4$ with silanol groups in hydroxyl nests that had been created during the dealumination. The tin thus predominately occupied sites with tetrahedral coordination. The tin content that could be achieved by this postsynthesis method was as high as 6.2 wt%, corresponding to a Sn/(Si+Sn) ratio of 0.034.

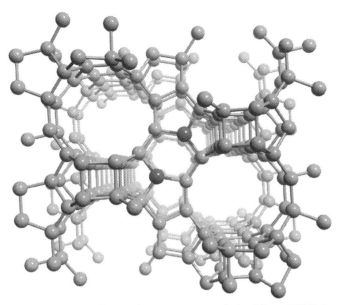

Figure 1.3 Representation of the Sn-beta structure as derived from EXAFS data, viewed along the *b*-axis (for clarity the oxygen atoms are not shown). The only Sn distribution consistent with the experimental EXAFS data is one where pairs of Sn atoms occupy opposite vertices of the six-member rings. A pair of T5 sites (red), with the required 5.1 Å separation is shown, representing one possible tin pair within the beta structure. This tin pair distorts two of the 12-membered ring channels as viewed from the [100] direction and all four 12-membered ring channels as viewed from the [010] direction. This distortion is either direct by the replacement of silicon by tin or by the expansion of the neighboring SiO₄ tetrahedra. *Reproduced with permission from Ref. (152) Copyright 2005, The American Chemical Society. (See the color plate.)*

Mal *et al.* *(66a)* obtained aluminum-free Sn–MTW by primary synthesis. The authors characterized the material by XRD, adsorption, and FTIR and Sn–NMR spectroscopies. In this material, SnO_x units appeared to be linked with SiO_4 tetrahedra through a common edge, and the tin sites can be described as structural defects. The authors used the material to investigate the catalytic oxidation of one- and two-ring aromatics with aqueous hydrogen peroxide as the oxidant. Both hydroxylation of the aromatic ring and oxidation of methyl substituents were observed. The product distributions demonstrated that the Sn–MTW-silicalite is generally capable of catalyzing the oxidation of bulky molecules such as naphthalene and 2-methylnaphthalene.

Sn–MCM–41 with tin in the framework has also been synthesized *(150,153)*. The material was found to catalyze a number of BV oxidation reactions. The activity for these transformations was not ascribed solely to the framework tin sites; it was postulated that an associated "basic" oxygen that stabilizes the transition states through hydrogen bonding is also required.

Corma *et al. (154)* synthesized Sn-MCM-41 grafted onto mesoporous MCM-41 by using various R_nSnX_{4-n} precursors, where R represents an alkyl group and X a halide. The catalyst was only active for BV oxidation reactions after careful calcination to remove the alkyl groups. Both the Lewis acidity and the catalytic performance of the grafted Sn(IV) centers in BV oxidation reactions were found to be quite similar to the corresponding properties of tin centers in directly synthesized Sn-MCM-41, but the Lewis acidity (as measured by the position of the carbonyl stretching frequency of adsorbed cyclohexanone) and activity were lower than those of reference Sn-beta catalysts on an equivalent-weight basis.

In summary, the incorporation of tin into framework positions of various molecular sieves was a breakthrough, as the incorporated tin sites constitute Lewis acid sites that are of a particular strength and are in a particular environment. These Lewis acid sites have been found to activate a number of reactants, particularly those with a carbonyl group, and can serve as catalytic sites for the associated transformations.

2.5. Iron-containing zeotype materials

A large number of papers have reported the synthesis of framework iron-containing zeotype materials. The syntheses of Fe-MFI *(82,155)*, Fe-beta *(84)*, Fe-MTT *(156)*, Fe-MOR *(157)*, Fe-TNU-9/-10 *(158)*, Fe-MCM-41 *(159)*, and Fe-SBA-15 *(160)* have been described, and the properties of these materials have been characterized.

Iron can be introduced into the frameworks of zeotype materials during their hydrothermal syntheses. Iron isomorphously substitutes for some silicon atoms in the framework if this route is chosen. The Si, Al, and Fe contents depend on the composition of the synthesis gel. The nature and distribution of the iron species in a particular sample strongly depend on the activation treatments that are applied after the synthesis.

There are several papers dedicated to the characterization of iron-containing zeotype materials, with an emphasis on determining the locations of the iron. For example, in an investigation of Fe-MFI, Milanesio *et al. (161)* used synchrotron X-ray powder diffraction in an attempt to determine the location of the Fe(III) in the MFI lattice. The data provided "a rather convincing indication" of the preferential location of iron at T9 and T10 sites (Figure 1.2). However, these results contradicted an earlier investigation by Hijar et al. (115b), who concluded that iron preferentially occupies the T8 sites. Calis et al. (162) and Fejes et al. (65) inferred, on the basis of ^{57}Fe Mössbauer, EPR, and ^{29}Si MAS NMR (162) spectra, that the fraction of Fe^{3+} incorporated at tetrahedral sites in the framework after calcination is limited. The iron framework positions and the formation of α-sites, defined as the oxygen species created when Fe-ZSM-5 reacts with N_2O during an activation procedure, have been discussed (163).

There have also been a few reports of the incorporation of iron into mesoporous materials. Huang et al. discussed the synthesis of Fe-SBA-15, whereby the iron was incorporated by incipient wetness impregnation following the hydrothermal synthesis of the SBA-15 (164). Jiang et al. synthesized Fe-MCM-41 using both direct hydrothermal and postsynthesis methods and claimed that tetrahedral Fe species are present in the MCM-41 framework (159); and Zhao et al. reported on the synthesis of Fe-MCM-48 using a mixed surfactant method (165).

There is still a dispute as to whether the catalytic activity of iron-containing zeotype materials, for example, Fe-ZSM-5, should be attributed to isomorphously substituted framework iron or to extra-framework iron oxide or iron hydroxide species that are highly dispersed in the material. These extra-framework iron species are present for two reasons, either because they were not incorporated into the framework during the synthesis or because they were ejected from the framework during postsynthesis treatments (such as calcination or other heat treatments). The unresolved issue of the origin of catalytic activity continues to be the subject of research, whereby state-of-the-art characterization techniques are being applied.

2.6. Gallium-containing zeotype materials

The success of syntheses designed to incorporate gallium into zeotype frameworks depends on the synthesis conditions, as is true for other framework metal-containing zeotype materials. Several researchers have claimed gallium incorporation into the frameworks of zeotype structures (5b, 166).

Ga-ZSM-5 that is synthesized (167) in the presence of methylamine is characterized by a uniform distribution of gallium throughout the framework. When both aluminum and gallium are present in the gel precursor, mixed complexes with methylamine are formed that are incorporated into the crystal lattice at different rates, while some unreacted gallium amine complex forms a coating on the outer surfaces of the metallosilicate crystals. Gallosilicates synthesized in fluoride media are characterized by a fairly homogeneous gallium incorporation, with an increased gallium concentration on the surface because of the formation of an overlayer from residual gallium–fluoro complexes in the synthesis medium (167).

Kofke et al. (168) determined the concentration of gallium in the framework of MFI-type materials by adsorption methods. Klik et al. (76) investigated the gallium incorporation by using high-field ^{71}Ga MAS NMR spectroscopy, showing that gallium preferentially occupies the framework positions in Ga-MFI materials that are prepared by direct synthesis from a gel.

Gabelica et al. (167) reported maximum incorporation of four gallium atoms per MFI unit cell, corresponding to a ratio of $Ga/(Ga+Si)$ of 0.041. Postsynthesis thermal treatments result in a partial degalliation of the framework. The nature, mobility, and location of the extra-framework gallium species depend markedly on the calcination conditions. During treatment in humid air, the homogeneous distribution of both Ga_2O_3 and residual Ga^{3+} framework ions is maintained.

In summary, the incorporation of gallium into zeotype materials, especially into the MFI structure, can be achieved by regular hydrothermal syntheses as well as by syntheses in fluoride media. Clear evidence has been presented that when incorporated during synthesis, gallium occupies framework positions. High gallium concentrations in the synthesis medium result in an overcoat of gallium species on the crystal surface. Extensive characterization has shown that the ratio of framework to non-framework gallium is sensitive to both the synthesis and calcination conditions. Degalliation of the framework leads to the formation of gallium oxide. The stability of framework gallium during catalysis depends on both the testing and regeneration conditions.

2.7. Further metal framework-containing zeotype or ordered mesoporous materials

In addition to the elements discussed above, there have been claims in the literature for the framework incorporation of other elements, viz germanium (145,169), chromium, copper, zirconium, and zinc. However, except for germanium, the incorporation of the metals into the framework is viewed critically by the authors. To the best of our knowledge, none of these materials have found commercial application, and they are not reviewed further.

3. CATALYTIC REACTIONS

3.1. Types of catalytic transformations grouped by nature of metal

In this section, the catalytic chemistry of selected framework metal-containing zeotype materials is reviewed, with an emphasis on commercial applications. The catalytic activities of framework metal-containing zeotype materials, especially those containing titanium, vanadium, or tin, have been investigated extensively. The enormous interest in these materials is attributed to their remarkable catalytic activities and especially their selectivities in oxidation reactions. Because hydrogen peroxide is generally used as the oxidant, water is formed as a by-product. Hence, oxidation reactions carried out with these catalysts can be considered environmentally clean processes. Several review articles have been published that summarize the catalytic reactions *(2a,3b–d,89)*. In this section, the focus is on selected industrially relevant reactions.

A review of the extensive literature indicates a diverse array of reactions catalyzed by framework metal-containing zeotype materials. The key reactions are illustrated in Figure 1.4.

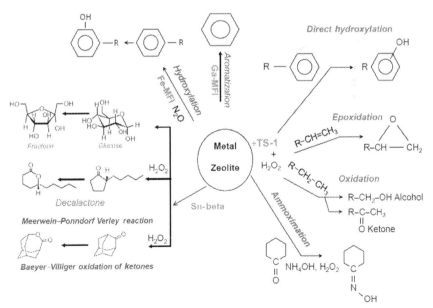

Figure 1.4 Oxidation reactions catalyzed by various metal-containing zeotype materials, with oxidizing agents specified.

The principal reactions for each of the framework metal-containing zeotype materials are the following:

Titanium-containing zeotype materials

- Epoxidation of olefins (e.g., of propylene)
- Hydroxylation of aromatics (e.g., of benzene and phenol)
- Ammoximation of ketones (e.g., of cyclohexanone) to oximes
- Oxidation of alkanes to alcohols and ketones
- Desulfurization

Tin-containing zeotype materials

- BV oxidation (e.g., of adamantanone)
- Meerwein–Ponndorf–Verley reaction
- Isomerization of sugars (e.g., glucose)

Gallium-containing zeotype materials

- Aromatization of hydrocarbons

Iron-containing zeotype materials

- Benzene hydroxylation with N_2O

Furthermore, the following reactions catalyzed by framework metal-containing zeotype materials have been reported; however, there are currently no major commercial applications:

- Oxidation of alkanes to alcohols and ketones with hydrogen peroxide *(52)*
- Oxidation of amines *(170)*
- Oxidation of sulfur-containing compounds *(171)*
- Hydroxyl-assisted chemically and stereochemically selective oxidations *(172)*
- Oxidative cleavage of hydrazones to carbonyl compounds *(173)*
- Cycloaddition and transesterification reactions to give polycarbonate precursors *(174)*
- Transesterification *(175)*

3.2. Epoxidation of olefins

3.2.1 Overview: Oxidants and effect of olefin constitution

Nucleophilic oxygen-containing compounds, for example peroxides, are efficient reagents for the epoxidation of carbon–carbon double bonds. Most epoxides are generated by treating alkenes with peroxide-containing reagents, which donate a single oxygen atom. Typical peroxide reagents include hydrogen peroxide, peroxycarboxylic acids, and alkyl hydroperoxides. A large number of papers have described the epoxidation of olefins with hydrogen peroxide or organic peroxide using framework metal-containing zeotype materials such as TS-1 *(176)*, TS-2 *(177)*, Ti-beta *(178)*, Ti-MWW

(41b), Ti-MCM-41 *(179)*, and Ti-MCM-36 *(180)*. Saxton *et al.* of ARCO have also specifically patented *(181)* the use of titanium catalysts for 1-hexene epoxidation.

Clerici *et al.* *(176a)* investigated the epoxidation of olefins (with chain length longer than C3) catalyzed by titanium silicalite (TS-1) under mild conditions. The authors performed the reaction at near room temperature in diluted alcoholic or aqueous solution of hydrogen peroxide. In methanol, the linear olefins (C4–C8) and allyl chloride and allyl alcohol showed fast reaction rates and high selectivities to epoxides. By-products originated from solvolysis of the oxirane ring or oxidation of the solvent. The yield and kinetic rate decreased with increasing chain length of the olefin. The rate of epoxidation also depended on the position of the double bond and on branching. The catalytic activity of TS-1 decreased when base was added, whereas it increased when acids were added.

3.2.2 Propylene oxide

3.2.2.1 Epoxidation of propylene to propylene oxide with hydrogen peroxide

Following the invention of TS-1, Clerici and Romano of Eniricerche filed a patent for the epoxidation of propylene with hydrogen peroxide *(182)*. Propylene is selectively epoxidized to propylene oxide (PO) in the presence of a solvent, and all by-products are secondary products formed through ring opening reactions. For example, propylene glycol and its monomethyl ethers and trace amounts of formaldehyde were reported as by-products. No allylic oxidation products were detected. Methanol was identified as the preferred solvent. The reported selectivity to PO is about 98% at high hydrogen peroxide conversion (95%).

A number of reports have sought to address the issues of selectivity and deactivation. Clerici *et al.* investigated *(176b)* the deactivation of TS-1 by repeatedly using the same sample in batch experiments. The authors stated that the catalytic activity decreased with time of usage, while the yield and product distribution were nearly stable. They found that the catalyst could be regenerated by calcination in air or a solvent wash. The deactivation was postulated to be caused by slowly diffusing by-products, which block the channels and hinder access to the sites. The selectivity was further improved, and the hydrolysis of the desired epoxides to the troublesome by-products was almost suppressed, when the residual acidity of the catalyst was completely eliminated. Wang *et al.* *(183)*, in an investigation focused on regeneration, arrived at a conclusion similar to that of Clerici *et al.* The deactivation of the TS-1 catalyst for epoxidation was ascribed to blocking of micropores by bulky organic by-products, which are formed through secondary reactions of the target product, for example, through dimerization or oligomerization of PO.

There are various recent developments in this area. Forlin et al. (184), Goebbel et al. (185), and Hofen and Thiele (186) developed continuous epoxidation processes with hydrogen peroxide. The incorporation of an additional metallic element improved the activity of TS-1, as demonstrated with doubly-substituted silicalites containing the combinations titanium–vanadium (131) or titanium–tin (187). The hydrophobicity of the catalysts was found to be important for the epoxidation activity (188). The preparation of extruded catalysts based on TS-1 for application in propylene epoxidation (26a) was described, and titanosilicate beads with hierarchical porosity (99a) were investigated.

The mechanism of epoxidation of propylene to PO with H_2O_2 on TS-1 has been thoroughly investigated and is thought to be well understood. The epoxidation of the olefin was reported to occur by a coordinative mechanism (189). In early mechanistic considerations, it was proposed that the oxidant species is a titanium peroxide in which the peroxo oxygens bind in a side-on mode to titanium, thus resulting in a Ti (η^2-O_2) configuration (as illustrated in Figure 1.5A, first structure). A second proposed mechanism is based on the formation of a titanium hydroperoxide, Ti–OOH, as the active species, with either an η^2 or η^1 configuration (190) (Figure 1.5A, second and third structures, respectively), or an η^2-coordinated –OOH stabilized by coordination of one molecule of the alcoholic solvent (Figure 1.5A, fourth structure). While the principal idea of formation and relevance of a Ti–OOH species received nearly unanimous consent, many disputes followed regarding specific details of the structure—for example, it was debated whether the hydroperoxide is η^2- or η^1-coordinated, and whether a protic molecule coordinates to atoms of the hydroperoxide and simultaneously to the titanium. In the original proposal, the hydroperoxide group is monodentate and is stabilized in a cyclic structure by the co-coordination of a protic molecule, generally water or alcohol. Reaction schemes for the epoxidation of propylene involving the Ti(η^1–OOH) (Figure 1.5B) and the Ti(η^2–OOH) and the Ti(η^2–OOH)(ROH) active species (Figure 1.5C) are also shown.

The TS-1-catalyzed propylene epoxidation has attracted significant interest from groups working on theoretical modeling, and numerous papers address the mechanism of epoxidation of propylene (191). Wells et al. (191b)

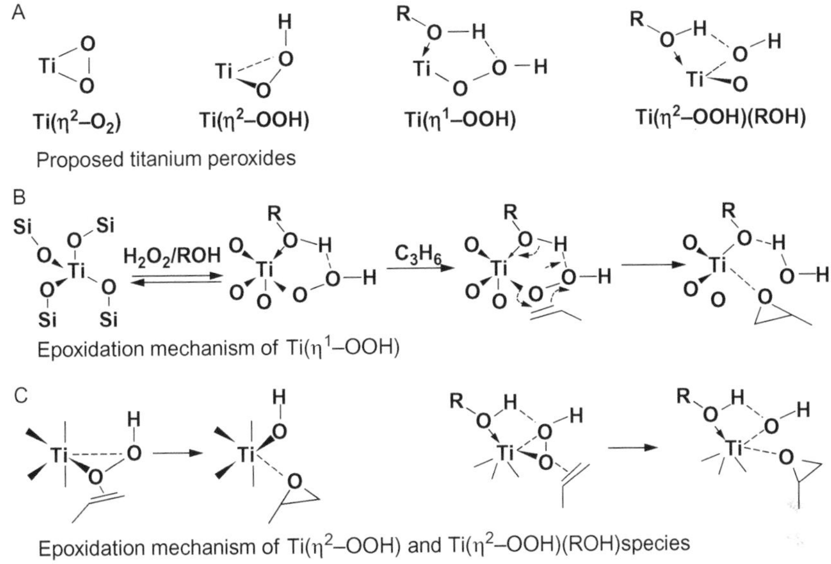

Figure 1.5 Structure and reactivity of various proposed titanium peroxides in TS-1. (A) Proposed titanium peroxide structures. (B and C) Mechanisms for the epoxidation of propylene formulated with three of the structures presented in A.

compared the following proposed mechanisms for the TS-1 catalyzed propylene epoxidation with H_2O_2 by using a consistent DFT method:

- Sinclair and Catlow *(192)* mechanism on tripodal sites through Ti–OOH species
- Vayssilov and van Santen *(193)* mechanism on tetrapodal site without Ti–OOH formation
- Munakata *et al. (191g)* mechanism involving peroxy (Ti–O–O–Si) species
- Defect site mechanism with a partial silanol nest (a titanium site with an adjacent Si–OH neighboring group)
- Defect site mechanism with a full silanol nest (i.e., three adjacent Si–OH groups; a silanol nest "heals" a Si defect site in the silicalite structure by replacing the missing Si atom with OH terminating groups, see Figure 1.6B.)

The use of a consistent level of theory allowed, for the first time, a meaningful comparison of the energetics associated with each of these pathways. The authors rigorously identified the important reaction intermediates and transition states and carried out a detailed thermochemical analysis. A summary of transition-state parameters for all of the mechanisms proposed

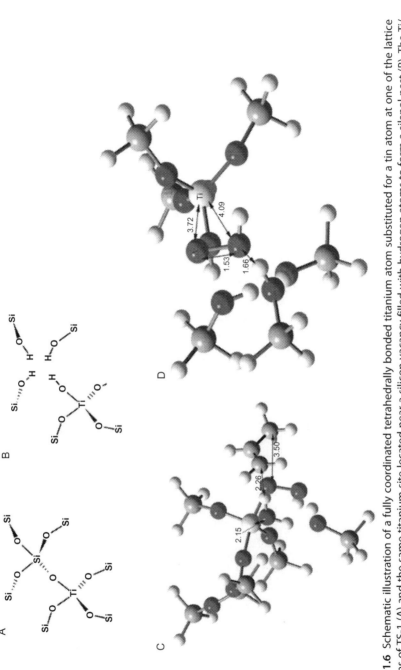

Figure 1.6 Schematic illustration of a fully coordinated tetrahedrally bonded titanium atom substituted for a tin atom at one of the lattice positions of TS-1 (A) and the same titanium site located near a silicon vacancy filled with hydrogen atoms to form a silanol nest (B). The Ti/defect mechanism for the partial silanol nest model showing the preadsorbed complex of propylene on the hydroperoxy intermediate (C) and the Ti/defect mechanism for the full silanol nest model showing the preadsorbed complex of H_2O_2 on the titanium site. Distances in Å. Color coding: small white spheres, H atoms; red spheres, O atoms; gray spheres, C atoms; large white spheres, Ti atoms; green spheres, Si atoms. *Adapted from Ref.* (191b), *with permission from The American Chemical Society.* (See the color plate.)

for the epoxidation step is given in Table 1.8, and two of the key structures are displayed in Figure 1.6C and D.

On the basis of the Gibbs energy of activation, the Sinclair and Catlow mechanism is the energetically most favorable mechanism for titanium sites on the *external* surface of TS-1 because of the tripodal nature of the titanium site in the model. The defect site mechanism involving a full silanol nest (proposed by Wells *et al. (191b)*) is characterized by an even lower Gibbs energy of activation and competes with the Sinclair and Catlow mechanism. The defect site mechanism likely represents the chemistry occurring inside the TS-1 pores in the liquid-phase epoxidation (H_2O_2/TS-1) process. If *only* the propylene epoxidation step is considered, then the Munakata peroxo intermediate (Si–O–O–Ti) is the most reactive intermediate, which can epoxidize propylene with a negligible activation barrier. However, formation of the Munakata intermediate is a highly activated step with $\Delta G_{act} = 19.8$ kcal/mol. Delgass and coworkers explained variations in the activation barriers for the individual mechanisms that arise from geometric and electronic parameters such as the relative orientation of adsorbed H_2O_2 and propylene, hydrogen bonding, O1—Ti bond distance in the Ti–O1–O2-H intermediate, and O1–O2 stretching in the transition state.

Table 1.8 Comparison of transition-state parameters for various mechanisms proposed for the epoxidation step in the synthesis of propylene epoxide on TS-1

| Mechanism | TS energy barrier (kcal/mol) | Octahedral complex[a] | | | Nearest H-bond | | Nearest |
		O1–Ti	O2–Ti	O–O	O	H	O–C
Munakata	Negligible	1.97	1.99	1.65	NA	NA	2.62
Defect (full nest)	4.62	1.88	2.12	1.83	1.53	1.27	2.12
Sinclair and Catlow	7.91	1.99	2.17	1.72	2.45	2.83	2.32
Defect (partial nest)	14.26	2.11	2.21	1.69	1.57	2.45	2.37
Vayssilov and van Santen	19.16	3.71	3.95	2.04	2.03	1.67	1.65
Gas phase (noncatalytic)	22.40	NA	NA	2.00	NA	1.75	1.75

[a]Note that the hydroperoxy intermediate is represented as Ti–O1–O2-H. Therefore, O1 is the proximal (to the Ti atom) oxygen atom and O2 is the distal oxygen atom.
Reprinted (adapted) with permission from *(191b)*. Copyright (2006) American Chemical Society.

Implications of various titanium site models for the activation barrier are also discussed, with consideration of external and internal, and nondefect and defect sites on TS-1 *(191a,c–f)*.

In summary, the TS-1 catalyzed epoxidation of propylene with H_2O_2 to PO is a thoroughly investigated epoxidation reaction. The oxidation chemistry is well known by now, and the catalyst, and the process parameters have been optimized. Currently, the titanium-based catalyst is in the process of being commercialized. Computational chemistry has been applied to elucidate the details of the surface chemistry and to identify the important reaction steps; at this point, various competing mechanisms, proposed by different authors, can be found in the literature.

3.2.2.2 Epoxidation of propylene to PO with organic hydroperoxides

The epoxidation of propylene to PO using organic hydroperoxides instead of H_2O_2 as oxidants is also facile, but a complication in this case is the coproduct formation, that is, the formation of stoichiometric amounts of the reduced form of the oxidant. *Tert*-butyl hydroperoxide (TBHP) *(194)*, ethylbenzene hydroperoxide *(195)*, and cumene hydroperoxide (CHP, systematic name: hydroperoxide, 1-methyl-1-phenylethyl) *(196)* are the most frequently used organic hydroperoxides for the oxidation of propylene to PO.

The discovery of mesoporous MCM-41 has opened up new avenues for the oxidation of larger substrates and the use of bulkier oxidants. The cumene hydroperoxide-based chemistry is characterized by a high selectivity and, consequently, very low by-product formation. The principal steps of the selective oxidation of propylene with cumene hydroperoxide are shown in Figure 1.7. When using mesoporous silicates with isomorphous titanium substitution, alkylaromatic hydroperoxides such as CHP become the more attractive oxidants than TBHP because of the ease of reuse.

Several framework titanium-substituted mesoporous silicates, including Ti-MCM-41 *(42,43)*, Ti-HMS *(198)*, Ti-MCM-36 *(180)*, Ti-MCM-48 *(199)*, and Ti-SBA-15 *(200)*, have shown promising activity for the epoxidation of bulky olefins with alkyl hydroperoxides as oxidants. Unfortunately, compared with the microporous MFI-type titanium silicates, the mesoporous materials exhibit low activity for epoxidation reactions. The hydrophilic nature of mesoporous silica catalysts with isomorphous titanium substitution is considered to be one of the major reasons for the low activity *(179)*. Various attempts have been made to improve the activity. Using a different synthetic procedure, titanium species have been grafted onto

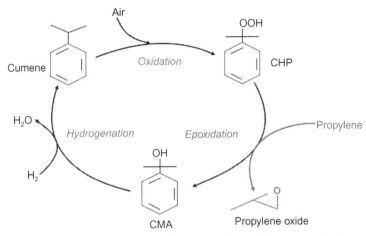

Figure 1.7 Propylene oxide formation from propylene with cumene hydroperoxide (CHP) as a recyclable oxidizing agent. After hydrogenolysis of the alcohol (which is the reduction product) to cumene, cumene hydroperoxide is generated again through oxidation of the cumene in air. *Adapted from Ref. (197a), with permission from the Royal Society of Chemistry.*

MCM-41 to generate more accessible titanium sites for selective epoxidation *(201)*. In another approach, a modified synthesis procedure was developed to obtain Ti-MCM-41 with high titanium content *(202)*. Indeed, the activity (on a per catalyst weight basis) for the epoxidation of olefins was improved significantly as a result of incorporation of a large amount of titanium *(202)*. The catalytic epoxidation requires nonaqueous oxidants to obtain satisfactory activity when using regular MCM-type materials, because MCM-41 has a limited stability in aqueous media. A number of recent reports *(43, 125a)* have described the advantages of silylation or other organic functionalization of the surface of Ti-MCM-41 materials; the functionalization makes the materials more hydrophobic and improves their activity for epoxidation.

Corma *et al. (203)* used high-throughput methods and genetic algorithms assisted by artificial neural networks to optimize the olefin epoxidation catalyst. This work is an excellent example of the use of state-of-the-art experimental techniques to design a high-performance catalyst. The strategy for the experimental design made it possible to explore the multidimensional space and optimize the catalyst properties and the reaction conditions for the highest epoxidation activity. This approach allowed the authors to simultaneously investigate multiple synthesis variables (e.g., concentration of OH$^-$, titanium, and surfactant) and their resulting nonlinear correlation to catalyst performance. High throughput synthesis and catalytic testing were accompanied by conventional characterization (e.g., XRD and diffuse reflectance UV spectroscopy), allowing the researchers to relate the catalytic results to material properties like titanium coordination, nature of the active sites, and hydrophobicity of the catalyst surface. Through clustering and factorial analysis of the multiparametric results, the gel composition could be correlated with catalytic performance, titanium silicate structure, and coordination of the titanium sites. The mesoporous molecular sieves Ti–MCM–41 and MCM–48 were among the most active catalysts. The best performing catalysts were tested for epoxidation of various linear olefins.

In conclusion, the epoxidation of propylene with bulky oxidants (such as cumene or TBHP) can be successfully achieved using titanium-containing mesoporous materials as catalysts. The catalytic chemistry of the active sites can be controlled via the synthesis conditions and postsynthesis modifications. The hydrophobicity of the catalyst is of great importance to achieve a highly selective catalyst. The Ti–MCM–41–based heterogeneous catalyst has demonstrated excellent performance in the commercial process for PO manufacture.

3.3. Hydroxylation of aromatic compounds
3.3.1 Overview: Target products and oxidants
Hydroxylation of aromatics to products such as phenol, hydroquinone, or catechol in a by-product-free, environmentally clean process is important in the fine chemicals industry. The selective partial oxidation of hydrocarbons has always been a challenging field of catalytic chemistry; the major challenge is the discovery of selective catalysts. Such catalysts should ideally perform the dual functions of activating the oxidant to generate a reactive catalyst-bound oxygen-containing species and activating the organic reactant appropriately so as to direct the reaction to the desired products. A variety of oxygen sources including O_2, H_2O_2, O_3, and N_2O are used. A reaction scheme for the hydroxylation of aromatics using framework metal-containing zeotype materials as catalysts is shown in Figure 1.8.

3.3.2 Oxidation of benzene
In the commercially relevant reports of benzene oxidation catalyzed by framework metal-containing zeotype materials hydrogen peroxide or nitrous oxide (N_2O) are used as oxidants. Thus phenol is produced without major by-product formation.

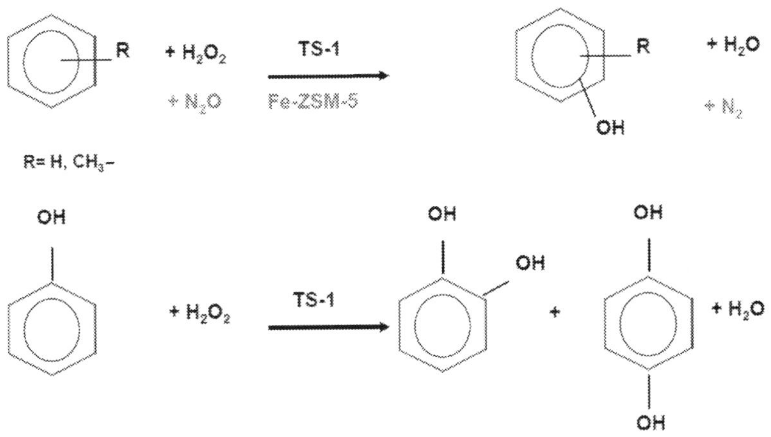

Figure 1.8 Hydroxylation of aromatic compounds with various combinations of catalyst and oxidation agent: TS-1 in combination with H_2O_2 or Fe-ZSM-5 in combination with N_2O.

3.3.2.1 Oxidation of benzene with hydrogen peroxide

Titanium silicalites (e.g., TS-1) catalyze the direct hydroxylation of aromatics with aqueous hydrogen peroxide as the oxidant. For example, hydrogen peroxide and TS-1 have been investigated for hydroxylation of benzene (87,204) to make phenol. The reactions were carried out with various cosolvents to homogenize the hydrophobic substrate and the aqueous hydrogen peroxide in the liquid phase. Titanium silicalite (made by secondary synthesis) (205), Ti-mordenite (206), and vanadium-beta (207) have been reported to be efficient catalysts. The benzene conversion, phenol selectivity, and hydrogen peroxide efficiency (defined as amount substrate hydroxylated (mol)/total amount of reacted H_2O_2 (mol), expressed as percentage) are usually moderate, and all in the same range of 80–95%. Potassium-exchanged titanium silicalite obtained by secondary synthesis is more active and selective for phenol formation than TS-1 made by primary synthesis.

3.3.2.2 Oxidation of Benzene with N_2O

Phenol can be produced directly from benzene by use of N_2O as an oxidizing agent in the gas phase. Catalysts are modified ZSM-5 or ZSM-11 materials, containing elements such as antimony, arsenic, beryllium, boron, cobalt, chromium, copper, gallium, indium, iron, nickel, scandium, vanadium or zinc. (208)

Fe-MFI is among the reported (209) catalysts for the oxidation of benzene to phenol with reactants in the gas phase. At a benzene conversion of 38%, the selectivity to phenol was 98%; these values are dependent on the catalyst and reaction conditions. The properties of Fe-beta *(210)*, Fe-TNU-9, Fe-TNU-10, and Fe-IM-5 *(158)* have also been investigated.

Zeotype materials containing a low iron concentration can be synthesized by hydrothermal synthesis and by ion exchange. The hydrothermally incorporated iron may partially migrate from the framework into extra-framework locations; this process can be induced by controlled heat treatment *(155a,211)*. Probably only a small fraction of the iron is active in the selective oxidation of benzene to phenol when N_2O is used as the oxidant. The active catalytic material is thought to consist of isolated iron ions in a pseudooctahedral configuration. These sites are positioned in hydroxyl nests (i.e., defect sites in the framework) and are selective for phenol formation as a result of *in situ* reduction during catalysis. Catalysts with iron incorporated during hydrothermal synthesis are characterized by lower rates of deactivation than those with iron introduced by ion exchange. However, the ion-exchanged catalysts have phenol productivities (related to catalyst mass) comparable to those of hydrothermally synthesized materials although the amount of Fe in extra-framework positions is about 20–30 × lower than in the catalysts with iron introduced during hydrothermal synthesis. The presence of aluminum in the zeolite framework is also beneficial for reducing the rate of deactivation *(209b)*.

The possible formation and activity of α-oxygen, defined as the active oxygen that is formed when N_2O reacts with ZSM-5 *(212)*, in Fe-MFI was proposed by Yang *et al.* *(213)* On the basis of DFT calculations, Fellah *et al.* suggested that phenol formation on binuclear (Fe (μ–O) Fe)$^{2+}$ sites in ZSM-5 is facilitated *(214)* in the presence of water. These computational findings are consistent with experimental observations and allow a rationalization of the findings at a molecular level.

3.3.3 Oxidation of phenol with hydrogen peroxide

The activation of phenol is much easier than that of benzene because of the high aromaticity of benzene and the corresponding charge delocalization. TS-1 has been used primarily as the catalyst for phenol hydroxylation, and the conditions have been thoroughly optimized *(215)*. TS-2 *(216)* and Ti-MCM-41 *(217)* were also evaluated, but they do not perform as well as TS-1. Ramaswamy *et al.* *(52)* compared various framework metal-containing zeotype materials with MEL structure for phenol hydroxylation; the incorporated metals were aluminum, tin, titanium, and vanadium. The efficiency of utilization of hydrogen peroxide in the hydroxylation reaction with these catalysts decreased in the following order: TS-2 > VS-2 > SnS-2 > Al-S-2.

With TS-1 as the catalyst, the oxidation products of phenol are hydroquinone and catechol (*para*- and *ortho*-hydroxyphenol), with minor yields of water and tar formed as by-products. Numerous early papers are concerned with this reaction *(218)*, and patents *(219)* have been filed. In the reaction catalyzed by TS-1, the conversion of phenol and the selectivity to dihydroxy products are significantly higher than achievable by either radical-initiated oxidation or acidic catalysts. The catechol/hydroquinone molar ratio is within the range of 0.5–1.3 and depends on the solvent. When the reaction occurs in aqueous acetone, the ratio is close to 1.3. It is believed that the product ratio is the result of restricted transition-state selectivity as well as mass transport shape selectivity associated with the different diffusivities of the *ortho* and *para* products. Hydroxylation at the *para*-position of phenol should be less hindered relative to that at the *ortho*-position, and hydroquinone has a smaller kinetic diameter than catechol, facilitating diffusion. Tuel and Taarit *(220)* proposed that catechol is mainly produced at the external surface of TS-1 crystals. Thus, the different catechol/hydroquinone ratios obtained when methanol or acetone is used as a solvent could be explained by either rapid or very slow poisoning of external sites by organic deposits, respectively. Accordingly, the authors were able to show that tars were easily dissolved by acetone (i.e., external sites for catechol formation remained available in this solvent) while they were insoluble in methanol.

The yield and kinetics of phenol hydroxylation are strongly dependent on the purity, crystal size, and the concentration of TS-1, and on the temperature. High selectivity is directly correlated to framework titanium, whereas extra-framework titanium species initiate the decomposition of hydrogen peroxide and subsequent hydroxylation paths, which are far less selective.

The effect of incorporating tin into titanium silicalite-1 (TS-1) on the kinetics of phenol hydroxylation to dihydroxybenzenes with aqueous hydrogen peroxide has been investigated *(197)*. The pathways are illustrated in Figure 1.9. The hydroxylation reaction was modeled using the results obtained with a batch reactor, whereby mass transfer limitations were carefully excluded. The analysis of the kinetics indicated that under the same reaction conditions, titanium–tin silicalite-1 (Ti–Sn-S-1) gave a higher phenol conversion rate than TS-1. This difference was attributed to the presence of active tin sites. The incorporation of tin influences the initiation of

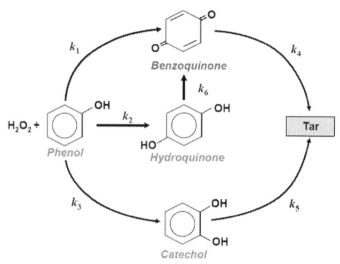

Figure 1.9 Reaction pathway for phenol hydroxylation with H_2O_2 as the oxidizing agent and TS-1 as the catalyst. The relative rate constants characterizing product (hydroquinone or catechol) formation and by-product (benzoquinone) formation and secondary reactions depend on the catalyst and are discussed in the text.

subsequent reactions of the products with hydrogen peroxide. Tin increases the rate of benzoquinone conversion to tar; however, it does not affect the rates of hydroquinone or catechol conversion. A Langmuir–Hinshelwood-type model was used to fit the observed rate data accounting for the occurrence of phenol hydroxylation and parallel reactions of the products. A surface-mediated reaction model yielded the best fit for reactions in the system. The model, however, failed to predict the outcome of the catechol reaction using TS-1, in which the catechol adsorption on the titanium active site was rate-limiting.

3.4. Ammoximation of cyclohexanone to cyclohexanone oxime

The ammoximation of cyclohexanone to cyclohexanone oxime using TS-1 was invented in 1987 *(221)*. The limited amount of published work focuses on the mechanism *(222)* and the kinetics *(223)* of this reaction on titanium silicate with dilute hydrogen peroxide as oxidant. There is no agreement thus far regarding the reaction pathway. The following two hypotheses have been formulated *(7,90,222b,224)*:

(i) The iminic route: First, the ammonia reacts with the cyclohexanone to form an imine. The iminic intermediate adsorbs on the catalyst surface and reacts with the oxygen donated by the hydrogen peroxide to give the oxime. The role of titanium is to activate the oxygen, which is then inserted into the iminic intermediate. This hypothesis, which was originally proposed by Thangaraj *et al. (225)*, is the most widely accepted for vapor phase catalysis.

(ii) The hydroxylamine route, proposed by Clerici *(2f)* and depicted in Figure 1.10: This mechanism involves the formation of hydroxylamine as the key step. The hydroxylamine is formed by the reaction between ammonia and hydrogen peroxide on titanium centers. In a second step, this intermediate reacts in the homogeneous phase with cyclohexanone to give the oxime. Clerici published this mechanism for the TS-1-catalyzed ammoximation of cyclohexanone *(2f)*. This mechanism is supported by spectroscopic investigations *(89b)* that demonstrate that the titanium sites in TS-1 are able to coordinate both NH_3 and H_2O_2 in an octahedral coordination, which is the coordination generally preferred by Ti^{4+}.

Several different by-products (e.g., 2-cyclohexylidene, cyclohexanone, and 2-(1-cyclohexen-1-yl)cyclohexanone) are formed during the reaction. These by-products deposit on the catalyst surface and deactivate the catalyst.

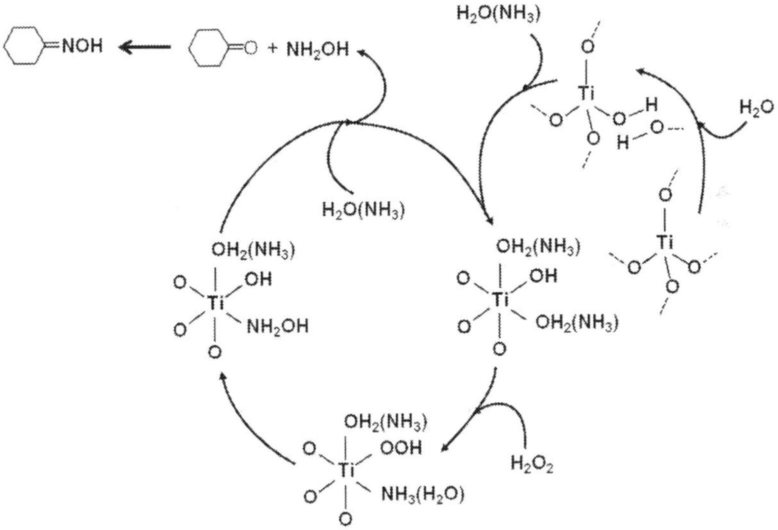

Figure 1.10 Proposed mechanism focusing on the chemistry at the active titanium site for the ammoximation reaction of cyclohexanone to cyclohexanone oxime catalyzed by TS-1. Hydroperoxide coordinated to the titanium oxidizes cocoordinated ammonia to give hydroxylamine. The hydroxylamine is released into solution and oxidizes cyclohexanone to cyclohexanone oxime. *Adapted from Ref. (2f), with permission from Enciclopedia Italiana.*

Careful analysis of the kinetics is compulsory to determine the optimized reaction conditions for such a complex network of parallel and consecutive reactions.

Several different titanium-containing zeotype structures have been investigated for the ammoximation of cyclohexanone, for example TS-1 *(221,224)*, TS-2 *(177,226)*, Ti-beta *(227)*, Ti-MOR *(228)*, Ti-MCM-41 *(229)*, and Ti-MCM-48 *(227a)*. Le Bars *et al. (227a)* evaluated zeotype materials with various pore sizes and concluded that the best performance and highest turnover in ammoximation is achieved with TS-1 as catalyst. The superiority of this material is believed to arise from its hydrophobicity, in combination with the three-dimensional framework of straight channels that facilitate the diffusion of reactants and products in the catalyst pores.

TS-1 *(230)* and TS-2 *(231)* have been tested as catalysts for the vapor phase Beckman rearrangement of cyclohexanone oxime to ε-caprolactam. However, as these framework metal-containing zeotypes are outperformed by metal-free high silica MFI *(232)*, this catalytic chemistry is not reviewed further.

3.5. Oxidations and reductions catalyzed by weak Lewis acids (with emphasis of tin-containing materials)

3.5.1 Overview of catalysts and reactions

Zeotype materials containing metal cations, for example ions of titanium, vanadium, chromium, iron, or tin, in the tetrahedral positions of their frameworks have been explored as solid Lewis acid catalysts *(1g)*. Such materials have been shown to be active in the Meerwein–Ponndorf–Verley reduction of carbonyl compounds *(151,233)*, and the BV oxidation (also called BV rearrangement) *(234)*.

For the oxidation chemistry, it is critical to selectively activate the carbonyl group, allowing the H_2O_2 to attack. Lewis acids are suitable for this task. The ideal solid catalyst would have isolated, single, uniform Lewis acid sites. A large number of potential catalysts, among them titanium-, vanadium-, chromium-, iron-, and tin-containing molecular sieves with microporous and mesoporous structures have been evaluated. The first three metals in the list are well known catalysts for activating H_2O_2 and selectively epoxidizing olefins, whereas iron primarily decomposes the H_2O_2 into H_2O and oxygen. Sn-beta has emerged as a suitable solid Lewis acid catalyst; it has been thoroughly evaluated and mechanistic investigations have been conducted.

Sn-beta *(68,151,235)* shows extraordinary chemoselectivity for the BV oxidation of unsaturated ketones using aqueous hydrogen peroxide as

Table 1.9 Comparison of the selectivities obtained for the Baeyer–Villiger oxidation of bicyclo[3.2.0]hept-2-en-6-one with various oxidants

	Reactant/ conversion/%	Products selectivity/%		
Sn-beta	>95	100	0	0
Meta-chloroperbenzoic acid (mCPBA)	>95	29	34	37
Enzymes *(236)*		100	0	0

For Sn-beta catalyst, the oxidant was hydrogen peroxide, whereas there was no catalyst for the mCPBA. Adapted from Ref. *(68)*, with permission from the Nature Publishing Group.

oxidant. Table 1.9 summarizes the selectivities toward formation of different lactones resulting from the use of Sn-beta compared with those achievable with traditional acid and enzymatic catalysts. Sn-beta was tested for the BV oxidation of adamantanone, cyclohexanone, and bicyclo [3.2.0] hept-2-en-6-one and an extraordinarily high selectivity, unique among heterogeneous catalysts, was demonstrated. The oxidation of bicycle-[3.2.0] hept-2-en-6-one using Sn-beta gave exclusively lactones, with selectivity comparable to that of an enzyme catalyst. *(68)* Another finding is that only Sn-beta was selective for the lactones, whereas Ti-beta produced epoxides, and the corresponding diols. In comparison when peracid is used as oxidant, a mixture of lactones, epoxides, and epoxylactone are formed.

The Sn-beta catalyst *(68,151,234g)* has also been synthesized, developed, and patented *(237)* for the formation of lactones from ketones and H$_2$O$_2$ via the BV rearrangement. The catalyst retains its activity and can be recycled without regeneration.

Several other oxidation reactions, such as the BV oxidation of aromatic aldehydes *(234g,238)*, and Melonal *(70)* fragrance synthesis have also been investigated and associated patents have been issued.

3.5.2 Mechanism of BV oxidation

To better understand the Sn-beta chemistry, a detailed experimental investigation into the mechanism of the BV oxidation was conducted *(68)*. It was established by IR spectroscopy that the catalytic cycle starts with the coordination and activation of the carbonyl group at the Sn–site. A suitable substrate, 2-methylcyclohexanone was labeled with oxygen-18. After the BV oxidation, labeled oxygen was found only in the carbonyl group of the lactone, and not in the ring. This result clearly demonstrated that the reaction mechanism of the BV oxidation with H_2O_2 catalyzed by Sn-beta proceeds via a "Criegee" adduct of H_2O_2 to the ketone. Therefore, the formation of dioxiranes or carbonyl oxides as intermediates can be excluded. The complete mechanistic cycle is depicted in Figure 1.11. While a coordination of the hydrogen peroxide to the tin center involving activation by deprotonation cannot be excluded there was no indication of such a Sn–O–O–H species in diffuse reflectance UV–vis spectra.

Sever and Root used DFT methods to investigate the reaction mechanism of Sn-catalyzed BV oxidation reactions (239). Their approach was to determine the mechanisms for a generic active tin center (modeled as $Sn(OH)_4$ cluster) rather than for tin in a specific zeotype structure. The

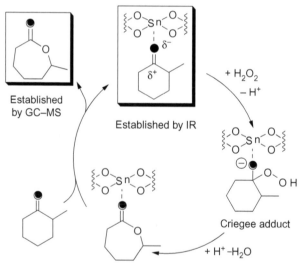

Figure 1.11 Proposed mechanism of Baeyer–Villiger oxidation catalyzed by Sn-beta. First, the ketone is coordinated to the Lewis acid center and, thereby, the carbonyl group is activated. This structure of the complex with the coordinated ketone was demonstrated by *in situ* FTIR spectroscopy. In the next step, the hydrogen peroxide attacks the electrophilic carbonyl carbon atom. After the following rearrangement step, the lactone product is replaced by a new substrate molecule. This mechanism is analogous to the mechanism observed when peracids are the oxidizing agents and has been proven by labeling the carbonyl oxygen in the substrate with ^{18}O, as indicated by the filled, "black oxygen" atoms in the figure. *Adapted from Ref. (68), with permission from Nature Publishing Group.*

"catalyst" was modeled with unconstrained, single-coordination sphere clusters using a B3LYP/ECP methodology. The BV reaction mechanism involves two principal steps: (i) addition of the hydrogen peroxide oxidant and the ketone substrate to form the Criegee intermediate, and (ii) BV rearrangement of the Criegee intermediate to yield the ester product. The activation barriers for the addition and rearrangement steps in the non-catalyzed mechanism were found to be 39.8 and 41.7 kcal/mol, respectively. In the absence of solvent coordination, the reaction of tin and hydrogen peroxide to produce a tin hydroperoxo (SnOOH) intermediate is characterized by an activation barrier of 15.4 kcal/mol. The tin-catalyzed mechanism proceeds through a Criegee intermediate containing a five-membered ring chelating the tin. This intermediate can be generated via two degenerate reaction pathways, one of which involves participation of the tin hydroperoxo intermediate. The Lewis-acidic tin center assists with the leaving of the hydroxyl group in the transition state of the BV rearrangement. Increased branching of the migrating carbon atom in the ketone substrate reduces the activation barrier for the rate-determining step by promoting nucleophilic attack of the peroxo bond by the migrating carbon.

The same authors also compared the relative performance of a titanium active site versus a tin active site for model epoxidation and BV reactions *(239b)*. The calculated activation energy for formation of the hydroperoxo intermediate on the tin active site is similar to that calculated for the titanium analogue. However, the stronger Lewis acidity of tin in comparison with titanium reduces the activation barrier for the rate-determining step in both the epoxidation and BV reaction mechanisms. For example, the Gibbs activation barriers for epoxidation of ethylene and dimethyl-1-butene are 3.4–4.4 kcal/mol lower for tin than for titanium, and for the BV oxidation of 2-methyl-3-pentanone the value is 7.2 kcal/mol lower for tin than for titanium. The overall reaction kinetics are similar for epoxidation processes using $Ti(IV)-H_2O_2$ or $Sn(IV)-H_2O_2$. In the BV reaction, however, the $Sn(IV)-H_2O_2$ combination substantially outperforms the $Ti(IV)-H_2O_2$ combination. The results obtained from this theory-based investigation aid in understanding experimental results obtained previously for the oxidation of unsaturated ketones with titanium- and tin-containing molecular sieves as catalysts.

The mechanism of the BV oxidation of cyclohexanone with hydrogen peroxide catalyzed by tin in a zeotype structure, Sn-beta, has been investigated by computational chemistry *(240)* and compared with that of the uncatalyzed gas-phase reaction. Based on the results obtained, a kinetic model was proposed for the Sn-beta catalyzed process and validated through experiments. It was found that a certain degree of flexibility in the frame-work is needed to allow the interaction between the tin center and the mol-ecules present in the reaction medium to occur; this flexibility can be provided by Sn–OH defects present in the framework. The computational chemistry investigation indicates that the active site in the zeotype material consists of two catalytic centers: a Lewis-acidic tin atom that activates the cyclohexanone carbonyl group and an adjacent basic oxygen atom prone to forming hydrogen bonds. Adsorption of hydrogen peroxide onto this sec-ond center (i.e., the basic oxygen) results in the formation of a complex (ter-med R1) in which the orientation of the two reactants is very favorable for the BV oxidation.

The complete mechanism *(240)* for the Sn-beta-catalyzed oxidation of cyclohexanone with hydrogen peroxide involves conversion of the complex termed R1 into a Criegee intermediate and subsequent rearrangement of this intermediate into ε-caprolactone and a water molecule, with an appar-ent activation energy not higher than 6 kcal/mol. In contrast, activation of hydrogen peroxide with formation of a tin hydroperoxo species is endother-mic by 4 kcal/mol and requires an activation energy of almost 13 kcal/mol, which makes this mechanism less probable. These activation barriers are slightly different from those calculated for the generic active tin site, dis-cussed above, and the difference is likely related to the incorporation of the tin in the beta structure in the latter work.

3.5.3 Structure–activity relationships

Shetty *et al. (241)* and Kulkarni *et al. (242)* evaluated the structure, bonding, and acidity of tin-substituted BEA-type material using a periodic approach based on DFT. The results demonstrate that the incorporation of tin into the BEA framework decreases the cohesive energy and is an endothermic pro-cess. Hence, it is clear that the amount of tin that can be incorporated into the BEA structure is limited. This limitation is the likely reason for the decrease in the turnover number in the BV oxidation reaction as the tin con-tent is increased in BEA samples; namely, at some concentration, additional tin will be located in extra-framework positions and these sites have dis-tinctly different Lewis acid strength than the framework tin sites.

Boronat *et al. (243)* investigated the differences between Sn-beta and Ti-beta molecular sieves using DFT methods. The authors performed a quantum-chemical analysis of the electronic properties of the active sites and sought to correlate the results with experimentally determined activities for various reactions. It was found that the best predictors for activity are the intrinsic Lewis acid strength of the modeled isolated active site, and the degree of backbonding from the catalyst to the organic reactant.

In summary, the Sn-beta catalyst has been shown to be highly selective and active for a large number of catalytic reactions, for example BV oxidation, glucose isomerization, and Meerwein–Ponndorf–Verley reactions. It is believed that the extraordinarily high chemoselectivity that is observed in some of these reactions is the result of a single type of catalytic site in Sn-beta.

3.6. Isomerizations and esterifications for the synthesis of chemicals and fuels from renewable feedstocks

Recent attempts to produce chemicals from biomass have renewed the interest in carbohydrate chemistry. One important intermediate step in a possible process scheme for the production of biofuels is the isomerization of glucose to fructose. It was discovered that Sn-beta is a suitable catalyst for this reaction and provides some advantages over the traditional enzyme catalyst. Sn-beta catalyzes glucose isomerization reactions with an activity that is comparable to that observed in biological processes. Davis and coworkers (244) reported that the active sites for the isomerization reaction in Sn-beta are tin atoms incorporated into the framework of the beta structure. Neither $SnCl_4*5H_2O$ nor SnO_2 were found to show isomerization activity. The mechanism by which a hydrogen atom is formally transferred from the C-2 to the C-1 position of the α-hydroxy aldehyde to form the corresponding α-hydroxy ketone remains unproven. Using deuterated solvents, it was shown that enzymes and bases perform the aldose–ketose transformation through a mechanism involving a cis 1,2-enediol intermediate, whereby the bonding electron pair at the C-2 carbon moves through the carbon skeleton to the C-1 carbon (shown in Figure 1.12). Analysis of the kinetics of these isomerization reactions reveals that certain acids and metals are able to catalyze a direct hydride shift between carbons C-2 and C-1. It is therefore conceivable that tin incorporated into the BEA framework performs the isomerization reaction via an intramolecular hydride shift between the carbonyl-containing C-1 and the hydroxyl-bearing C-2 carbon of glucose, with intermediate formation of a chelate complex characterized by a five-membered ring involving the tin (as illustrated in Figure 1.12). Important factors in the Sn-beta-catalyzed isomerization of glucose in water include the solvent, the confinement and polarity effects

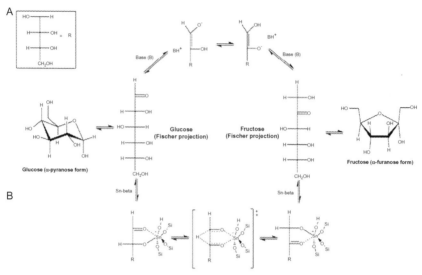

Figure 1.12 Pathways characterizing the isomerization of glucose to fructose catalyzed by base (A) or by Sn-beta (B). The base-catalyzed process is initiated by proton abstraction, whereas in Sn-beta, the glucose forms a chelating complex with the tin. *Adapted from Ref. (244b), with permission from Wiley.*

within the micropores of the molecular sieve, and the coordination state of the tin atoms in the framework, where they can be present in partially hydrolyzed form or fully framework coordinated. The Sn–beta catalyst proved to be superior to enzyme catalysts for this chemistry. The zeotype catalyst is not only stable but also robust; it is unaffected by impurities that harm enzyme activity and can operate in aggressive environments, for example, in acidic solutions. Consequently, Sn-beta can be employed as a catalyst for one-pot processes with several reactions. Glucose was successfully converted directly to 5-hydroxymethylfurfural (HMF) in high yields by combining a dehydration reaction in hydrochloric acid solution with the isomerization on Sn-beta.

Similar catalytic chemistry, also motivated by the interest in renewable feedstocks, has been reported. Taarning *et al. (245)* investigated lactic acid synthesis via isomerization of trioses (C3 sugars), specifically dihydroxyacetone and glyceraldehyde. Lewis–acidic zeotype catalysts such as Sn-beta show surprisingly high activity and selectivity for the isomerization of trioses to lactic acid, and for the isomerization–esterification to form lactate esters. Holm *et al. (246)* described a catalytic process for the direct formation of methyl lactate from common sugars. Sn-beta catalyzes the conversion of mono- and disaccharides that are dissolved in methanol to methyl lactate. With sucrose as the substrate, the methyl lactate yield reaches 68%. Additional contributions in this area have been published by Lew *et al. (247)*, Vlachos and coworkers *(248)*, and Lobo *(249)*.

Assary and Curtiss *(250)* presented the results of an examination of the isomerization of glyceraldehyde to dihydroxyacetone on Sn-beta by DFT. Various structural models were analyzed to understand the binding modes of glyceraldehyde with the active site, and the detailed free energy landscape was computed for the isomerization process. It was shown that the rate-limiting step for the isomerization is the 1,2-hydride shift, which is facilitated by the active participation of the hydroxyl group in the hydrolyzed Sn-beta. According to the computed activation barriers for isomerization on Sn, Zr, Ti, and Si ions in the framework, the activity of the catalytic sites in aqueous dielectric media declines in the following order $Sn > Zr > Ti > Si$.

3.7. Aromatization of light petroleum gas

The zeotype material Ga-MFI is a bifunctional material and therefore an effective catalyst for aromatization of light hydrocarbons *(76,251)*. The gallium catalyzes the dehydrogenation of light hydrocarbons, while the acid sites catalyze the aromatization of olefinic species. However, there is still much controversy over the specific gallium sites and the location of the gallium responsible for the catalytic activity. It is also generally known that in the aromatization of propylene, metallosilicates show a higher selectivity to aromatics than the purely protic catalyst H–ZSM-5.

4. INDUSTRIAL APPLICATIONS

4.1. Overview of catalysts and oxidants

4.1.1 Commercially employed catalyst materials and their availability

In this section, the commercial applications involving framework metal-containing zeotype catalysts are summarized. Notwithstanding the intriguing properties, only a limited number of framework metal-containing zeotype materials, among them TS-1, Ga-MFI, and some mesoporous catalysts, are currently used for commercial applications. Framework metal-containing catalysts are no different than other catalysts in that the activity, stability, lifetime, manufacturability, and catalyst cost must meet certain requirements for a new catalytic process to be implemented.

TS-1 as a new commercial catalyst of ENI company was reported by Perego *et al. (218)* Polimeri Europa's Web site (operated by ENI) *(114)* gives more details, from the properties of TS-1 to its applications. The catalyst capacity that is being newly installed amounts to more than 100 t/yr. The commercial products come in the form of extrudates for use in fixed-bed reactors or as microspheres for use in slurry reactors. The average bulk density of the extrudates is 650 kg/m^3, and that of the microspheres is 600 kg/m^3.

Other manufacturers offering TS-1 for sale include PQ, which is offering its PQ-TS-1, and CleanScience PUNE, India *(252)*. Both companies manufacture TS-1 in the form of extrudates and microspheres. The listed process applications are: aromatic hydroxylation, oxidation of primary alcohols to aldehydes, olefin epoxidation, and alkane oxidations. Presumably, there are more companies, especially in China, producing framework metal-containing zeotype materials.

Large pore materials, hydrophobic framework metal-containing zeotype materials, hybrid materials, and materials with unique Lewis acidity are in the current pipeline for commercialization. Some new developments that are close to commercialization will be also be reviewed.

4.1.2 Assessment of commercial availability of H_2O_2 for oxidation processes catalyzed by framework metal-containing zeotype materials

Many of the oxidations catalyzed by framework metal-containing zeotype materials require H_2O_2 as oxidant (see Sections 4.2.2 and 4.3.1). The cost of hydrogen peroxide is the economic limitation for process developments, and the possibility to use inexpensive H_2O_2 is the key driver for commercialization. In a Chemsystems Report *(253)* published in 2009, every currently existing process involving the use of H_2O_2 as oxidant was evaluated. The authors of the report calculated and reported the cost of production and return on capital. The report concluded that the anthraquinone process is the most attractive process for the manufacture of H_2O_2 from an economic standpoint. One potential solution is to integrate anthraquinone-based inexpensive "crude grade" hydrogen peroxide into the production of PO (e.g., a Dow Chemical PO plant with H_2O_2 supplied by pipeline from a Shell plant, thereby reducing shipping costs). Another solution is the *in situ* production of hydrogen peroxide from a mixture of H_2 and O_2, but there are both economic *(253)* and safety *(254–256)* issues related to this process at commercial scale.

4.2. Epoxidation of olefins: Production of PO through epoxidation of propylene

4.2.1 Comparison of existing and new technologies for propylene epoxidation

Key players in the development and commercial application of technology for the manufacture of PO are Dow–BASF, Degussa–Uhde, ENI–Sumitomo, and ARCO/Lyondell, as listed in Table 1.10. Based on a Chinese patent search, there is also a focused effort in China in this area. Efforts to contact several of the authors did not elicit a response, and thus further information could not be included in this review.

PO has historically been produced by the chlorohydrin method or the organic peroxide method. While these processes generate large amounts of by-products (namely, calcium chloride) they also deliver coproducts such as *tert*-butanol, which may be used for the production of methyl tertiary-butyl ether (MTBE), or styrene. Key producers of PO are Sumitomo, Repsol, and Huntsman *(275)*. Shell's SMPO (Propylene Oxide with

Table 1.10 List of processes for propylene oxide manufacture with metal framework-containing zeotype or other porous materials as catalysts

Technology	Assignee(s)	Oxidant	Catalyst	References
HPPO	Dow–BASF–Solvay	H_2O_2	TS-1	*(184,257–260,260b)*
HPPO	Degussa–UHDE	H_2O_2	TS-1	*(257,261a,262)*
PO/Cumene HP	Sumitomo	Cumene-HP	Titanium-molecular sieve	*(263,264)*
PO[a]	Lyondell, BASF, Degussa, Headwater, and others	*In situ* H_2O_2	Pd-Pt/TS-1	*(2c,265–268)*
PO[a]	Sumitomo and others	Gas-phase *in situ* H_2O_2	Noble metal/titanium heterogeneous catalyst	*(269,270)*
PO[a]	Dow, Nippon Shokubai, BASF, Bayer	*In situ* generated H_2O_2	Au/titanium silicalites	*(271–274)*

[a]Under development.

Styrene Monomer) process *(275)* has operated continuously using a silica-supported titania. The coproduct is styrene, which is positive for the process economics.

ChemSystems *(257)* published a detailed report on the technology and economics of existing PO processes including PO/MTBE, Chlorohydrin, and PO/Styrene. The present review provides an up-to-date summary of the technologies that rely on framework metal-containing zeotype catalysts. Processes using hydrogen peroxide as oxidant (e.g., those of Dow–BASF and Degussa–Uhde) and processes using cumene hydroperoxide (e.g., that of Sumitomo) are presented.

4.2.2 PO synthesis using hydrogen peroxide and TS-1
4.2.2.1 The HPPO process of Dow Chemical and BASF

Dow entered the business of PO manufacture with the hydrogen peroxide-based technology through its 2001 acquisition of ENIChem's polyurethane business, which included ENI's PO technology. In 2003, Dow and BASF announced the intent to construct the "world's largest scale" PO plant in Amsterdam that would use the hydrogen peroxide route. The "hydrogen peroxide-based propylene oxide" (HPPO) joint venture partners broke ground for the production facility in September 2006. The largest plant using this new process, with a capacity of 300,000 t/yr PO, was constructed. A mega-sized 230,000 t/yr plant, $3 \times$ larger than the largest ever built, for the synthesis of H_2O_2 using Solvay's anthraquinone technology was built close to the BASF/Dow HPPO plant. The integration of the H_2O_2 and PO processes allows a significant reduction of transportation costs, and in general offers opportunities for the reduction of H_2O_2 production costs. In 2009, Dow and BASF announced *(258)* that they completed and started up the commercial unit (with a capacity of 300,000 t/yr). Dow and BASF received the Presidential Green Challenge Award, Innovation and Excellence Award in Core Engineering *(259)*, for their jointly developed hydrogen peroxide to propylene oxide (HPPO) technology that vastly improves the production process of a key chemical intermediate, PO. The economics of the HPPO process have been analyzed by ChemSystems. *(257,265)*

After the HPPO technology was successfully implemented in Europe, SCG-Dow Group, a joint venture between Dow and Siam Cement Group, started-up a new PO facility in Thailand, and has successfully completed its full 390,000 t/yr capacity performance test *(276)* in 2012. The plant operation exceeded expectations for all quality and yield performance parameters. The technology was invented by Forlin *et al. (184)* and Göbbel *et al.*

(184,277) from Dow Company, who developed a process for the continuous production of epoxides, such as PO, by direct catalytic oxidation of an olefin with aqueous hydrogen peroxide. The process is comprised of a sequence of process steps, which encompass reaction, distillation, decomposition, phase separation, condensation, and further distillation with recycle of various streams to improve catalyst lifetime and reaction selectivity. The flow diagram of the continuous process from Dow is presented in Figure 1.13.

The catalysis occurs in isothermally operated continuously stirred tank reactors. The propylene stream (which includes fresh and recycled propylene) is introduced into the reaction stage at controlled conditions, and in excess, to maximize the conversion and selectivity toward PO and to maintain the reaction pressure.

According to one embodiment of the invention of the process, the fresh olefin is purified in a distillation column before it is introduced into the main reactors (sequentially connected R1–R3, shown as a single reactor in Figure 1.13 for simplification). The feed stream of hydrogen peroxide is divided into two portions such that the ratio of the streams to the reactors R1 and R2 is 50:50. The reactors R1 and R2 are operated under similar conditions; the temperature is about 55–75 °C and the pressure is 13 bar. The reactor R3, which serves as the finishing reactor, is operated at 70–90 °C and at a pressure of 8 bar. The distillation (flash) columns operate largely under the same operating conditions and separate the top streams in vapor phase comprising unreacted olefin, epoxide, and inert gases. Methanol is used as the solvent for the epoxidation reaction, and by-products are methyl formate and dimethoxymethane. The decomposition reactors (R-4A/B), which serve to decompose unreacted peroxides, are fixed-bed tubular reactors arranged in series.

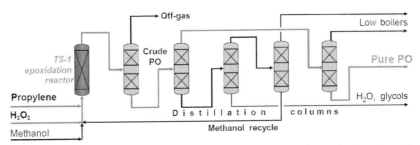

Figure 1.13 Simplified process flow diagram for propylene oxide production via oxidation of propylene catalyzed by TS-1 with H_2O_2 as the oxidizing agent (Dow–BASF HPPO process). *Adapted from US 7,138,534.*

The process and performance parameters, as specified in the claims of the patent *(184)*, are the following:

- Reactor R1 and R2 operated at a temperature of 55–75 °C and a pressure of 13 bar
- Reactor 3 operated at a temperature of 70–90 °C and a pressure 8 bar
- Minimum purity of propylene feed: 96%
- Hydrogen peroxide concentration in feed 35 wt.% in methanol
- Catalyst: titanium silicalite as described in US Patent 4,937,216
- 60 ppm ammonium salt in feed, pH 6.5
- Propylene and hydrogen peroxide conversion >95%
- Purity of PO >99.8%

Compared with conventional PO process technologies, the HPPO process offers unique benefits in three areas: economics, protection of the environment, and opportunities for future growth.

The economic benefits include:

(i) significantly lower capital cost to build

(ii) elimination of the need for additional infrastructure or markets for coproducts, as the process produces only PO and water, and

(iii) simple raw materials integration—only hydrogen peroxide and propylene are needed as raw materials.

Environmental benefits include:

(i) reduction of wastewater by 70–80%, as compared with existing PO technologies,

(ii) reduction of energy usage by 35%, as compared with existing PO technologies,

(iii) reduction of infrastructural needs and physical footprint with simpler raw material integration and avoidance of coproducts.

4.2.2.2 Degussa/UHDE PO process

Degussa, the world's second largest producer of H_2O_2, has developed a HPPO process together with Uhde, and licensed this technology to SKC, who built a plant with 100,000 t/yr capacity in Ulsan, South Korea. In this plant, titanium silicalite is used as the catalyst. Hydrogen peroxide is supplied from a conventional anthraquinone plant that is also located in Ulsan and belongs to Degussa/Headwater.

Degussa's *(261)* continuous process comprises reaction between a gas and a liquid stream in countercurrent flow mode in a fixed-bed reactor. The gas phase, which contains the propylene, is brought into contact with a liquid phase, which contains the hydrogen peroxide, in the presence of titanium silicalite. In this process *(186,278–280)*, the reactor is operated at a temperature of 60 °C. To some extent, the high selectivity is reached because of the moderate reaction temperature. The product PO stream is depressurized; the resulting propylene-rich gas phase is recirculated. The expanded liquid phase is separated into a raw PO stream and a stream consisting of methanol and water. Water and methanol are separated, and the methanol is recirculated. The raw PO is purified by extractive distillation using propylene glycol which effectively removes impurities such as acetaldehyde. The final purity of the PO is 99.97 wt%. The main by-product is propylene glycol. The catalyst deactivates as a result of the formation of PO oligomers, and is regenerated by treatment with H_2O_2 solution or calcination. Substances that can deactivate the catalyst are continuously removed from circulation, which brings about a considerable improvement in catalyst performance and longevity.

4.2.3 PO synthesis using cumene hydroperoxide

Sumitomo developed a recirculation process for manufacture of PO using CHP as oxidant *(196)*. The company developed both a new catalyst and a new process for PO production. The production method is fundamentally similar to known methods involving organic peroxides as oxidants; the major difference is that cumene is used as the reaction medium and hence the process is referred to as the "cumene PO-only" process. Laboratory tests started in 2000 and pilot plant testing in 2001. A plant was completed in 2002 and started up in 2003. This commercial plant was the first PO-only plant in Japan, producing PO by oxidation of propylene with cumene hydroperoxide without a significant formation of coproducts. The plant is located in the Chiba prefecture, operated by a joint venture between Nihon Oxirane Co. and Lyondell, and produces around 200,000 t of PO/year. A second plant was started in May 2009 in Saudi Arabia, as a joint project of Sumitomo with Saudi Arabian Oil Co.

Sumitomo developed a titanium-based epoxidation catalyst, and in the new plant, a high-performance proprietary catalyst that allows a more compact plant configuration is employed. Development of this catalyst made possible the design of a process without coproduction of styrene monomer.

In US Patent 6,211,388 *(196a)*, Tsuji *et al.* described the catalyst as a titanium-containing silicon oxide, characterized by the following properties and protocol of preparation: one peak showing an inter-planar spacing (*d*) larger than 18 Å in XRD, an average pore size of 10 Å or more, a pore size distribution such that the size of the pores that make up 90% of the total pore volume is 5–200 Å, a specific pore volume of 0.2 cm^3/g, a quaternary ammonium ion (e.g., cetyltrimethylammonium bromide) used as SDA, a band at 960 cm^{-1} detected in FTIR spectra, and a silylation treatment applied to the catalyst. *(281)*

Tsuji *et al.* summarized *(263)* the key properties of a titanium-containing silica catalyst (referred to as "Sumitomo Ti catalyst"). This catalyst was the first commercially used mesoporous heterogeneous catalyst for epoxidation. It was designed to exhibit a high level of activity for large molecules like the oxidant cumene hydroperoxide. There are three properties, illustrated schematically in Figure 1.14, that are believed to be essential for the performance: (i) tetrahedral titanium, (ii) meso–macro pores allowing for the diffusion of large molecules, and (iii) hydrophobicity. The pore size distribution in particular distinguishes this catalyst from materials such as TS-1 (also shown in Figure 1.14).

Cavani and Gaffney *(271)* published a flow diagram of the Sumitomo process (see Figure 1.15), and Sumitomo process patents *(196b,282)* describe the following process steps: (1) An oxidation process where cumene is oxidized in air to obtain CMP, (2) An epoxidation process where α,α-dimethyl benzyl-alcohol (CMA) is obtained from CMP and propylene in the presence of an epoxidation catalyst, (3) A hydrogenation process where CMA is hydrogenated, and cumene is obtained in the presence of a hydrogenation catalyst and hydrogen, (4) A cumene purification process where the cumene

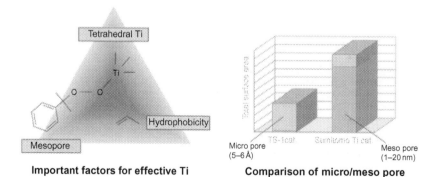

Important factors for effective Ti epoxidation catalyst   **Comparison of micro/meso pore structure between TS-1 catalyst and sumitomo Ti catalyst**

Figure 1.14 Critical properties of Sumitomo's titanium molecular sieve. (Left) Essential chemical and morphological properties. (Right) Relative contributions of mesopores to total surface area for a regular TS-1 sample and for the Sumitomo catalyst. *Adapted from Ref. (263), with permission from Sumitomo Chemical Co. Ltd.*

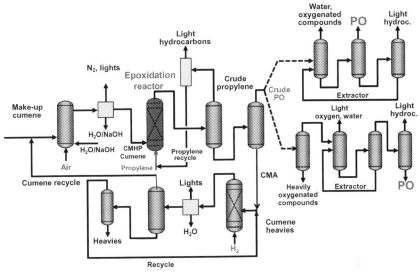

Figure 1.15 Process flow diagram for the oxidation of propylene with cumene hydroperoxide as the oxidizing agent and titanium-containing mesoporous material as the heterogeneous catalyst (Sumitomo process). The process involves the following steps: (1) A process for oxidation of cumene with air to obtain CMHP, (2) a process for epoxidation of propylene in the presence of a catalyst whereby α,α-dimethyl benzyl-alcohol (CMA) is concomitantly obtained from CMHP, (3) a process for the hydrogenation of CMA with H_2 in the presence of a catalyst to obtain cumene, (4) a process for purification of the cumene, followed by recycle of cumene to the oxidation process, and (5) a process for the purification of PO. *Adapted from Ref. (271), with permission from Wiley-VCH.* (See the color plate.)

obtained is purified and recycled to the oxidation process, and (5) A PO purification process where PO is purified.

4.2.4 PO synthesis with *in situ generation of H_2O_2 from H_2 and O_2*

A process for the epoxidation of propylene with *in situ* generation of hydrogen peroxide was proposed in the 1990s by the Tosoh Corporation *(283)*. The company suggested that PO could be made in a flow system via a direct reaction between H_2 and O_2 in the presence of propylene by using a catalyst made of palladium supported on crystalline titanium silicate. A propylene conversion of 0.8% was reported, with a selectivity to PO of 99%. ARCO (now Lyondell) described catalysts that produce PO with better selectivity and yield as compared with those reported earlier *(284)*. BASF has also claimed the use of framework metal-modified TS-1 catalysts for this catalytic chemistry *(266)*. Various catalyst compositions were described that can be grouped into titanium- or vanadium-substituted silicalites containing rare earth ions, and titanium- or vanadium-substituted silicalites containing late transition metals (Re, Ru, Os, Rh, Ir, Pd, Pt, Ag, Au). Hoechst *(285)* investigated the epoxidation of propene with *in situ* generated H_2O_2, using a titanium silicate support and platinum and palladium metals. The catalyst was made of 1 wt% Pd and 0.02 wt% Pt supported on TS-1.

The direct synthesis of hydrogen peroxide has been developed jointly by Degussa-Evonik and Headwaters (now Nanokinetix) *(286)*. A demonstration unit for the production of several thousand metric tons per year of dilute hydrogen peroxide in methanol went on stream in 2006, in Hanau–Wolfgang, Germany. The process is operated at conditions below the explosion limit, and palladium and platinum nanocatalysts are used. A consortium including Degussa, Uhde, and academic teams investigated the reaction of propylene with *in situ* produced hydrogen peroxide vapors. In the presence of a catalyst, milder reaction conditions could be used *(287)*. The catalyst developed by Headwaters Technology Innovation has now been named NxCat.

The epoxidation of gaseous propylene can be carried out on catalysts that are largely similar to those used for ethylene epoxidation *(271,288)* if small amounts of either a nitrogen oxide species (i.e., NO, NO_2) or a volatile organic chloride (e.g., ethyl chloride) are added as initiator. The catalytic epoxidation with reactants in the gas phase has been investigated in depth by ARCO (now Lyondell), and many patents have been issued on this subject *(289)*. The catalysts have some analogies with those originally reported by Dow, who cited best values of 3.7% for propylene conversion and 47.2% selectivity to PO at a reaction temperature of 180 °C (with an Ag/Mg–SiO_2 catalyst) *(290)*.

Table 1.10 summarizes the PO processes that are either already commercialized or under development.

To arrive at new, simplified processes, *in situ* H_2O_2 synthesis is the preferred approach. Catalysts that are structurally stable, exhibit stable activity, are easily regenerated, and characterized by high selectivity are needed, and candidate materials must demonstrate high performance in the development stage, prior to commercialization. Several problems remain to be solved with respect to these fascinating processes that directly oxidize propylene with green oxidants (*in situ* generated H_2O_2 or O_2) in a surface-mediated gas-phase reaction. These problems include: Safety issues (potentially explosive mixtures), rapid deactivation of the catalyst, the need for gas-phase promoters (e.g., NO, chlorocarbons), and lower space velocities than desirable.

4.3. Hydroxylation of aromatic compounds: Oxidation of benzene or phenol

4.3.1 Direct oxidation of benzene to phenol with H_2O_2

The single-step oxidation of benzene to phenol without coproduct formation is a great industrial challenge. Currently, phenol is produced by the Cumene process, which also yields stoichiometric amounts of acetone. However, there are recent developments in the oxidation of benzene to phenol with aqueous hydroperoxides as oxidants, TS-1 as the catalyst, and sulfolane (thiophene, tetrahydro, 1,1-dioxide) as a cosolvent *(204b)*.

The selective oxidation of benzene to phenol with hydrogen peroxide is achieved by using a new zeotype catalyst (referred to as "TS-1B"), obtained by modification of titanium silicalite with NH_4HF_2 and H_2O_2. This new catalyst is used in the presence of a particular cosolvent (sulfolane) and in this way, the produced phenol is protected from over oxidation and the selectivity is enhanced substantially. Clerici and Rivetti *(2f)* published a simplified flow diagram of the process (see Figure 1.16). The oxidation of benzene is carried out continuously in a fixed-bed reactor charged with modified TS-1B, at temperatures between 95 and 110 °C, and at 6 bar total pressure. Unconverted benzene (which also acts as solvent), water, and phenol are removed via distillation, while the sulfolane solvent is purified by extracting the by-products (which are hydroquinone, catechol, and tar) with aqueous sodium hydroxide, and then recirculated. The process is run at low benzene conversion, and the selectivity of benzene conversion to phenol is 97%. Hydrogen peroxide is converted to phenol with a selectivity of 71%. The process is in the development stage, and may be an alternative to the cumene-based phenol process in the near future.

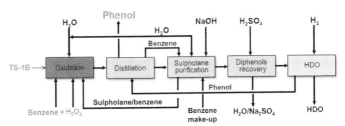

Figure 1.16 Process scheme for oxidation of benzene to phenol catalyzed by TS-1B with hydrogen peroxide as the oxidizing agent. *Adapted from Ref. (2f), with permission from Enciclopedia Italiana.*

4.3.2 Direct oxidation of benzene with N₂O to phenol

Another strategy for the direct oxidation of benzene to phenol is based on the use of N_2O as oxidant. While the catalytic chemistry is feasible, commercialization hinges on the availability of appropriate amounts of N_2O at a nonprohibitive cost. Some process designs thus seek to use N_2O from waste streams.

Benzene is oxidized directly to phenol using N_2O and a framework iron-containing or otherwise iron-modified ZSM-5 catalyst. The reaction produces phenol and N_2 as initial products. The Institute of Catalysis of Novosibirsk were first to report this new route to phenol *(209a,c,291)*, which was enabled by an Fe-ZSM-5 catalyst. The authors synthesized the ferrisilicate *(292)* via hydrothermal synthesis to incorporate iron into the framework; this catalyst was reported to exhibit high selectivity.

A possible source of N_2O is adipic acid production. N_2O is the main component of the NO_x emitted during the oxidation of cyclohexanol to adipic acid with nitric acid as oxidant. Since phenol is used to make cyclohexanol, the processes could be coupled as shown in Figure 1.17. The molar amount of N_2O is equivalent to that of reacted cyclohexanol.

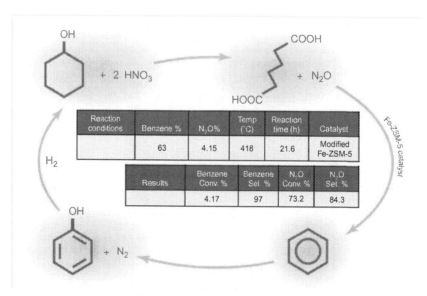

Figure 1.17 Reaction steps, reaction conditions, and typical results for the direct hydroxylation of benzene to phenol with modified Fe-ZSM-5 as the catalyst and N₂O as the oxidizing agent according to the Solutia process. *Adapted from Chemical Engineering, September 2004, p. 1, with permission from Chemical Engineering.*

Therefore, the balance is only assured in the case of 100% selectivity. Indeed, Solutia (formerly the chemical division of Monsanto) *(293)* has claimed a new process for the production of adipic acid without formation of acetone (which is normally obtained as a coproduct during cleavage of cumene hydroperoxide to phenol). Solutia patents *(293,294)* claim that the benzene-to-phenol selectivity is high while the N_2O selectivity is lower than 90%. These numbers seem to indicate that some N_2O has to be added, and the possibility for successful industrial application will depend on the amount of extra N_2O required.

Solutia and the Boreskov Institute of Catalysis (Russia) have developed a new technology, called the AlphOx Process, to produce phenol from benzene in a single step. In this process, which eliminates cumene as an intermediate and acetone as a coproduct, gaseous benzene and nitrous oxide (N_2O) react on an Fe-MFI catalyst to give phenol. Solutia is planning to commercialize the process, using an offgas stream from its adipic acid plant as the nitrous oxide source. A pilot plant employing AlphOx has been in operation since 1996, and the first commercial AlphOx facility was scheduled for start-up by 2000. A ChemSystems technical and economic evaluation of a plant using an off-gas stream from adipic acid production indicates that the total fixed capital investment for AlphOx is lower than that for a comparable cumene-based phenol process. The net production cost for AlphOx is also competitive and is not affected by fluctuations in the acetone market. However, the phenol plant capacity is limited by the availability of N_2O recovered from the adipic acid process. Unfortunately, a relatively small phenol plant requires a world-scale adipic acid plant for its N_2O supply. Therefore, the AlphOx technology would seem to appeal to producers looking to build a phenol plant with small capacity that would not be economical with the conventional cumene-based process.

Chemical Engineering Journal published a flow diagram of the AlphOx process (see Figure 1.18) in 2004 *(295)*. GTC Technology Inc., working with Solutia, stated "the investment cost for 300 Mlb/year plant is about 55–60% of an integrated cumene-to-phenol plant." The AlphOx technology has been extensively tested at Solutia's demonstration facility in Pensacola, FL. GTC is currently negotiating the details with petrochemical firms for the first commercial plant. Unfortunately, no more information is provided by GTC.

In 2001 Solutia Inc., St. Louis, Mo., and JLM Industries Inc., Tampa, Fla., announced the end of an agreement to build a benzene-to-phenol plant at Solutia's Pensacola, Fla., facility. The companies agreed to terminate the

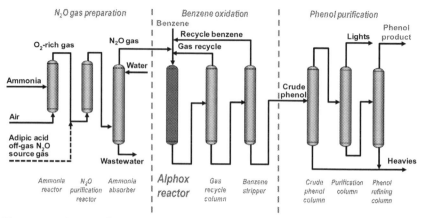

Figure 1.18 Process flow diagram for the direct hydroxylation of benzene to phenol with modified Fe-ZSM-5 as the catalyst and N$_2$O as the oxidizing agent. N$_2$O is obtained from the waste stream of adipic acid production. This scheme depicts the Solutia Alphox process. (See the color plate.)

planned project because of "current dynamics in the global market for phenol." They plan to pursue separate strategies in the phenol market.

GE has also patented a process for the direct hydroxylation of benzene to phenol with N$_2$O as oxidant *(296)*.

4.3.3 Hydroxylation of phenol using TS-1

The hydroxylation of phenol to hydroquinone and catechol was the first commercial application of TS-1 by ENI *(218,219b)*. The reaction, which occurs in aqueous or aqueous–organic solutions, produces a mixture of hydroquinone (*p*-dihydroxybenzene) and catechol (*o*-dihydroxybenzene). Water and tarry compounds are the major by-products.

Both the conversion and selectivity are significantly higher than those achieved by radical and acidic catalysts *(218)*. EniChem developed the synthesis of *o*- and *p*-dihydroxybenzenes by this route in the early 1980's and proceeded to the industrial scale, in substitution of the Brichima process. A plant producing 10,000 t/yr of diphenols has been in operation in Ravenna, Italy, since 1986. In comparison to radical catalysts, TS-1 excels with a lower consumption of hydrogen peroxide, and the number of phenol separation and recycle steps can be decreased, thus significantly reducing the production costs. The current production is 24,000 t/yr and is managed by Borregaard Italia. A process flow diagram *(2b)* is shown in Figure 1.19.

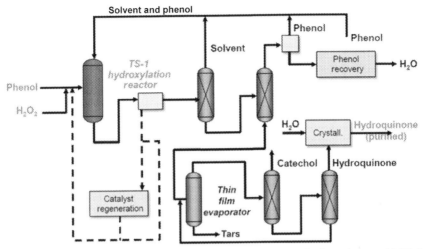

Figure 1.19 Process flow diagram for phenol hydroxylation to hydroquinone with TS-1 as the catalyst and hydrogen peroxide as the oxidizing agent. *Adapted from Ref. (297), with permission from Kluwer Academic.*

NCL Puna also developed a patented synthesis of a titanium–substituted catalyst *(298)* that is used for oxidation chemistry including direct hydroxylation of phenol. CleanScience *(252)* is offering derivatives of hydroquinone in commercial quantities that are produced via phenol hydroxylation and alkylation.

4.4. Production of cyclohexanone oxime through cyclohexanone ammoximation

Cyclohexanone oxime is produced by the reaction of cyclohexanone with a hydroxylamine derivative. Several commercial processes are available for this step, and they differ mainly in the manufacture of the hydroxylamine derivative. At the beginning of the 1990s, a "once-through" synthesis of cyclohexanone oxime was claimed by EniChem *(224,299)*.

The direct ammoximation of cyclohexanone has been performed with hydrogen peroxide and ammonia using a catalyst based on TS-1 *(5a)*. Good performance was obtained, with water as the only side product. EniChem built a 12,000 t/yr demonstration plant *(300)*, which has been running successfully since 1994. Sumitomo built a unit with a capacity of 60,000 t/yr in 2003; and both companies' plants have operated without any major problems. *(232)* Sumitomo *(301)* commercialized its process, in combination with a zeolite-catalyzed Beckmann rearrangement of cyclohexanone oxime to ε-caprolactone. Izumi *et al.* *(232)* have provided more details of the Beckmann rearrangement process.

According to the detailed information provided by Clerici and Rivetti *(2f)* (Figure 1.20) and Centi and Trifiro *(2b)*, the ammoximation is carried out continuously in the liquid phase in a stirred reactor. The reaction medium is a slurry, with the catalyst dispersed at a concentration of 2–3% by weight. The reaction is typically conducted at a temperature of 80–90 °C and at a slight overpressure. Cyclohexanone, NH_3, and aqueous H_2O_2 are supplied in a molar ratio of 1.0:2.0:1.1, the solvents are water and *tert*-butanol, and the residence time is 1.5 h. The reaction mixture is separated from the catalyst, and the unconverted excess of ammonia and the solvent (water/*tert*-butanol azeotrope) are recycled after separation from the reaction product in a column. The aqueous solution of crude oxime is withdrawn from the bottom of the column and is conveyed to the purification unit for recovery of cyclohexanone oxime with the required purity. During the process, the catalyst must undergo periodic purging, washing, and make-up operations since, in the presence of ammonia, the siliceous structure of TS-1 slowly dissolves, with loss of weight of the catalyst accompanied by migration of the titanium to the outside surface of the solid and subsequent loss of catalytic activity. If the process is operated under the conditions

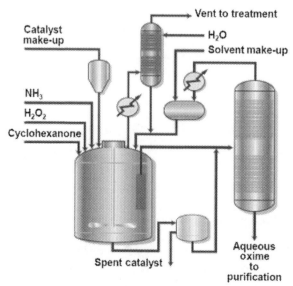

Figure 1.20 Process scheme for cyclohexanone ammoximation catalyzed by TS-1. *Adapted from Ref. (2f), with permission from Enciclopedia Italiana.*

described above, the conversion of the cyclohexanone is nearly complete at 99.9%, the selectivity to the oxime (based on cyclohexanone) is higher than 98%, and the yield of oxime related to H_2O_2 is about 94%. The main inorganic by-products, derived from the oxidation of ammonia or the decomposition of H_2O_2, are N_2, N_2O, O_2, nitrites, and ammonium nitrates.

4.5. Potential industrial applications in the BV oxidation for production of fragrances and flavors

Sn-beta, combined with hydrogen peroxide as oxidant, has the potential to substitute classical oxidants, for example peracids, in BV oxidations. As an example, the oxidation of delfone (2 pentylcyclopentanone) to d–decalactone (tetrahydro-6-pentyl-2H-pyran-2-one) is currently achieved using the corrosive peracetic acid. The resultant lactones have a creamy coconut- and peach-like aroma and are important flavor constituents of many types of fruit, and cheese and other dairy products. The lactones are also used in fragrances; the two enantiomers have different aromas. Sn-beta was tested for this BV transformation in a stirred reactor. The desired lactone product was obtained in 86% yield in the presence of the Sn-beta catalyst. (302) This result demonstrated clearly that the combination of Sn-beta and hydrogen peroxide is an environmentally friendly alternative to the commonly used organic peracids, even in asymmetric synthesis. Instead of a stoichiometric amount of carboxylic acid waste, water is produced as a side product from the oxidant.

4.6. Aromatization of light petroleum gas

The BP-UOP's Cyclar™ process converts low-value propane and butanes, or light feedstocks containing olefins and paraffins, to higher value petrochemical-grade benzene, toluene, and xylenes (303). A simplified block flow diagram of the Cyclar process is shown in Figure 1.21. A Cyclar unit is divided into three major sections. The reactor section includes the radial flow reactor stack, combined feed exchanger, charge heater, and inter-heaters. The regenerator section includes the regenerator stack and catalyst transfer system. The product recovery section includes the product separators, compressors, stripper, and gas recovery equipment. The flow scheme is similar to the UOP continuous catalytic regeneration (CCR) Platforming process, which is used widely throughout the world for reforming petroleum naphtha. BP commissioned the first commercial-scale Cyclar unit at its refinery in Grangemouth, Scotland, in January 1990. In

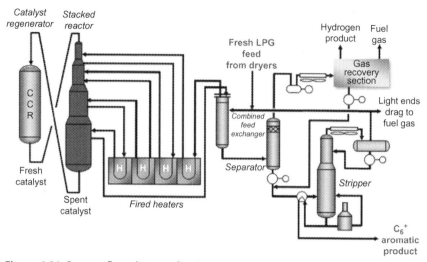

Figure 1.21 Process flow diagram for the conversion of light alkanes and alkenes to benzene, toluene, and xylene catalyzed by a gallium-modified zeolite (UOP-BP Cyclar™ process). *Adapted from Ref.* (303).

1995, UOP licensed the first Cyclar-based aromatics complex in the Middle East. This Cyclar unit is a low-pressure design that is capable of converting 1.3 million metric tons per annum (MTA) of light petroleum gas (LPG) to aromatics. The associated aromatics complex is designed to produce 350,000 MTA of benzene, 300,000 MTA of *para*-xylene, and 80,000 MTA of *ortho*-xylene. This new aromatics complex was started in August 1999 and continues to operate at present.

Although a variety of reactions need to be catalyzed, a single gallium-modified zeotype catalyst is used in the aromatization process developed by BP and UOP. The aromatization unit is operated in conjunction with UOP's CCR system. Acidic sites catalyze dehydrogenation, oligomerization, and cyclization. The shape selectivity of the cavities promotes the cyclization reactions and limits the size of the rings. Reportedly, an improved, second generation Ga-MFI catalyst is employed in this process.

5. FUTURE DIRECTIONS AND OPPORTUNITIES

Commercial zeolite catalysis, as a field in catalysis science, is now more than 50 years old, following the great inventions (*304*) by Union Carbide (*305*) and Mobil researchers (*306*), and it has reached some maturity. There have been continuous developments in the synthesis, manufacture, and commercial application of zeolites. However, there continues to be a steady and continuous stream of reports of synthesis of new zeolitic micro- and macroporous structures.

Compared to traditional zeolite catalysis, the catalysis by zeotype materials with incorporated metallic elements such as titanium, gallium, iron, and tin with atomic dispersion into framework position(s) is a relatively new area. The incorporation of these elements into molecular sieves and zeolites has been well-studied and a large number of synthesis methods are described in literature. Additionally, many unique chemical and catalytic properties are associated with these framework metal-containing zeotype materials. The degree of control of material properties that is nowadays exerted in the preparation of metal-containing solids is already impressive, but it can be expected that knowledge and skill will further increase. We believe that in this area, there is a vast amount of unpublished knowledge and trade secrets that have been developed during the commercialization of these materials.

A detailed understanding of both the solid state and surface chemistry has been greatly assisted by improved characterization techniques (including X-ray absorption spectroscopy, high resolution electron microscopy, 3D-NMR spectroscopy, and others) and new molecular modeling methods.

Thirty years after the invention of ENI's TS-1, new environmentally friendly processes have been commercialized for the epoxidation of propylene and the ammoximation of cyclohexanone that have no major by-products or coproducts. The BASF–Dow and Degussa–Uhde processes claim to be environmentally friendly, have low capital investment costs, and have no significant quantities of by-products when compared to either the chlorohydrin or the PO with styrene monomer processes. The Sumitomo PO process is also beneficial for the environment, because it needs only a cumene-to-cumene hydroperoxide reactivation process step.

The successful incorporation of gallium into ZSM-5 resulted in a cornerstone technology developed by UOP-BP, called the Cyclar process. This catalyst is used for aromatization of light paraffins to aromatics.

Iron incorporation into MFI resulted in the development of another important application for framework metal-containing zeotype materials; that is, the ability to synthesize phenol directly from benzene using N_2O as oxidant.

Tin incorporation into the beta structure has produced a new exiting material for reactions catalyzed by weak Lewis acids. The Sn–beta catalyst has already demonstrated very high selectivity in BV oxidations and in glucose isomerization.

However, there is still opportunity for the commercialization of new processes using framework metal-containing zeotype materials. It is remarkable that in this particular research field the successful commercialization has been achieved partly as a result of cooperation between scientists in academic groups and industrial researchers. For example, such cooperation played a role in the development of the PO processes. However, the costs involved in the syntheses of metal-containing zeotype materials (mainly due to SDAs and other reagents), and frequently the costs associated with both scale-up and manufacture, have tended to inhibit wide application of these new materials. As a result, new materials are finding initial catalytic applications only in specialty applications (i.e., production of high value-added products) or "niche" areas.

We believe that future developments of environmentally friendly processes using framework metal-containing zeotype materials will only be achieved when the following challenges are met by academia and industry.

Primarily for Academia:
• Better understanding of the chemistry and limitations of framework metal-containing zeotype materials
• New materials with close to 100% selectivity
• New materials for biomass conversion
• New selective uniform catalysts for direct oxidation with air
• Development of new, high-productivity, tools for catalyst synthesis and catalytic evaluation
• New analytical tools that provide understanding of the molecular level chemistry
• New molecular modeling tools

And for Industry:
• Improved stability and extended lifetime of framework metal-containing zeotype materials
• Use new processes to intensify and improve the framework metal-containing zeotype materials synthesis (thus reducing cost)
• Scale up more selective and cheaper catalysts
• Use new laboratory tools to speed up the catalyst commercialization
• Cheaper oxidants (and also safer hydrogen peroxide from water)
• Solve engineering and safety aspects of *in situ* hydrogen peroxide production from hydrogen and oxygen
• Lower manufacturing cost of catalyst and capital investment for simplified process
• Process optimization (energy, safety, etc.)

Both academic and industrial research and development, as well as cooperation between industries, has resulted in a large effort in the discovery, synthesis, and catalytic application of framework metal-containing zeotype materials. These collaborations have led to new applications of metal-containing zeotype materials for the large-scale production of chemicals without major by-product formation. New environmentally friendly processes such as propylene epoxidation, phenol hydroxylation, ammoximation of cyclohexanone, and aromatization of light paraffins have been commercialized. Many new developments are in the pipeline, and they will likely be commercialized when both the economics and the environmental requirements become favorable.

ACKNOWLEDGMENTS

The authors would like to express their appreciation to UOP LLC, a Honeywell company, for permission to publish this review, and to our colleagues and friends for fruitful discussions that contributed to this paper. Raweewan Klaewkla, a former Ph.D student, summarized the synthesis of framework metal-containing zeotype materials. Jean Blaho (UOP) provided support on literature search and visualization of literature. Robert Jensen, Rick Rosin, Greg Lewis, Mike Gatter, Bob Broach, and Jaime Moscoso provided great collaboration over the years. Special thanks to Avelino Corma (University of Valencia) for successful cooperation, to Giuseppe Bellussi (ENI), Billy Bardin (Dow), and Pam Klipstein (PQ) for providing information related to commercial applications.

REFERENCES

1. (a) Ertl, G.; Knoezinger, H.; Schueth, F.; Weitkamp, J.; Eds. In *Handbook of Heterogeneous Catalysis*; Vols. 1–8; VCH, a Wiley Company, Weinheim, 2008, 719 pp; (b) Bartholomew, H.; Farrauto, R. J. *Fundamentals of Industrial Catalytic Processes*; Vol. 47, 2nd ed., John Wiley & Sons, 2008; (c) Weitkamp, J.; Puppe, L. Eds. In: *Catalysis and Zeolites: Fundamentals and Applications*, 1st ed.; Springer-Verlag, Heidelberg, 2006 (August 15, 1999); (d) Cejka, J.; van Bekkam, H.; Corma, A.; Schutz, F. *Introduction to Zeolite Science and Parctice*; Elsevier, 2007; pp 1–1058; (e) Rase, Howard F. *Handbook of Commercial Catalysts: Heterogeneous Catalysts*; Vol. 98, 2002; p 81, Procelli, J. V; (f) Corma, A. *Chem. Rev.* **1995**, *95* (3), 559–614; (g) Corma, A.; García, H. *Chem. Rev.* **2002**, *102* (10), 3837–3892; (h) Crabtree, R. H. Ed.; *Handbook of Green Chemistry, Volume 2: Heterogeneous Catalysis*; 2009, 338 pp; (i) Tanabe, K.; Hölderich, W. F. *Appl. Catal. A Gen.* **1999**, *181* (2), 399–434; (j) Corma, A. *J. Catal.* **2003**, *216* (1–2), 298–312.
2. (a) Bellussi, G.; Perego, C. *Handbook of Heterogeneous Catalysis*; In: Ertl, G.; Knözinger, H.; Schuth, F.; Weitkamp, J. Eds.; Vol. 5, Wiley, 1997; (b) Centi, G.; Cavani, F.; Trifiro, F. *Selective Oxidation by Heterogeneous Catalysis*; Kluger Academic: New York, NY, 1998; pp 1–505; (c) Cavani, F.; Teles, J. H. *ChemSusChem* **2009**, *2* (6), 508–534; (d) Sheldon, R. A. *Stud. Surf. Sci. Catal.* **1990**, *55*, 1–32, New Developments in Selective Oxidation; (e) Sheldon, R. A.; Arends, I. W. C. E.; Lempers, H. E. B. *Spec. Publ. R. Soc. Chem.* **1998**, *216*, 37–46, Supported Reagents and Catalysts in Chemistry; (f) Clerici, M. G.;

Ricci, M.; Rivetti, F. *Encyclopedia of Hydrocarbons*; In: Beccani, M.; Romano, U. Eds.; Vol. 3, Instituto Della Enciclopedia Italiana Roma: Italy, 2006; pp 661–686.

3. (a) Bellussi, G.; Fattore, V. *Studies in Surface Science and Catalysis*; In: Jacobs, P. A.; Jaeger, N. I.; Kubelková, L.; Wichterlov, B.; Jacobs, P. A.; Jaeger, N. I.; Kubelková, L.; Wichterlov, B. Eds.; Vol. 69; Elsevier, 1991; pp 79–92; (b) Bruno, N. *Advances in Catalysis*; In: Eley, D. D.; Haag, W. O.; Bruce, G. Eds.; Vol. 41; Academic Press, 1996; pp 253–334; (c) Ratnasamy, P.; Kumar, R. *Studies in Surface Science and Catalysis*; Vol. 97; Elsevier, 1995; pp 367–376; (d) Ratnasamy, P.; Srinivas, D.; Knözinger, H. *Advances in Catalysis*; Vol. 48; Academic Press, 2004; pp 1–169; (e) Ingallina, P.; Clerici, M. G.; Rossi, L.; Bellussi, G. *Stud. Surf. Sci. Catal.* **1995**, *92*, 31–39, Science and Technology in Catalysis 1994; (f) Notari, B. *Catal. Today* **1993**, *18* (2), 163–172; (g) Ratnasamy, P.; Srinivas, D.; Knoezinger, H. *Adv. Catal.* **2004**, *48*, 1–169.

4. Skeels, G. W.; Flanigen, E. M. *Studies in Surface Science and Catalysis*; In: Jacobs, P. A.; van Santen, R. A. Eds.; Vol. 49; Elsevier, 1989; pp 331–344.

5. (a) Taramasso, M.; Perego, G.; Notari, B. Preparation of Porous Crystalline Synthetic Material Comprised of Silicon and Titanium Oxides. 4410501, **1983**; (b) Choudhary, V. R.; Kinage, A. K.; Choudhary, T. V. *Appl. Catal. A Gen.* **1997**, *162* (1–2), 239–248.

6. Thangaraj, A.; Kumar, R.; Mirajkar, S. P.; Ratnasamy, P. *J. Catal.* **1991**, *130* (1), 1–8.

7. Padovan, M.; Leofanti, G.; Roffia, P. Method for the Preparation of Titanium-Silicalite Catalysts. 88-116870 311983, 19881011, 1989.

8. (a) van der Pol, A. J. H. P.; Verduyn, A. J.; van Hooff, J. H. C. *Appl. Catal. A Gen.* **1992**, *92* (2), 113–130; (b) Van der Pol, A. J. H. P.; Van Hooff, J. H. C. *Appl. Catal. A Gen.* **1992**, *92* (2), 93–111.

9. Dwyer, J.; Zhao, J.; Rawlence, D.; et al. In: *Proc. 9th Int. Conf. Zeolites*, Butterworth and Heinemann, London, 1992; p 155.

10. Tuel, A.; Taarit, Y. B. *Zeolites* **1994**, *14* (7), 594–599.

11. Zhang, H.; Liu, Y.; Jiao, Z.; He, M.; Wu, P. *Ind. Eng. Chem. Res.* **2009**, *48* (9), 4334–4339.

12. Jorda, E.; Tuel, A.; Teissier, R.; Kervennal, J. *Zeolites* **1997**, *19* (4), 238–245.

13. Senderov, E.; Hinchey, R. J.; Halasz, I. TS-PQ Titano-Silicate Molecular Sieves and Methods for Synthesis and Use Thereof. 20030152510, **2003**.

14. Xu, L.; Ren, Y.; Wu, H.; Liu, Y.; Wang, Z.; Zhang, Y.; Xu, J.; Peng, H.; Wu, P. *J. Mater. Chem.* **2011**, *21* (29), 10852–10858.

15. Peng, H.; Xu, L.; Wu, H.; Wang, Z.; Liu, Y.; Li, X.; He, M.; Wu, P. *Microporous Mesoporous Mater.* **2012**, *153*, 8–17.

16. Sanz, R.; Serrano, D. P.; Pizarro, P.; Moreno, I. *Chem. Eng. J.* **2011**, *171* (3), 1428–1438.

17. Zhou, J.; Hua, Z.; Cui, X.; Ye, Z.; Cui, F.; Shi, J. *Chem. Commun.* **2010**, *46* (27), 4994–4996.

18. Serrano, D. P.; Sanz, R.; Pizarro, P.; Moreno, I. *Top. Catal.* **2010**, *53* (19–20), 1319–1329.

19. Wang, X.; Li, G.; Wang, W.; Jin, C.; Chen, Y. *Microporous Mesoporous Mater.* **2011**, *142* (2–3), 494–502.

20. Reichinger, M.; Gies, H.; van den Berg, M.; Gruenert, W.; Kirschhock, C. *Stud. Surf. Sci. Catal.* **2007**, *170A*, 276–281, From Zeolites to Porous MOF Materials.

21. Wang, Y.; Lin, M.; Tuel, A. *Microporous Mesoporous Mater.* **2007**, *102* (1–3), 80–85.

22. Ok, D.-Y.; Jiang, N.; Prasetyanto, E. A.; Jin, H.; Park, S.-E. *Microporous Mesoporous Mater.* **2011**, *141* (1–3), 2–7.

23. Au, L. T. Y.; Chau, J. L. H.; Ariso, C. T.; Yeung, K. L. *J. Membr. Sci.* **2001**, *183* (2), 269–291.

24. Sebastian, V.; Motuzas, J.; Dirrix, R. W. J.; Terpstra, R. A.; Mallada, R.; Julbe, A. *Sep. Purif. Technol.* **2010**, *75* (3), 249–256.

25. Nemeth, L. T.; Malloy, T. P.; Jones, R. R. Epoxidation of Olefins Using a Titania-Supported Titanosilicate. 5354875, **1994**.

26. (a) Serrano, D. P.; Sanz, R.; Pizarro, P.; Moreno, I.; de Frutos, P.; Blázquez, S. *Catal. Today* **2009**, *143* (1–2), 151–157; (b) Strebelle, M.; Catinat, J.-P. Epoxidation Catalyst Based on Titanium Zeolite. 1998-EP7527 9928029, 19981118, **1999**.

27. Muller, U.; Senk, R.; Harder, W.; Rudolf, P.; Rieber, N. Manufacture of Zeolite-Based Catalysts, Especially for Epoxidation of Olefins. 2001-10110139, 20010302, **2002**.

28. Wang, X.; Zhang, X.; Wang, Y.; Liu, H.; Wang, J.; Qiu, J.; Ho, H. L.; Han, W.; Yeung, K. L. *Chem. Eng. J.* **2011**, *175*, 408–416.

29. Reddy, J. S.; Sivasanker, S.; Ratnasamy, P. *J. Mol. Catal.* **1992**, *71* (3), 373–381.

30. Serrano, D. P.; Uguina, M. A.; Sanz, R.; Castillo, E.; Rodrı́guez, A.; Sánchez, P. *Microporous Mesoporous Mater.* **2004**, *69* (3), 197–208.

31. Reddy, J. S.; Sivasanker, S. *Catal. Lett.* **1991**, *11* (2), 241–244.

32. Bellussi, G.; Carati, A.; Clerici, M. G.; Esposito, A. *Stud. Surf. Sci. Catal.* **1991**, *63*, 421–429, Prep. Catal. 5.

33. Nemeth, L.; Lewis, G.J.; Rosin, R.R. Titanostannosilicates: Epoxidation of Olefins. 5780654, **1998**.

34. Tuel, A.; Ben Taârit, Y. *Zeolites* **1995**, *15* (2), 164–170.

35. Camblor, M. A.; Corma, A.; Martínez, A.; Pérez-Pariente, J.; Primo, J. *Studies in Surface Science and Catalysis*; In: Guisnet, M.; Barbier, J.; Barrault, J.; Bouchoule, C.; Duprez, D.; Pérot, G.; Montassier, C. Eds.; Vol. 78; Elsevier, 1993; pp 393–399.

36. Davis, R. J.; Liu, Z.; Tabora, J. E.; Wieland, W. S. *Catal. Lett.* **1995**, *34* (1,2), 101–113.

37. (a) Kuznicki, S.M.; Jacubinas, R.M.; Langner, T.W. Polymorph-Enriched ETS-4. 2001-US24722 2002014219, 20010807, **2002**; (b) Nastro, A.; Hayhurst, D. T.; Kuznicki, S. M. *Studies in Surface Science and Catalysis*; In: Karge, H. G.; Weitkamp, J. Eds.; Vol. 98; Elsevier, 1995; pp 22–23; (c) De Luca, P.; Kuznicki, S.; Nastro, A. *Studies in Surface Science and Catalysis*; In: Laurent, B.; Serge, K. Eds.; Vol. 97; Elsevier, 1995; pp 443–445.

38. Kuznicki, S.M.; Curran, J.S.; Yang, X. Crystalline titanium Silicate Molecular Sieves ETS-14, Their Manufacture, and Cathodes Containing the ETS-14. 98-US596 9832695, 19980107, **1998**.

39. Corma, A.; Diaz, U.; Fornes, V.; Jorda, J. L.; Domine, M.; Rey, F. *Chem. Commun.* **1999**, *9*, 779–780.

40. Diaz-Cabanas, M.-J.; Villaescusa, L. A.; Camblor, M. A. *Chem. Commun.* **2000**, *9*, 761–762.

41. (a) Wu, P.; Tatsumi, T. *Chem. Commun.* **2001**, *10*, 897–898; (b) Wu, P.; Komatsu, T.; Yashima, T.; Tatsumi, T. *Stud. Surf. Sci. Catal.* **2001**, *135*, 4375–4383, Zeolites and Mesoporous Materials at the Dawn of the 21st Century.

42. Corma, A.; Navarro, M. T.; Pérez-Pariente, J. *J. Chem. Soc. Chem. Commun.* **1994**, *2*, 147–148.

43. Corma, A.; Domine, M.; Gaona, J. A.; Jorda, J. L.; Navarro, M. T.; Rey, F.; Pérez-Pariente, J.; Tsuji, J.; McCulloch, B.; Nemeth, L. T. *Chem. Commun.* **1998**, *20*, 2211–2212.

44. (a) Pena, M. L.; Dellarocca, V.; Rey, F.; Corma, A.; Coluccia, S.; Marchese, L. *Microporous Mesoporous Mater.* **2001**, *44–45*, 345–356; (b) Dellarocca, V.; Marchese, L.; Peña, M. L.; Rey, F.; Corma, A.; Coluccia, S. *Studies in Surface Science and Catalysis*; In: Gamba, A.; Colella, C.; Coluccia, S. Eds.; Vol. 140; Elsevier, 2001; pp 209–220.

45. (a) Balkus, K. J., Jr.; Khanmamedova, A.; Gabrielov, A. G.; Zones, S. I. *Studies in Surface Science and Catalysis*; In: Hightower, J. W.; Nicholas Delgass, W.; Alexis, T. B.; Iglesia, E. Eds.; Vol. 101; Elsevier, 1996; pp 1341–1348; (b) Balkus, K. J., Jr.; Gabrielov, A. G.; Zones, S. I. *Studies in Surface Science and Catalysis*; In: Laurent, B.; Serge, K. Eds.; Vol. 97; Elsevier, 1995; pp 519–525.

46. Wu, P.; Tatsumi, T.; Komatsu, T.; Yashima, T. *Chem. Mater.* **2002**, *14* (4), 1657–1664.
47. Shan, Z.; Jansen, J. C.; Marchese, L.; Maschmeyer, T. *Microporous Mesoporous Mater.* **2001**, *48* (1–3), 181–187.
48. Thomas, J. M.; Raja, R. *Microporous Mesoporous Mater.* **2007**, *105* (1–2), 5–9.
49. Rigutto, M. S.; Van Bekkum, H. *Appl. Catal.* **1991**, *68* (1), L1–L7.
50. Sen, T.; Ramaswamy, V.; Ganapathy, S.; Rajamohanan, P. R.; Sivasanker, S. *J. Phys. Chem.* **1996**, *100* (9), 3809–3817.
51. Rao, P. R. H.; Ramaswamy, A. V. *Appl. Catal. A Gen.* **1993**, *93* (2), 123–130.
52. Ramaswamy, A. V.; Sivasanker, S.; Ratnasamy, P. *Microporous Mater.* **1994**, *2* (5), 451–458.
53. Sudhakar Reddy, J.; Liu, P.; Sayari, A. *Appl. Catal. A Gen.* **1996**, *148* (1), 7–21.
54. Bhaumik, A.; Dongare, M. K.; Kumar, R. *Microporous Mater.* **1995**, *5* (3), 173–178.
55. Reddy, K. R.; Ramaswamy, A. V.; Ratnasamy, P. *J. Catal.* **1993**, *143*, 275–285.
56. Kumar, R. In: *Catalysis: Modern Trends, [Papers presented at the National Symposium on Catalysis], 12th Bombay, Dec. 19-22, 1994,* 1996; pp 48–56.
57. Sayari, A.; Reddy, K. M.; Moudrakovski, I. *Stud. Surf. Sci. Catal.* **1995**, *98*, 19–21 Zeolite Science 1994: Recent Progress and Discussions.
58. Zhang, Q.; Wang, Y.; Ohishi, Y.; Shishido, T.; Takehira, K. *J. Catal.* **2001**, *202* (2), 308–318.
59. Pena, M. L.; Dejoz, A.; Fornes, V.; Rey, F.; Vazquez, M. I.; Lopez Nieto, J. M. *Appl. Catal. A Gen.* **2001**, *209* (1,2), 155–164.
60. Tuel, A.; Ben Taarit, Y. *Zeolites* **1994**, *14* (1), 18–24.
61. Chien, S.-H.; Ho, J.-C.; Mon, S.-S. *Zeolites* **1997**, *18* (2–3), 182–187.
62. Corma, A.; Domine, M. E.; Valencia, S. *J. Catal.* **2003**, *215* (2), 294–304.
63. Mal, N. K.; Ramaswamy, V.; Rajamohanan, P. R.; Ramaswamy, A. V. *Microporous Mater.* **1997**, *12* (4–6), 331–340.
64. Skeels, G.W.; Chapman, D.M.; Flanigen, E.M. Fluoride Treatment for Aluminum Substitution by Tin in the Frame of Microporous Aluminosilicates. US Patent 5401488, **1995**.
65. Fejes, P.; Nagy, J. B.; Kovács, K.; Vankó, G. *Appl. Catal. A Gen.* **1996**, *145* (1–2), 155–184.
66. (a) Mal, N. K.; Bhaumik, A.; Ramaswamy, V.; Belhekar, A. A.; Ramaswamy, A. V. *Stud. Surf. Sci. Catal.* **1995**, *94*, 317–324, Catalysis by Microporous Materials; (b) Mal, N. K.; Ramaswamy, V.; Ganapathy, S.; Ramaswamy, A. V. *J. Chem. Soc. Chem. Commun.* **1994**, *17*, 1933–1934.
67. Mal, N. K.; Bhaumik, A.; Kumar, R.; Ramaswamy, A. V. *Catal. Lett.* **1995**, *33* (3,4), 387–394.
68. Corma, A.; Nemeth, L. T.; Renz, M.; Valencia, S. *Nature* **2001**, *412* (6845), 423–425.
69. Li, P.; Liu, G.; Wu, H.; Liu, Y.; Jiang, J.-g; Wu, P. *J. Phys. Chem. C* **2011**, *115* (9), 3663–3670.
70. Corma, A.; Iborra, S.; Mifsud, M.; Renz, M. *J. Catal.* **2005**, *234* (1), 96–100.
71. Choudhary, V. R.; Kinage, A. K.; Belhekar, A. A. *Zeolites* **1997**, *18* (4), 274–277.
72. Machado, F. J.; López, C. M.; Goldwasser, J.; Méndez, B.; Campos, Y.; Escalante, D.; Tovar, M.; Ramírez-Agudelo, M. M. *Zeolites* **1997**, *19* (5–6), 387–394.
73. Ma, D.; Shu, Y.; Zhang, C.; Zhang, W.; Han, X.; Xu, Y.; Bao, X. *J. Mol. Catal. A Chem.* **2001**, *168* (1–2), 139–146.
74. Dooley, K. M.; Price, G. L.; Kanazirev, V. I.; Hart, V. I. *Catal. Today* **1996**, *31* (3–4), 305–315.
75. Degnan, T.F.; Klocke, D.J.; Rubin, M.K. Manufacture of Improved ZSM-5 Catalyst. 1994-325838 5554274, 19941019, **1996**.
76. Klik, R.; Bosáček, V.; Kubelková, L.; Freude, D.; Michel, D. *Zeolites* **1997**, *19* (5-6), 343–348.
77. Camblor, M. A.; Pérez-Pariente, J.; Forne´s, V. *Zeolites* **1992**, *12* (3), 280–286.

78. Sasidharan, M.; Hegde, S. G.; Kumar, R. *Microporous Mesoporous Mater.* **1998**, *24* (1–3), 59–67.
79. Fechete, I.; Caullet, P.; Dumitriu, E.; Hulea, V.; Kessler, H. *Appl. Catal. A Gen.* **2005**, *280* (2), 245–254.
80. Kumar, N.; Lindfors, L. E. *Appl. Catal. A Gen.* **1996**, *147* (1), 175–187.
81. Milanesio, M.; Artioli, G.; Gualtieri, A. F.; Palin, L.; Lamberti, C. *J. Am. Chem. Soc.* **2003**, *125* (47), 14549–14558.
82. Centi, G.; Perathoner, S.; Arrigo, R.; Giordano, G.; Katovic, A.; Pedula, V. *Appl. Catal. A Gen.* **2006**, *307* (1), 30–41.
83. Testa, F.; Pasqua, L.; Crea, F.; Aiello, R.; Lázár, K.; Fejes, P.; Lentz, P.; Nagy, J. B. *Microporous Mesoporous Mater.* **2003**, *57* (1), 57–72.
84. Kumar, R.; Thangaraj, A.; Bhat, R. N.; Ratnasamy, P. *Zeolites* **1990**, *10* (2), 85–89.
85. Hamdy, M. S.; Mul, G.; Jansen, J. C.; Ebaid, A.; Shan, Z.; Overweg, A. R.; Maschmeyer, T. *Catal. Today* **2005**, *100* (3–4), 255–260.
86. Bellussi, G.; Rigutto, M. S. *Studies in Surface Science and Catalysis*; In: Jansen, J. C.; Stöcker, M.; Karge, H. G.; Weitkamp, J. Eds.; Vol. 85; Elsevier, 1994; pp 177–213.
87. Peregot, G.; Bellussi, G.; Corno, C.; Taramasso, M.; Buonomot, F.; Esposito, A. *Studies in Surface Science and Catalysis*; In: Murakami, A. I. Y.; Ward, J. W. Eds.; Vol. 28; Elsevier, 1986; pp 129–136.
88. Skeels, G.W.; Ramos, R. Substituted Aluminosilicate Compositions and Process for Preparing Same. 5098687, **1992**.
89. (a) Bellussi, G.; Rigutto, M. S. *Studies in Surface Science and Catalysis*; In: van Bekkum, H.; Flangen, E. M.; Jacobs, P. A.; Jansen, J. C. Eds.; Vol. 137; Elsevier, 2001; pp 911–955; (b) Clerici, M. G. *Oil Gas Eur. Mag.* **2006**, *32* (2), 77–82.
90. Padovan, M.; Genoni, F.; Leofanti, G.; Petrini, G.; Trezza, G.; Zecchina, A. *Stud. Surf. Sci. Catal.* **1991**, *63*, 431.
91. Notari, B. *Studies in Surface Science and Catalysis*; In: Grobet, P. J.; Mortier, W. J.; Vansant, E. F.; Schulz-Ekloff, G. Eds.; Vol. 37; Elsevier, 1988; pp 413–425.
92. (a) Millini, R.; Perego, G. *Gazz. Chim. Ital.* **1996**, *126* (3), 133–140; (b) Millini, R.; Previde Massara, E.; Perego, G.; Bellussi, G. *J. Catal.* **1992**, *137* (2), 497–503.
93. Bellussi, G.; Buonomo, F.; Esposito, A.; Clerici, M.; Romano, U.; Notari, B. Catalyst on the Basis of Silicon and Titanium having High Mechanical Strength and a Process for Its Preparation. 4954653, **1990**.
94. Bordiga, S.; Damin, A.; Bonino, F.; Lamberti, C. *Top. Organometal. Chem.* **2005**, *16*, 37–68, Surface and Interfacial Organometallic Chemistry and Catalysis.
95. Phonthammachai, N.; Krissanasaeranee, M.; Gulari, E.; Jamieson, A. M.; Wongkasemjit, S. *Mater. Chem. Phys.* **2006**, *97* (2–3), 458–467.
96. Guo, X.; Li, G.; Zhang, X.; Wang, X. *Stud. Surf. Sci. Catal.* **1997**, *112*, 499–508, Spillover and Migration of Surface Species on Catalysts.
97. Kooyman, P. J.; Jansen, J. C.; van Bekkum, H. In: Proceedings from the Ninth International Zeolite Conference, Vol. 1, 1993; pp 505–512.
98. Drago, R. S.; Dias, S. C.; McGilvray, J. M.; Mateus, A. L. M. L. *J. Phys. Chem. B* **1998**, *102* (9), 1508–1514.
99. (a) Lin, K.; Li, L.; Sels, B. F.; Jacobs, P. A.; Pescarmona, P. P. *Catal. Today* **2011**, *173* (1), 89–94; (b) Serrano, D.; Sanz, R.; Pizarro, P.; Moreno, I. *Chem. Commun.* **2009**, *11*, 1407–1409.
100. Ravishankar, R.; Kirschhock, C.; Schoeman, B. J.; De Vos, D.; Grobet, P. J.; Jacobs, P. A.; Martens, J. A. In: *Proceedings of the 12th International Zeolite Conference, Baltimore, July 5-10, 1998*, Vol. 3, 1999; pp 1825–1832.
101. (a) Wang, X.; Zhang, X.; Liu, H.; Yeung, K. L.; Wang, J. *Chem. Eng. J.* **2010**, *156* (3), 562–570; (b) Maira, A. J.; Lau, W. N.; Lee, C. Y.; Yue, P. L.; Chan, C. K.; Yeung, K. L. *Chem. Eng. Sci.* **2003**, *58* (3–6), 959–962.

102. Wang, X. S.; Li, G.; Yan, H. S.; Guo, X. W. *Stud. Surf. Sci. Catal.* **2001**, *135*, 3923–3929, Zeolites and Mesoporous Materials at the Dawn of the 21st Century.
103. Balducci, L.; Ungarelli, R.; Tonini, C. Manufacture of Spherical Silica-Zeolite Composite for Use as Catalysts. 1997-101858, 19970206, **1997**.
104. Onimus, W. H.; Cooker, B.; Morales, E. Catalyst Preparation. 7182932, **2007**.
105. (a) Teles, J. H.; Rehfinger, A.; Bassler, P.; Wenzel, A.; Rieber, N.; Mueller, U.; Rudolf, P. Regeneration of Zeolite Catalysts for Epoxidation. 2001-EP10490 2002022259, 20010911, **2002**; (b) Mueller, U.; Teles, J. H.; Wenzel, A.; Harder, W.; Rudolf, P.; Rehfinger, A.; Bassler, P.; Rieber, N. Regeneration of Zeolite Catalyst for Epoxidation of Propene to Propylene Oxide. 2001-EP10489 2002022260, 20010911, **2002**.
106. Arca, V.; Boscolo Boscoletto, A.; Fracasso, N.; Meda, L.; Ranghino, G. *J. Mol. Catal. A Chem.* **2006**, *243* (2), 264–277.
107. Fang, Y.; Hu, H. *Catal. Commun.* **2007**, *8* (5), 817–820.
108. Meng, L.; Jiang, H.; Chen, R.; Gu, X.; Jin, W. *Appl. Surf. Sci.* **2010**, *257* (6), 1928–1931.
109. Cheng, S.; Min, E.; Wu, W.; Jin, Z.; Zhu, L.; Shan, H.; Sun, B.; Zhang, S.; Wang, E. Preparation of Ti-Si Zeolite. 1998-117501 1245089, 19980818, **2000**.
110. Liu, Y.; Yang, J.; Yao, M.; Wang, L.; Xing, R.; Wu, P.; He, M. Modification Method of Titanium Silicalite Molecular Sieve. 2008-10204149 101417238, 20081208, **2009**.
111. Lin, M.; Shi, C.; Zhu, B.; Shu, X.; Mu, X.; Luo, Y.; Wang, X.; Ru, Y. Method for Treating Titanosilicate Zeolite with Noble Metal, Source. 2008-10119341 101665256, 20080904, **2010**.
112. Oyama, S. T.; Gaudet, J.; Zhang, W.; Su, D. S.; Bando, K. K. *ChemCatChem* **2010**, *2* (12), 1582–1586.
113. Lee, W.-S.; Cem Akatay, M.; Stach, E. A.; Ribeiro, F. H.; Nicholas Delgass, W. *J. Catal.* **2012**, *287*, 178–189.
114. Europe, E. P.; Licencing; http://www.eni.com/en_IT/attachments/azienda/attivita-strategie/petrolchimica/licensing/TS1-flyer-lug09.pdf, TS-1 Catalysts for licencing.
115. (a) Hijar, C. A.; Jacubinas, R. M.; Eckert, J.; Henson, N. J.; Hay, P. J.; Ott, K. C. *J. Phys. Chem. B* **2000**, *104*, 12157–12164; (b) Hijar, C. A.; Jacubinas, R. M.; Eckert, J.; Henson, N. J.; Hay, P. J.; Ott, K. C. *J. Phys. Chem. B* **2000**, *104* (51), 12157–12164.
116. Henry, P. F.; Weller, M. T.; Wilson, C. C. *J. Phys. Chem. B* **2001**, *105* (31), 7452–7458.
117. Lamberti, C.; Bordiga, S.; Zecchina, A.; Artioli, G.; Marra, G.; Spano, G. *J. Am. Chem. Soc.* **2001**, *123* (10), 2204–2212.
118. Gale, J. D. *Solid State Sci.* **2006**, *8* (3–4), 234–240.
119. Yuan, S.; Si, H.; Fu, A.; Chu, T.; Tian, F.; Duan, Y.-B.; Wang, J. *J. Phys. Chem. A* **2011**, *115* (5), 940–947.
120. Jentys, A. C.; Richard, C. A. *Catal. Lett.* **1993**, *22* (3), 251.
121. Vincent, C.; Joseph, E.R.; Clarence, D.V. Process for Producing Low-Bulk Density Silica. 3556725, **1971**.
122. (a) Vartuli, J. C.; Kresge, C. T.; Roth, W. J.; McCullen, S. B.; Beck, J. S.; Schmitt, K. D.; Leonowicz, M. E.; Lutner, J. D.; Sheppard, E. W. *Advanced Catalysts and Nanostructured Materials*; In: William, R. M. Ed.; Academic Press: San Diego, CA, 1996, pp 1–19; (b) Kresge, C.T.; Roth, W.J. Manufacture of Mesoporous Crystalline Materials. 92-964235 5300277 U S 5,198,203, 19921021, **1994**; (c) Kresge, C. T.; Leonowicz, M. E.; Roth, W. J.; Vartulli, J. C.; Beck, J. S. *Nature* **1992**, *359*, 710.
123. Corma Canos, A.; Navarro Villalba, M. T.; Pérez Pariente, J. Structure Material of the Zeolite Type with Ultralarge Pores and a Lattice Comprised of Silicone and Titanium Oxides; Its Synthesis and Utilization for the Selective Oxidation of Organic Products. 1994-ES59 9429022, 19940610, **1994**.

124. Maschmeyer, T.; Thomas, J. M.; Masters, A. F. *NATO ASI Ser.* C **1997**, *498*, 461–494, New Trends in Materials Chemistry.
125. (a) Yamamoto, K.; Nohara, Y.; Tatsumi, T. *Chem. Lett.* **2001**, *7*, 648–649; (b) Igarashi, N.; Kidani, S.; Rizwan, A.; Tatsumi, T. *Stud. Surf. Sci. Catal.* **2000**, *129*, 163–168, Nanoporous Materials II, Proceedings of the Conference on Access in Nanoporous Materials, 2000.
126. Morey, M. S.; O'Brien, S.; Schwarz, S.; Stucky, G. D. *Chem. Mater.* **2000**, *12* (4), 898–911.
127. (a) Luan, Z.; Maes, E. M.; Van der Heide, P. A. W.; Zhao, D.; Czernuszewicz, R. S.; Kevan, L. *Chem. Mater.* **1999**, *11* (12), 3680–3686; (b) Berube, F.; Kleitz, F.; Kaliaguine, S. *J. Phys. Chem.* C **2008**, *112* (37), 14403–14411.
128. Markowitz, M. A.; Jayasundera, S.; Miller, J. B.; Klaehn, J.; Burleigh, M. C.; Spector, M. S.; Golledge, S. L.; Castner, D. G.; Gaber, B. P. *Dalton Trans.* **2003**, *17*, 3398–3403.
129. Freyhardt, C. C.; Tsapatsis, M.; Lobo, R. F.; Balkus, K. J., Jr.; Davis, M. E. *Nature* **1996**, *381* (6580), 295–298.
130. Trong On, D.; Nguyen, S. V.; Hulea, V.; Dumitriu, E.; Kaliaguine, S. *Microporous Mesoporous Mater.* **2003**, *57* (2), 169–180.
131. Nemeth, L. T.; Lewis, G. J.; Rosin, R. R. Titanovanadosilicalites as Epoxidation Catalysts for Olefins. 5744619, **1998**.
132. Reddy, J. S.; Kumar, R.; Ratnasamy, P. *Appl. Catal.* **1990**, *58* (1), L1–L4.
133. (a) Xiong, C.; Chen, Q. *React. Kinet. Catal. Lett.* **2001**, *73* (1), 39–45; (b) Uguina, M. A.; Serrano, D. P.; Ovejero, G.; Van Grieken, R.; Camacho, M. *Zeolites* **1997**, *18* (5–6), 368–378; (c) de Lucas, A.; Rodri´guez, L.; Sánchez, P. *Appl. Catal. A Gen.* **1999**, *180* (1–2), 375–383.
134. Kuznicki, S. M.; Curran, J. S.; Yang, X. ETS-14 Crystalline Titanium Silicate Molecular Sieves, Manufacture and Use Thereof. 5882624, **1999**.
135. (a) Krijnen, S.; Sanchez, P.; Jakobs, B. T. F.; Van Hooff, J. H. C. *Microporous Mesoporous Mater.* **1999**, *31* (1–2), 163–173; (b) Blasco, T.; Camblor, M. A.; Corma, A.; Esteve, P.; Guil, J. M.; Martinez, A.; Perdigon-Melon, J. A.; Valencia, S. *J. Phys. Chem.* B **1998**, *102* (1), 75–88; (c) Corma, A.; Iglesias, M.; Sanchez, F. *Catal. Lett.* **1996**, *39* (3,4), 153–156; (d) Camblor, M. A.; Corma, A.; Martinez, A.; Pérez-Pariente, J. *J. Chem. Soc. Chem. Commun.* **1992**, *8*, 589–590; (e) Wilkenhöner, U.; Gammon, D. W.; van Steen, E. *Studies in Surface Science and Catalysis*; In: Aiello, R.; Giordano, G.; Testa, F. Eds.; Vol. 142; Elsevier, 2002; pp 619–626.
136. (a) Wu, P.; Tatsumi, T. *J. Catal.* **2003**, *214* (2), 317–326; (b) Wu, P.; Tatsumi, T.; Komatsu, T.; Yashima, T. *J. Phys. Chem.* B **2001**, *105* (15), 2897–2905.
137. (a) Corma, A.; Diaz, U.; Domine, M. E.; Fornes, V. *J. Am. Chem. Soc.* **2000**, *122* (12), 2804–2809; (b) Corma, A.; Diaz, U.; Domine, M. E.; Fornes, V. *Chem. Commun.* **2000**, *2*, 137–138.
138. Dartt, C. B.; Davis, M. E. *Appl. Catal. A Gen.* **1996**, *143* (1), 53–73.
139. Fan, W.; Wu, P.; Namba, S.; Tatsumi, T. *Angew. Chem. Int. Ed.* **2003**, *43* (2), 236–240.
140. (a) Rao, P. R. H. P.; Kumar, R.; Ramaswamy, A. V.; Ratnasamy, P. *Zeolites* **1993**, *13* (8), 663–670; (b) Bhaumik, A.; Kumar, R.; Ratnasamy, P. *Stud. Surf. Sci. Catal.* **1994**, *84*, 1883–1888, ZEOLITES AND RELATED MICROPOROUS MATERIALS, PT. C.
141. Weckhuysen, B. M.; Keller, D. E. *Catal. Today* **2003**, *78* (1–4), 25–46.
142. Dzwigaj, S.; Che, M. *Catal. Today* **2011**, *169* (1), 232–241.
143. Albuquerque, A.; Pastore, H. O.; Marchese, L. *Studies in Surface Science and Catalysis*; In: Čejka, J.; Žlkovzi, N.; Nachtigall, P. Eds.; Vol. 158; Elsevier, 2005; pp 901–908, Part A.
144. Gontier, S.; Tuel, A. *Microporous Mater.* **1995**, *5* (3), 161–171.

145. Zones, S. I.; Chen, C.-Y. Zeolite Me-UTD-1 as Hydrocarbon Conversion Catalysts, Reduction Catalysts, and Adsorbents and Method of Preparation. 97-US9900 9746486, 19970606, **1997**.

146. Tielens, F.; Calatayud, M.; Dzwigaj, S.; Che, M. *Microporous Mesoporous Mater.* **2009**, *119* (1–3), 137–143.

147. Valtchev, V. *God. Sofii. Univ. "Sv. Kliment Okhridski", Geol.-Geogr. Fak.* **1992**, *83* (1), 87–97.

148. Skeels, G. W.; Flanigen, E. M. *Secondary Synthesis* **1990**, 4933161.

149. Mal, N. K.; Ramaswamy, V.; Ganapathy, S.; Ramaswamy, A. V. *Appl. Catal. A Gen.* **1995**, *125* (2), 233–245.

150. Corma, A.; Navarro, M. T.; Nemeth, L.; Renz, M. *Chem. Commun.* **2001**, *21*, 2190–2191.

151. Corma, A.; Domine Marcelo, E.; Nemeth, L.; Valencia, S. *J. Am. Chem. Soc.* **2002**, *124* (13), 3194–3195.

152. Bare, S. R.; Kelly, S. D.; Sinkler, W.; Low, J. J.; Modica, F. S.; Valencia, S.; Corma, A.; Nemeth, L. T. *J. Am. Chem. Soc.* **2005**, *127* (37), 12924–12932.

153. Corma, A.; Renz, M. *Collect. Czechoslov. Chem. Commun.* **2005**, *70* (10), 1727–1736.

154. Corma, A.; Navarro, M. T.; Renz, M. *J. Catal.* **2003**, *219* (1), 242–246.

155. (a) Centi, G.; Giordano, G.; Fejes, P.; Katovic, A.; Lazar, K.; Nagy, J. B.; Perathoner, S.; Pino, F. *Stud. Surf. Sci. Catal.* **2004**, *154C*, 2566–2573, Recent Advances in the Science and Technology of Zeolites and Related Materials; (b) Fejes, P.; Kiricsi, I.; Lázár, K.; Marsi, I.; Rockenbauer, A.; Korecz, L.; Nagy, J. B.; Aiello, R.; Testa, F. *Appl. Catal. A Gen.* **2003**, *242* (2), 247–266; (c) Patarin, J.; Tuilier; Durr, J.; Kessler, H. *Zeolites* **1992**, *12* (1), 70–75.

156. Kumar, R.; Ratnasamy, P. *J. Catal.* **1990**, *121* (1), 89–98.

157. Chandwadkar, A. J.; Bhat, R. N.; Ratnasamy, P. *Zeolites* **1991**, *11* (1), 42–47.

158. Lee, J. K.; Kim, Y. J.; Lee, H.-J.; Kim, S. H.; Cho, S. J.; Nam, I.-S.; Hong, S. B. *J. Catal.* **2011**, *284* (1), 23–33.

159. Jiang, Y.; Lin, K.; Zhang, Y.; Liu, J.; Li, G.; Sun, J.; Xu, X. *Appl. Catal. A Gen.* **2012**, *445–446*, 172–179.

160. Li, Y.; Feng, Z.; Lian, Y.; Sun, K.; Zhang, L.; Jia, G.; Yang, Q.; Li, C. *Microporous Mesoporous Mater.* **2005**, *84* (1–3), 41–49.

161. Milanesio, M.; Lamberti, C.; Aiello, R.; Testa, F.; Piana, M.; Viterbo, D. *J. Phys. Chem. B* **2000**, *104* (43), 9951–9953.

162. Calis, G.; Frenken, P.; de Boer, E.; Swolfs, A.; Hefni, M. A. *Zeolites* **1987**, *7* (4), 319–326.

163. Dubkov, K. A.; Ovanesyan, N. S.; Shteinman, A. A.; Starokon, E. V.; Panov, G. I. *J. Catal.* **2002**, *207* (2), 341–352.

164. Huang, R.; Yan, H.; Li, L.; Deng, D.; Shu, Y.; Zhang, Q. *Appl. Catal. B Environ.* **2011**, *106* (1–2), 264–271.

165. Zhao, W.; Qin, M.; Wang, L.; Chu, J.; Qu, J.; Li, S.; Li, Q.; Qi, T. *J. Colloid Interface Sci.* **2012**, *384* (1), 81–86.

166. (a) Singh, A. P.; Reddy, K. R. *Zeolites* **1994**, *14* (4), 290–294; (b) Kosslick, H.; Richter, M.; Tuan, V. A.; Parlitz, B.; Szulzewsky, K.; Fricke, R. *Studies in Surface Science and Catalysis*; In: Jacobs, P. A.; Jaeger, N. I.; Kubelková, L.; Wichterlová, B. Eds.; Vol. 69; Elsevier, 1991; pp 109–117; (c) Bayense, C. R.; van Hooff, J. H. C. *Appl. Catal. A Gen.* **1991**, *79* (1), 127–140.

167. Gabelica, Z.; Demortier, G. *Nucl. Instrum. Methods Phys. Res. B* **1998**, *136–138*, 1312–1321.

168. Kofke, T. J. G.; Gorte, R. J.; Kokotailo, G. T. *Appl. Catal.* **1989**, *54* (1), 177–188.

169. (a) Moliner Marin, M.; Diaz Cabanas, M. J.; Cantin Sanz, A.; Corma Canos, A. Manufacture and Use of the Microporous Crystalline Material Zeolite ITQ-37.

2007-ES70036 2007099190, 20070222, **2007**; (b) Corma Canos, A.; Sabater Picot, M. J.; Valencia Valencia, S. Microporous Crystalline Zeolite Material, Zeolite ITQ-28, Production Method Thereof and Use of Same. 2004-ES70074 2005030646, 20040924, **2005**; (c) Corma Canos, A.; Diaz Cabanas, M. J. Zeolite ITQ-30. 2005-ES70072 2005118476, 20050525, **2005**; (d) Corma Canos, A.; Castaneda Sanchez, R.; Fornes Segui, V.; Rey Garcia, F. Porous Crystalline Material (Zeolite ITQ-24), Preparation Method Thereof and Use of Same in the Catalytic Conversion of Organic Compounds. 2004-ES70006 2004071956, 20040210, **2004**; (e) Corma Canos, A.; Rey Garcia, F.; Valencia Valencia, S.; Martinez Triguero, L. J. Microporous Crystalline Zeolite Material (zeolite ITQ-22), Synthesis Thereof and Use of Same as a Catalyst. 2003-ES246 2003099719, 20030523, **2003**. (f) Corma Canos, A.; Rey Garcia, F.; Diaz Cabanas, M. J. Porous Crystalline Material (Zeolite ITQ-21), the Preparation Method Thereof and the Use of Same in the Catalytic Conversion of Organic Compounds. 2002-ES223 2002092511, 20020510, **2002**; (g) Chu, C. C.; Kaeding, W. W. Zeolite Catalyst Modified With a Group IVB Metal. 1980-206820 4394300, 19801114, **1983**.

170. Sonawane, H. R.; Pol, A. V.; Moghe, P. P.; Biswas, S. S.; Sudalai, A. *J. Chem. Soc. Chem. Commun.* **1994**, (10), 1215–1216.

171. (a) Nemeth, L.; Bare, S. R.; Rathbun, W.; Gatter, M.; Low, J. *Studies in Surface Science and Catalysis*; In: Antoine Gédéon, P. M.; Florence, B. Eds.; Vol. 174; Elsevier, 2008; pp 1017–1020, Part B; (b) Kumar, R.; Reddy, J.S.; Reddy, R.S.; Kumar, P. Catalytic Sulfoxidation of Thioethers using TS-2 and TS-1 as Catalyst and Dilute Hydrogen Peroxide as Oxidizing Agent. Berichte - Deutsche Wissenschaftliche Gesellschaft fuer Erdoel, Erdgas und Kohle, Tagungsbericht **1992**, 9204 (Proc. DGMK-Conf. Sel. Oxid. Petrochem, 1992), pp 367–376.

172. (a) Kumar, P.; Kumar, R.; Pandey, B. *J. Indian Inst. Sci.* **1994**, *74* (2), 293–307; (b) Kumar, R.; Pais, G. C. G.; Kumar, P.; Pandey, B. *J. Chem. Soc. Chem. Commun.* **1995**, *13*, 1315–1316.

173. Kumar, P.; et al. *J. Chem. Soc. Chem. Commun.* **1993**, *20*, 1553.

174. Srivastava, R.; Srinivas, D.; Ratnasamy, P. *Catal. Lett.* **2003**, *91* (1–2), 133–139.

175. Srinivas, D.; Srivastava, R.; Ratnasamy, P. *Catal. Today* **2004**, *96* (3), 127–133.

176. (a) Clerici, M. G.; Ingallina, P. *J. Catal.* **1993**, *140* (1), 71–83; (b) Clerici, M. G.; Bellussi, G.; Romano, U. *J. Catal.* **1991**, *129* (1), 159–167.

177. Reddy, J. S.; Sivasanker, S. *Ind. J. Technol.* **1992**, *30* (2), 64–70.

178. (a) Saxton, R.J.; Crocco, G.L.; Zajacek, J.G.; Wijesekera, K.S. Epoxidation Process and Catalyst Therefor. 199500659685, **1995**; (b) Corma, A.; Esteve, P.; Martinez, A.; Valencia, S. *J. Catal.* **1995**, *152* (1), 18–24

179. Corma, A.; Jorda, J. L.; Navarro, M. T.; Rey, F. *Chem. Commun.* **1998**, *17*, 1899–1900.

180. Kim, S.-Y.; Ban, H.-J.; Ahn, W.-S. *Catal. Lett.* **2007**, *113* (3–4), 160–164.

181. Saxton, R. J.; Zajacek, J. G.; Crocco, G. L. Epoxidation of Olefins with Hydrogen Peroxide Catalyzed by Titanium-Containing Molecular Sieve Isomorphous to Zeolite b. 1993-172404 5412122, 19931223, **1995**.

182. Clerici, M. G.; Romano, U. Olefin Epoxidation Catalysts. 1987-100625 230949, 19870119, **1987**.

183. Wang, Q.; Wang, L.; Chen, J.; Wu, Y.; Mi, Z. *J. Mol. Catal. A Chem.* **2007**, *273* (1–2), 73–80.

184. Forlin, A.; Tegon, P.; Paparatto, G. Process for the Continuous Production of an Olefinic Oxide. 7138534, **2006**.

185. Goebbel, H.-G.; Bassler, P.; Teles, J. H.; Rudolf, P.; Mueller, U.; Forlin, A.; Schulz, M.; Weidenbach, M. Process for Epoxidizing Propene. 7786317, **2010**.

186. Hofen, W.; Thiele, G.A. Continuous and Cost-Effective Process for the Catalytic Epoxidation of Olefins for the Manufacture of Propylene Oxide from

Countercurrently Contacted Liquid and Gas-Phase Streams of Hydrogen Peroxide and Propene. 2000-102543 1122247, 20000207, **2001**.

187. Nemeth, L.T.; Lewis, G.J.; Rosin, R.R. Epoxidation of Olefins Using Catalysts Containing Titanostannosilicalites. 98-199271 2000026440, 19980714, **2000**.

188. Park, S.; Cho, K. M.; Youn, M. H.; Seo, J. G.; Jung, J. C.; Baeck, S.-H.; Kim, T. J.; Chung, Y.-M.; Oh, S.-H.; Song, I. K. *Catal. Commun.* **2008**, *9* (15), 2485–2488.

189. Huybrechts, D. R. C.; Bruycker, L. D.; Jacobs, P. A. *Nature* **1990**, *345*, 240.

190. Bellussi, G.; Carati, A.; Clerici, M. G.; Maddinelli, G.; Millini, R. *J. Catal.* **1992**, *133* (1), 220–230.

191. (a) To, J.; Sokol, A. A.; French, S. A.; Catlow, C. R. A. *J. Phys. Chem. C* **2007**, *111* (40), 14720–14731; (b) Wells, D. H., Jr.; Joshi, A. M.; Delgass, W. N.; Thomson, K. T. *J. Phys. Chem. B* **2006**, *110* (30), 14627–14639; (c) To, J.; Sherwood, P.; Sokol, A. A.; Bush, I. J.; Catlow, C. R. A.; van Dam, H. J. J.; French, S. A.; Guest, M. F. *J. Mater. Chem.* **2006**, *16* (20), 1919–1926; (d) Deka, R. C.; Nasluzov, V. A.; Ivanova Shor, E. A.; Shor, A. M.; Vayssilov, G. N.; Roesch, N. *J. Phys. Chem. B* **2005**, *109* (51), v4304–v24310; (e) Catlow, C. R. A.; French, S. A.; Sokol, A. A.; Thomas, J. M. *Philos. Trans. R. Soc. London, Ser. A* **1829**, *2005* (363), 913–936; (f) Gleeson, D.; Sankar, G.; Catlow, C. R. A.; Thomas, J. M.; Spano, G.; Bordiga, S.; Zecchina, A.; Lamberti, C. *Phys. Chem. Chem. Phys.* **2000**, *2* (20), 4812–4817; (g) Munakata, H.; Oumi, Y.; Miyamoto, A. *J. Phys. Chem. B* **2001**, *105*, 3493; (h) Munakata, H.; Oumi, Y.; Miyamoto, A. *Stud. Surf. Sci. Catal.* **1999**, *121*, 227–232, Science and Technology in Catalysis 1998.

192. Sinclair, P. E.; Catlow, C. R. A. *J. Phys. Chem. B* **1999**, *103*, 1084.

193. Vayssilov, G. N.; van Santen, R. A. *J. Catal.* **1998**, *175*, 170.

194. Chang, T. Process for Epoxidation of Propylene. EP0003966 199909901445, **1999**.

195. Dubner, W. S. Improved Process for the Co-Production of Propylene Oxide and Styrene Monomer. 199300569248, **1993**.

196. (a) Tsuji, J.; Yamamoto, J.; Canos, A. C.; Ray, G., Fernando Method for Producing Propylene Oxide. 6211388, **2001**; (b) Tsuji, J.; Yamamoto, J. Process for Producing Propylene Oxide. 200100170710, **2001**.

197. (a) Klaewkla, R.; Kulprathipanja, S.; Rangsunvigit, P.; Rirksomboon, T.; Rathbun, W.; Nemeth, L. *Chem. Eng. J.* **2007**, *129* (1–3), 21–30; (b) Klaewkla, R.; Kulprathipanja, S.; Rangsunvigit, P.; Rirksomboon, T.; Nemeth, L. *Chem. Commun.* **2003**, *13*, 1500–1501.

198. Zhang, W.; Froeba, M.; Wang, J.; Tanev, P. T.; Wong, J.; Pinnavaia, T. J. *J. Am. Chem. Soc.* **1996**, *118* (38), 9171.

199. Li, K.-T.; Lin, C.-C.; Lin, P.-H. *Prepr. Am. Chem. Soc. Div. Petrol. Chem.* **2007**, *52* (2), 233–236.

200. Srinivas, D.; Ratnasamy, P. *Microporous Mesoporous Mater.* **2007**, *105* (1–2), 170–180.

201. (a) Maschmeyer, T.; Thomas, J. M. *Phys. Chem. Chem. Phys.* **1999**, *1*, 585; (b) Oldroyd, R. D.; Thomas, J. M.; Maschmeyer, T.; MacFaul, P. A.; Snelgrove, D. W.; Ingold, K. U.; Wayner, D. D. M. *Angew. Chem. Int. Ed. Engl.* **1997**, *35* (23/24), 2787–2790.

202. Prasad, M. R.; Madhavi, G.; Rao, A. R.; Kulkarni, S. J.; Raghavan, K. V. *J. Porous. Mater.* **2006**, *13* (1), 81–94.

203. Corma, A.; Serra, J. M.; Serna, P.; Valero, S.; Argente, E.; Botti, V. *J. Catal.* **2005**, *229* (2), 513–524.

204. (a) Bianchi, D.; Balducci, L.; Bortolo, R.; D'Aloisio, R.; Ricci, M.; Spano, G.; Tassinari, R.; Tonini, C.; Ungarelli, R. *Adv. Synth. Catal.* **2007**, *349* (6), 979–986; (b) Balducci, L.; Bianchi, D.; Bortolo, R.; D'Aloisio, R.; Ricci, M.; Tassinari, R.; Ungarelli, R. *Angew. Chem. Int. Ed.* **2003**, *42* (40), 4937–4940.

205. Nemeth, L. T.; Hyatt, E. M.; Malloy, T. P. Oxidation of Aromatics to Hydroxyaromatics Using Aluminosilicates Containing Framework Titanium. 5233097, **1993**.
206. Wu, P.; Komatsu, T.; Yashima, T. *J. Phys. Chem. B* **1998**, *102* (46), 9297–9303.
207. Dimitrova, R.; Spassova, M. *Catal. Commun.* **2007**, *8* (4), 693–696.
208. Gubelmann, M.; Popa, J. M.; Tirel, P. J. Preparation of Phenols by Hydroxylation of Benzene Derivatives by Nitrous Oxide Over Modified Zeolite Catalysts. 1990-401668 406050, 19900615, **1991**.
209. (a) Sobolev, V. I.; Panov, G. I.; Kharitonov, A. S.; Romannikov, V. N.; Volodin, A. M.; Ione, K. G. *J. Catal.* **1993**, *139* (2), 435–443; (b) Sobolev, V. I.; Kharitonov, A. S.; Paukshtis, Y. A.; Panov, G. I. *J. Mol. Catal.* **1993**, *84* (1), 117–124; (c) Panov, G. I.; Kharitonov, A. S.; Sobolev, V. I. *Appl. Catal. A Gen.* **1993**, *98* (1), 1–20.
210. Pérez-Ramírez, J.; Groen, J. C.; Brückner, A.; Kumar, M. S.; Bentrup, U.; Debbagh, M. N.; Villaescusa, L. A. *J. Catal.* **2005**, *232* (2), 318–334.
211. Perathoner, S.; Pino, F.; Centi, G.; Giordano, G.; Katovic, A.; Nagy, J. B. *Top. Catal.* **2003**, *23* (1–4), 125–136.
212. Pirngruber, G. D.; Roy, P. K. *Catal. Today* **2005**, *110* (3–4), 199–210.
213. Yang, G.; Zhou, D.; Liu, X.; Han, X.; Bao, X. *J. Mol. Struct.* **2006**, *797* (1–3), 131–139.
214. Fellah, M. F.; Pidko, E. A.; van Santen, R. A.; Onal, I. *J. Phys. Chem. C* **2011**, *115* (19), 9668–9680.
215. (a) Tuel, A.; Moussa-Khouzami, S.; Ben Taarit, Y.; Naccache, C. *J. Mol. Catal.* **1991**, *68* (1), 45–52; (b) Bellussi, G.; Clerici, M.; Buonomo, F.; Romano, U.; Esposito, A.; Notari, B. Catalyst Based on Silicon and Titanium having High Mechanical Strength. 1986-200663 200260, 19860418, **1986**; (c) Romano, U.; Esposito, A.; Maspero, F.; Neri, C.; Clerici, M. G. *Chim. Ind. (Milan, Italy)* **1990**, *72* (7), 610–616.
216. Keshavaraja, A.; Ramaswamy, V.; Soni, H. S.; Ramaswamy, A. V.; Ratnasamy, P. *J. Catal.* **1995**, *157* (2), 501–511.
217. Corma Canos, A.; Navarro Villalba, T.; Pérez Pariente, J. Structure Material of the Zeolite Type With Ultralarge Pores and a Lattice Comprised of Silicone and Titanium Oxides: Its Synthesis and Utilization for the Selective Oxidation of Organic Products. 5783167, **1998**.
218. Perego, C.; Carati, A.; Ingallina, P.; Mantegazza, M. A.; Bellussi, G. *Appl. Catal. A Gen.* **2001**, *221* (1–2), 63–72.
219. (a) Esposito, A.; Neri, C.; Taramasso, M. Hydroxylation of Phenol. **1985**; (b) Esposito, A.; Neri, C.; Buonomo, F. Hydroxylation of Aromatic Hydrocarbons. 1983-3309669 3309669, 19830317, **1983**.
220. Tuel, A.; Taarit, Y. B. *Appl. Catal. A Gen.* **1993**, *102* (1), 69–77.
221. (a) Roffia, P.; Padovan, M.; Leofanti, G.; Mantegazza, M. A.; De Alberti, G.; Tauszik, G. R. Catalytic Manufacture of Oximes. 87-108577 267362, 19870612, **1988**; (b) Roffia, P.; Padovan, M.; Moretti, E.; De Alberti, G. Preparation of Cyclohexanone Oxime by Ammoximation of Cyclohexene in the Presence of a Titanium-Silicalite Catalyst. 86-109400 208311, 19860709, **1987**. (c) Roffia, P.; Padovan, M.; Moretti, E.; De Alberti, G. Ammoximation of Cyclohexanone to Cyclohexanone Oxime, **1987**.
222. (a) Dal Pozzo, L.; Fornasari, G.; Monti, T. *Catal. Commun.* **2002**, *3* (8), 369–375; (b) Zecchina, A.; Spoto, G.; Bordiga, S.; Geobaldo, F.; Petrini, G.; Leofanti, G.; Padovan, M.; Mantegazza, M.; Roffia, P. *Studies in Surface Science and Catalysis*; In: Guczi, L.; Solymosi, F.; Tétényi, P. Eds.; Vol. 75; Elsevier, 1993; pp 719–729.
223. Cai, F. f.; Wang,, S.-l; Zhou, J.-c. *Huaxue Yu Shengwu Gongcheng* **2005**, *22* (11), 29–30, 32.
224. Roffia, P.; Leofanti, G.; Cesana, A.; Mantegazza, M.; Padovan, M.; Petrini, G.; Tonti, S.; Gervasuiti, P. *Stud. Surf. Sci. Catal.* **1990**, *55*, 43–52, New Developments in Selective Oxidation.

225. Thangaraj, A.; Sivasanker, S.; Ratnasamy, P. *J. Catal.* **1991**, *131* (2), 394–400.
226. Reddy, J. S.; Sivasanker, S.; Ratnasamy, P. *J. Mol. Catal.* **1991**, *69* (3), 383–392.
227. (a) Le Bars, J.; Dakka, J.; Sheldon, R. A. *Appl. Catal. A Gen.* **1996**, *136* (1), 69–80; (b) Sudhakar Reddy, J.; Sayari, A. *Appl. Catal. A Gen.* **1995**, *128* (2), 231–242.
228. Wu, P.; Komatsu, T.; Yashima, T. *J. Catal.* **1997**, *168* (2), 400–411.
229. Oikawa, M.; Kitamura, M. Oximation Method for Producing Oxime From Ketones, Ammonia, and Organic Peroxide. 2009-251778 2145876, 20090710, **2010**.
230. (a) Palkovits, R.; Schmidt, W.; Ilhan, Y.; Erdem-Senatalar, A.; Schueth, F. *Microporous Mesoporous Mater.* **2008**, *117* (1–2), 228–232; (b) Thangaraj, A.; Sivasanker, S.; Ratnasamy, P. *J. Catal.* **1992**, *137* (1), 252–256.
231. Reddy, J. S.; Ravishankar, R.; Sivasanker, S.; Ratnasamy, P. *Catal. Lett.* **1993**, *17* (1–2), 139–140.
232. Izumi, Y.; Ichihashi, H.; Shimazu, Y.; Kitamura, M.; Sato, H. *Bull. Chem. Soc. Jpn.* **2007**, *80*, 1280–1287.
233. Boronat, M.; Corma, A.; Renz, M. *J. Phys. Chem. B* **2006**, *110* (42), 21168–21174.
234. (a) Li, G.; Ling, Z.; Yuan, X.; Wu, J.; Luo, H. *Huaxue Fanying Gongcheng Yu Gongyi* **2010**, *26* (2), 162–166; (b) Dutta, B.; Jana, S.; Bhunia, S.; Honda, H.; Koner, S. *Appl. Catal. A Gen.* **2010**, *382* (1), 90–98; (c) Suzuki, K. *Shokubai* **2009**, *51* (7), 563; (d) Nemeth, L.; Bare, S.; Corma, A. In: *Abstracts of Papers, 235th ACS National Meeting, New Orleans, LA, United States, April 6-10, 2008* 2008, CATL-019; (e) Raja, R.; Thomas, J. M. *Z. Anorg. Allg. Chem.* **2005**, *631* (13–14), 2942–2946; (f) Hao, X.; Yamazaki, O.; Yoshida, A.; Nishikido, J. *Green Chem.* **2003**, *5* (5), 524–528; (g) Renz, M.; Blasco, T.; Corma, A.; Fornes, V.; Jensen, R.; Nemeth, L. *Chem. Eur. J.* **2002**, *8* (20), 4708–4717.
235. Canos, A. C.; Nemeth, L. T.; Renz, M.; Moscoso, J. G. The Oxidation of Ketones to Esters Using a Tin Substituted Zeolite Beta. 200100181291, **2001**.
236. Kelly, D. R.; Wan, P. W. H.; Tang, J. In: Rehm, H.-J.; Reed, G., Eds.; *Biotechnology*; Kelly, D. R, Ed.; *Volume 8a, Biotransformations I;* Wiley-VCH: Weinheim, 1998, p. 535.
237. Canos, A. C.; Nemeth, L. T.; Renz, M.; Moscoso, J. G. The Oxidation of Ketones to Esters Using a Tin Substituted Zeolite Beta. 2002-00181291B, **2002**.
238. Corma, A.; Fornes, V.; Iborra, S.; Mifsud, M.; Renz, M. *J. Catal.* **2004**, *221* (1), 67–76.
239. (a) Sever, R. R.; Root, T. W. *J. Phys. Chem. B* **2003**, *107* (39), 10848–10862; (b) Sever, R. R.; Root, T. W. *J. Phys. Chem. B* **2003**, *107* (38), 10521–10530.
240. Boronat, M.; Corma, A.; Renz, M.; Sastre, G.; Viruela, P. M. *Chem. Eur. J.* **2005**, *11* (23), 6905–6915.
241. Shetty, S.; Pal, S.; Kanhere, D. G.; Goursot, A. *Chem. Eur. J.* **2006**, *12* (2), 518–523.
242. Kulkarni, B. S.; Krishnamurty, S.; Pal, S. *J. Mol. Catal. A Chem.* **2010**, *329* (1–2), 36–43.
243. (a) Boronat, M.; Corma, A.; Renz, M.; Viruela, P. M. *Chem. Eur. J.* **2006**, *12* (27), 7067–7077; (b) Boronat, M.; Corma, A.; Renz, M. *J. Phys. Chem. B* **2006**, *110* (42), 21168–21174.
244. (a) Nikolla, E.; Roman-Leshkov, Y.; Moliner, M.; Davis, M. E. *ACS Catal.* **2011**, *1* (4), 408–410; (b) Roman-Leshkov, Y.; Moliner, M.; Labinger, J. A.; Davis, M. E. *Angew. Chem. Int. Ed.* **2010**, *49* (47), 8954–8957, S8954/1–S8954/8; (c) Moliner, M.; Roman-Leshkov, Y.; Davis, M. E. *Proc. Natl. Acad. Sci. U. S. A.* **2010**, *107*, 6164–6168.
245. Taarning, E.; Saravanamurugan, S.; Spangsberg Holm, M.; Xiong, J.; West, R. M.; Christensen, C. H. *ChemSusChem* **2009**, *2* (7), 625–627.
246. Holm, M. S.; Saravanamurugan, S.; Taarning, E. *Science* **2010**, *328* (5978), 602–605.
247. Lew, C. M.; Rajabbeigi, N.; Tsapatsis, M. *Microporous Mesoporous Mater.* **2012**, *153*, 55–58.

248. (a) Choudhary, V.; Pinar, A. B.; Sandler, S. I.; Vlachos, D. G.; Lobo, R. F. *ACS Catal.* **2011**, *1* (12), 1724–1728; (b) Vlachos, D.; Caratzoulas, S. In: *Pacifichem 2010, International Chemical Congress of Pacific Basin Societies, Honolulu, HI, United States, December 15-20,* 2010, AETECH-141, 2010.

249. Lobo, R. F. *ChemSusChem* **2010**, *3* (11), 1237–1240.

250. Assary, R. S.; Curtiss, L. A. *J. Phys. Chem. A* **2011**, *115* (31), 8754–8760.

251. (a) Moissette, A.; Lobo, R. F.; Vezin, H.; Bremard, C. *J. Phys. Chem. C* **2011**, *115* (14), 6635–6643; (b) Guisnet, M.; Gnep, N. S.; Alario, F. *Appl. Catal. A Gen.* **1992**, *89* (1), 1–30.

252. http://www.cleanscience.co.in/TS-1%20Zeolite%20Catalyst.html, **2012**.

253. Program, C.-P. Hydrogen Peroxide 07/08-3; White Plains, NY 10601, **2009**; pp 1–110 + attachements.

254. Vanden Bussche, K. M.; Oroskar, A. R.; Bricker, J. C.; Nemeth, L. T.; Towler, G. P. Apparatus and Process for the Synthesis of Hydrogen Peroxide Directly from Hydrogen and Oxygen. 20060096869, **2006**.

255. Grey, R. A.; Le-Khac, B. Oxidation Process with In-Situ H_2O_2 Generation and Polymer-Encapsulated Catalysts Therefor. 7030255, **2006**.

256. Meiers, R.; Dingerdissen, U.; Holderich, W. F. *J. Catal.* **1998**, *176* (2), 376–386.

257. Chemsysten—PERP Report Propylene Oxide 07/08–6; White Plains, NY 10601, **2008-2009**; pp 1–97 + attachements.

258. Announcement, D.-B. P. http://www.dow.com/news/corporate/2009/20090305a. htm, Public Announcement P-09-154 New BASF and Dow HPPO Plant in Antwerp Completes Start-Up Phase [Online], **2009**.

259. Announcements, D. B. A. Dow and BASF Receive Presidential Green Chemistry Challenge Award for HPPO Technology. http://www.dow.com/propyleneoxide/news/20091105a.htm, http://www.dow.com/news/corporate/2010/20100622a. htm, http://www.dow.com/news/multimedia/media_kits/2010_06_21a/pdfs/green_chem_20100621.pdf. http://www.basf.com/group/corporate/en/news-and-media-relations/news-releases/P-10-327, Presidential Green Chemistry Award [Online], **2010**.

260. (a) Gobbel, H.-G.; Schultz, H.; Schultz, P.; Patrascu, R.; Schultz, M.; Weidenbach, M. Separation of Propylene Oxide from a Mixture Comprising Propylene Oxide and Methanol. 20060006054, **2006**; (b) Strebelle, M.; Catinat, J.-P. Titanium Zeolite Epoxidation Catalyst. 1998-EP7528 9928030, 19981118, **1999**.

261. (a) Thiele, G. F.; Roland, E. *J. Mol. Catal. A Chem.* **1997**, *117* (1–3), 351–356, Proceedings of the 6th International Symposium on the Activation of Dioxygen and Homogeneous Catalytic Oxidation, 1996; (b) Thiele, G. Process for Preparing Epoxides from Olefins. 1996-108370 757043, 19960525, **1997**. (c) Thiele, G. Regeneration of Epoxidation Catalyst and Manufacture of an Epoxy Compound using the Catalyst. 1995-19528220 19950801, **1997**.

262. (a) Hasenzahl, S.; Thiele, G. Process for the Preparation of Titanium Containing Molecular Sieves. 6054112, **2000**; (b) Hasenzahl, S.; Markowz, G.; Viandt, M.; Roland, E. E.; Thiele, G.; Goor, G.; Moller, A. Granulates which Contain Titanium Silicalite-1. 6106803, **2000**.

263. Tsuji, J.; Yamamoto, J.; Ishino, M.; Oku, N. *Sumitomo Kagaku* **2006**, *1*, 4.

264. Tsuji, J.; Yamamoto, J.; Canos, A. C.; Garcia, F. R. Method for Producing Propylene Oxide. 6211388, **2001**.

265. ChemSystems PERP Report Hydrogen Peroxide-Based Propylene Oxide 06/07S2; Nexant's ChemSystems PERP report, 06/07S2, Hydrogen Peroxide-Based Propylene Oxide, **2007**.

266. Muller, U.; Schulz, M.; Gehrer, E.; Grosch, G. H.; Harder, W.; Dembowski US Patent 5,859,265 and 6,008,389 BASF; Epoxidation, **1997**.

267. (a) Grey, R. A.; Le-Khac, B. Oxidation Process with In-Situ H2O2 Generation and Polymer-Encapsulated Catalysts. 2004-796680 20050202957, 20040309, **2005**; (b) Rueter, M. Direct Hydrogen Peroxide Production Using Staged Hydrogen Addition. 7067103, **2006**.

268. Dessau, R. M.; Kahn, A. P.; Grey, R. A.; Jones, C. A.; Jewson, J. D. Epoxidation process, 6008388, **1999**.

269. Kawabata, T. Preparation of Olefin Oxides by Epoxidation of Olefins Catalyzed by Titanosilicates and Precious Metal-Loading Activated Carbon. 2008-19030 2009179580, 20080130, **2009**.

270. Sugita, K.; Yagi, T. Process for Producing Olefin Oxides. 5573989, **1996**.

271. Cavani, F.; Gaffney, A. M. In: *Sustainable Industrial Processes*; Cavani, F.; Centi, G.; Perathoner, S.; Trifiro, F. Eds.; Wiley-VCH, 2009; pp 319–365.

272. (a) Zhan, G.; Du, M.; Huang, J.; Li, Q. *Catal. Commun.* **2011**, *12* (9), 830–833; (b) Lee, W.-S.; Cem Akatay, M.; Stach, E. A.; Ribeiro, F. H.; Nicholas Delgass, W. *J. Catal.* **2012**, *287*, 178–189.

273. Liu, T.; Luo, M.; Lu, J. *Shiyou Huagong* **2009**, *38* (3), 260–266.

274. (a) Haruta, M.; Angelov, K. J.; Tsubota, T.; Hayashi, T.; Wada, M. Catalysts Containing Titanium and Gold for Partial Oxidation of Hydrocarbons and Preparation of Oxygen-Containing Organic Compounds Using the Catalysts. 1997-302833 11128743, 19971105, **1999**; (b) Haruda, M.; Angelov, K. J.; Tsubota, T.; Hayashi, T. Catalysts for Partial Oxidation of Hydrocarbons and Method for Manufacture of Oxygen-Containing Hydrocarbons. 1997-240612 11076820, 19970905, **1999**.

275. Buijink, J. K. F.; Lange, J.-P.; Bos, A. N. R.; Horton, A. D.; Niele, F. G. M. In: *Mechanisms in Homogeneous and Heterogeneous Epoxidation Catalysis*; Oyama, S. T. Ed.; Elsevier: Amsterdam, 2008; pp 355–371.

276. Announcement, D. B. P. http://www.dow.com/polyurethane/news/ 2012/20120104a.htm Public Announcement. A Joint Venture Between Dow and Siam Cement Group, Has Finalized the Start-Up of Its New Propylene Oxide (PO) Facility in Thailand by Successfully Completing Its Full Capacity Performance Test [Online], **2012**.

277. Gobbel, H.-G.; Bassler, P.; Teles, J. H.; Rudolf, P.; Muller, U.; Forlin, A.; Schulz, M.; Weidenbach, M. Process for Epoxidizing Propene. 20080306290, **2008**.

278. (a) Haas, T.; Brasse, C.; Stochniol, G.; Woell, W.; Hofen, W.; Thiele, G. A Continuous Process for Epoxidation of Olefins with Hydrogen Peroxide. 2002-21966 1403259, 20020930, **2004**. (b) Haas, T.; Hofen, W.; Sauer, J.; Thiele, G. Process for the Epoxidation of Olefins. 6600055, **2003**.

279. Haas, T.; Hofen, W.; Thiele, G.; Kampeis, P. *DGMK Tagungsbericht* **2001**, *2001-4*, 127–130, Proceedings of the DGMK-Conference "Creating Value from Light Olefins–Production and Conversion", 2001.

280. Bredemeyer, N.; Langanke, B.; Ullrich, N.; Haas, T.; Hofen, W.; Jaeger, B. *Proceedings of the DGMK/SCI on "Oxidation and Functionalization: Classical and alternative routes and sources", October 2005*; DGMK Tagungsbericht: Milan, 2005–2; p. 170.

281. Corma, A.; Lorda, J. L.; Navarro, M. T.; Pérez-Pariente, J.; Rey, F.; Tsuji, J. *Stud. Surf. Sci. Catal.* **2000**, *129*, 169–178, Nanoporous Materials II, Proceedings of the Conference on Access in Nanoporous Materials.

282. Oku, N.; Seo, T. Process for Producing Propylene Oxide. 20020151730, **2002**.

283. Sato, A.; Miyake, T. JP Koka 4-352771 Epoxidation of Propylene with H2O2, **1992**.

284. Dessau, R. M.; Kahn, A. P.; Grey, R. A. US Patent 6,005,123 BS US Patent 6,008,388; Epoxidation, **1999**.

285. Hoelderich, W. Epoxidation, 1998 DE 98-19845975 and J. Mol. Catal. A, 141, 215.

286. Brasse, C.; Jaeger, B.; Epoxidation, D. A. O.; http://www.degussa-award.com/degussa/en/press/news/details. Degussa Announcement on Epoxidation, **2008**.

287. Zhou, B., Rueter, M. and Parasher, S. Nano Catalyst US Patent 7,011,807—7,045,479—7,144,565 and 7,045,481. US Patent 7,011,807—7,045,479—7,144,565 and 7,045,481, **2006**.

288. Hayden, P.; Sampson, R. J. 4,007,135, **1977**.

289. Rangasamy, P.; Kahn, A.P.; Gaffney, A.M. Epoxidation US Patent 5,625,084, 5,686,380, 5,703,254, 5,864,047. 5,861,51. US Patent 5,625,084, 5,686,380, 5,703,254, 5,864,047. 5,861,51, **1997-1999**.

290. Bowman, R. G. Epoxidation Eur Patent 318,815, **1989**.

291. Panov, G. I.; Sobolev, V. I.; Kharitonov, A. S.; Paukshtis, E. A. *Chem. Ind.* **1995**, *62*, 525–530, Catalysis of Organic Reactions.

292. Kharitonov, A. S.; Sheveleva, G. A.; Panov, G. I.; Sobolev, V. I.; Paukshtis, Y. A.; Romannikov, V. N. *Appl. Catal. A Gen.* **1993**, *98* (1), 33–43.

293. (a) McGhee, W. D. Selective Introduction of Active Sites for Hydroxylation of Benzene. 5977008, **1999**; (b) McGhee, W. D. Selective Introduction of Active Sites for Hydroxylation of Benzene. 5808167, **1998**.

294. Bellussi, G.; Perego, C. *CATTECH* **2000**, *4*, 4–16.

295. www.che.com, Benzene to Phenol Ready for Commercialization.

296. Kustov, L. M.; Tarasov, A. L.; Tyrlov, A. A.; Bogdan, V. I. Method for the Oxidation of Benzene and/or Toluene to Phenol and/or Cresols. 6388145, **2002**.

297. Centi, G.; Cavani, F.; Trifirò, F. *Selective Oxidation by Heterogeneous Catalysis*; Kluwer Academic: New York, NY, 2000 p.118.

298. Kumar, R.; Raj, A.; Kumar, S. B.; Ratnasamy, P. Process for the Preparation of Titanium Silicates. 5885546, **1999**.

299. Roffia, P.; Paparatto, G.; Cesana, A.; Tauszik, G. Direct Preparation of Ketoximes From Secondary Alcohols. 88-112060 301486, 19880726, **1989**.

300. Petrini, G.; Leofanti, G.; Mantegazza, M. A.; Pignataro, F. *Green Chemistry: Designing Chemistry for the Environment*; Anastas, P. T.; Williamson, T. C. Eds.; ACS Symposium Series, 626, American Chemical Society: Washington D.C., 1996; p. 33.

301. (a) Oikawa, M.; Fukao, M. Method for Producing Cyclohexanone Oxime. 7067699, **2006**; (b) Fukao, M.; Oikawa, M. Continuous Catalytic Ammoxidation Reaction for Producing Cycloalkanone Oximes from Cycloalkanones and Hydrogen Peroxide and Ammonia in the Presence of a Titanosilicate and a C1-6 Alcohol Solvent. 2006-8306 1717222, 20060421, **2006**; (c) Fukao, M.; Kawase, S. Method for Producing Cyclohexanone Oxime. 2006232774, **2006**. (d) Oikawa, M.; Fukao, M. Oximation Method and Catalysts for Producing Cyclohexanone Oxime from Cyclohexanone and Ammonia and Hydrogen Peroxide. 2003-457383 2004002619, 20030610, **2004**.

302. Corma, A.; Iborra, S.; Mifsud, M.; Renz, M.; Susarte, M. *Adv. Synth. Catal.* **2004**, *346* (2–3), 257–262.

303. Jeanneret, J. J. *Handbook of Petroleum Refining Processes 2.5. UOP-BP Cyclar Process.* McGrawHIll, 1996.

304. John, N. A. *Catal. Today* **2011**, *163* (1), 3–9.

305. (a) Rabo, J. A. *Appl. Catal. A Gen.* **2002**, *229* (1–2), 7–10; (b) Rabo, J. A.; Schoonover, M. W. *Appl. Catal. A Gen.* **2001**, *222* (1–2), 261–275; (c) Rabo, J. A. *Catal. Today* **1994**, *22* (2), 201–233; (d) Pujadó, P. R.; Rabó, J. A.; Antos, G. J.; Gembicki, S. A. *Catal. Today* **1992**, *13* (1), 113–141.

306. Haag, W. O. *Studies in Surface Science and Catalysis*; In: Weitkamp, J.; Karge, H. G.; Pfeifer, H.; Hölderich, W. Eds.; Vol. 84; Elsevier, 1994; pp 1375–1394.

Identification and Characterization of Surface Hydroxyl Groups by Infrared Spectroscopy

Konstantin Hadjiivanov

Institute of General and Inorganic Chemistry, Bulgarian Academy of Sciences, Sofia 1113, Bulgaria

Contents

☆Dedicated to the memory of Helmut Knözinger.

Advances in Catalysis, Volume 57
ISSN 0360-0564
http://dx.doi.org/10.1016/B978-0-12-800127-1.00002-3

Abstract

Surface hydroxyl groups are active centers in many catalytic reactions and can play an important role during catalyst preparation. In this chapter, the application of infrared spectroscopy for identification and characterization of surface OH groups is reviewed. The potential of other techniques is also briefly described. The vibrational signature of various types of hydroxyls is discussed; however, the amount of information that can be gathered from the hydroxyl spectra itself is limited. In contrast, application of probe molecules allows a profound characterization of hydroxyl species. Two scenarios are considered: the formation of H-bonds between hydroxyl groups and probe molecules, and chemical reactions between hydroxyl groups and molecules or ions (such as protonation of basic probe molecules, exchange reactions, redox processes). Means to explore the accessibility and location of hydroxyl groups are introduced. The properties of OD and OH groups are compared, and the application of H/D exchange as a diagnostic reaction is discussed. Finally, the hydroxyl population on materials of practical interest is analyzed.

ABBREVIATIONS

2D two-dimensional
ACI accessibility index
AEL structure code of a zeolite framework
AFI structure code of a zeolite framework
AlPO porous aluminophosphate
BEA structure code of a zeolite framework
BHW Bellamy–Hallam–Williams
CHA structure code of a zeolite framework
CoAPO-*n* porous cobalt aluminophosphate, $n = 5, 11, 18$
DAY dealuminated Y (FAU) zeolite
DFT density functional theory
DMP 2,6-dimethylpyridine, lutidine
DTBP 2,6-di-*tert*-butylpyridine
EMT structure code of a zeolite framework
ERI structure code of a zeolite framework
FAU structure code of a zeolite framework
FER structure code of a zeolite framework
FTIR Fourier transform infrared
FWHM full width at half maximum
INS inelastic neutron scattering
ITQ Instituto de Tecnologia Quimica Valencia; zeolites with different structures
KFI structure code of a zeolite framework
LTA structure code of a zeolite framework
MAS NMR magic angle spinning nuclear magnetic resonance
MAZ structure code of a zeolite framework
MCM-*n* Mobil composition of matter, ordered porous materials with different structures:
  $n = 22, 49, 58, 69$ (zeolites); $n = 41, 48$ (mesoporous materials)
MFI structure code of a zeolite framework
MOF metal-organic framework
MOR structure code of a zeolite framework
MTW structure code of a zeolite framework
NCL-1 a type of zeolite
NMR nuclear magnetic resonance
PA proton affinity
Py pyridine
SAPO-*n* porous silicoaluminophosphate, $n = 4, 5, 18, 34, 37, 40, 44$
SBA Santa Barbara; ordered porous materials with different structures
SSZ-*n* Standard Oil Synthetic Zeolite; porous alumosilicates with different structures
  (SSZ-13, CHA; SSZ-24, AFI)
SUZ Sunbury Zeolite; SUZ-4, SZR structure
SZR structure code of a zeolite framework
TMA trimethylamine
TON structure code of a zeolite framework
USY ultrastable Y (FAU) zeolite
X zeolite with a FAU structure

Y zeolite with a FAU structure
ZSM Zeolite Socony Mobil; porous materials with different structures: ZSM-5, MFI; ZSM-12, MTW

GREEK SYMBOLS

ν vibrational mode
$\nu(OH)$ and ν_{OH} OH stretching frequency
$\tilde{\nu}$ wavenumber in cm^{-1}
γ out-of-plane OH mode
δ deformation mode; in-plane OH deformation mode

1. INTRODUCTION

Because of the humidity of the atmosphere, most of the solid materials in our surrounding are, at least in part, covered by surface hydroxyl groups or adsorbed water molecules. This situation also applies to heterogeneous catalysts: on most surfaces that have been exposed to the ambient, OH groups are present. Very often, these groups are active sites in catalytic reactions. For example, it is well established that those hydroxyl groups in zeolites that exhibit Brønsted acidity act as catalytically active sites for a variety of reactions. In many cases, the catalysts are bifunctional, and protonic acidity is one of the functions required for the reaction. Surface hydroxyls are often intermediates in catalytic processes. Undoubtedly, the OH groups are spectator species in some catalytic reactions or may even block some catalytically active sites. However, the role of the hydroxyl groups is not restricted to the catalytic reaction process. Hydroxyl groups are frequently involved in the catalyst preparation process by acting as functional groups in surface chemical reactions. For instance, metal (and other) cations can be incorporated into zeolites by exchanging the protons of the so-called bridging hydroxyls with the desired cations. Many processes of grafting, for example, onto silica, involve surface OH groups. It is well established that hydroxyls participate in the formation of the active phase, also when preparation techniques are applied that do not rely on the presence of OH groups, such as impregnation.

For the reasons detailed above, surface OH groups attract the continuous attention of the catalysis community. A search of the Scopus database with the keywords "hydroxyls," "catalysis," and "infrared spectroscopy" resulted in 304 documents in 1996 and in 830 documents in 2012. Many efforts have been made to characterize the properties of surface hydroxyls in detail. Attention has been paid to the geometry, acidity or basicity, stability, redox behavior, location, etc., and attempts have been made to connect these properties with the reactivity of the solid exhibited during catalyst preparation steps or in catalytic processes. Without doubt, the most convenient and most frequently used technique for detection and characterization of surface hydroxyl groups is infrared spectroscopy. For this reason, this review is focused on the possibilities presented by this technique. However, there are many other techniques that give valuable information on surface hydroxyls; among these, NMR spectroscopy is perhaps the most important. These techniques and their potential will also be briefly described and discussed.

Exchange of protium in hydroxyl groups with deuterium is an often used approach to analyze properties of OH groups. Here, the kind of information that can be obtained from such experiments will be described, and the differences (albeit small) between the properties of OH groups and the respective OD groups will be emphasized.

The most commonly used quantity for describing an OH group is the frequency of the associated O–H stretching vibration. It will be shown that, although a basic parameter in spectroscopy, the frequency of the OH bands cannot be used to unambiguously determine the type of site to which an OH is bound (thus distinguishing Si–OH, Fe–OH, etc.). The cause of this ambiguity is the fact that even weak H-bonding (that can occur with oxygen from the solid) can strongly affect the frequency. Therefore, we will consider the problem in its entire complexity and will give, when available, information on other IR-active modes (including combination and overtone modes) and on acid–base, redox, or other properties.

In almost all investigations of powdered samples with IR spectroscopy, the authors have detected and usually reported the existence of OH groups. This fact makes the number of papers addressing one or another aspect of surface hydroxyls enormous. Therefore, we are not able to cover all published data in this review. We have tried to consider papers presenting essential or new (at the time of publication) information on surface hydroxyl groups.

This chapter is organized as follows. We start with answering general questions regarding surface OH groups: why they exist, how stable they are, what types of OH groups are distinguished, what their vibrational properties are, and what techniques are available for detection and characterization. In the two subsequent sections, we focus on the interaction of OH groups with various reagents. This topic includes the reversible interaction of OH groups with probe molecules to determine specific properties such as acidity, basicity, or type. Special attention is paid to IR data revealing the location of OH groups, mainly in porous materials. Chemical transformation of OH groups constitute the second part of this topic. The next section is dedicated to the properties of isotopically substituted OH groups. Thereafter, some hints on conducting experiments and avoiding misinterpretations are provided. The last part summarizes the data on various materials of practical importance. We believe this information may help researchers to characterize unknown materials and interpret the data. This part starts with a general description of the OH population on different classes of materials and then considers OH groups bound to a particular chemical element. At the end, we briefly summarize the current status of the field and outline some unresolved problems. We hope that this work will be useful not only for the catalysis community but for a broader circle of scientists and engineers.

2. TYPES OF SURFACE OH GROUPS AND THEIR SPECTRAL SIGNATURE

The hydroxyl group is defined as a functional group that consists of an oxygen atom and a hydrogen atom joined by a single bond. When located on a solid surface, it is called a surface hydroxyl group.

2.1. O–H stretching frequency

The free OH^- group is a negatively charged species and, therefore, tends to be chemically bound to a positively charged site. However, the free hydroxyl anion OH^- can be generated and detected in the gas phase or by matrix isolation. As a diatomic species, the hydroxyl group possesses only one vibration, which is the O–H stretching mode. The O–H bond order is about 1, and the O–H stretching frequency is related to this bond order.

In 1986, the fundamental O–H stretching absorption frequency of the isolated hydroxyl anion in the gas phase was reported to be 3555.6 cm^{-1} (1). Previously, scientists had used a reference value for the frequency proposed by Herzberg (2), which was 3700 cm^{-1}. This value led to a number of incorrect conclusions, also about surface hydroxyls. Interestingly, the ν(OH) mode of the isolated hydroxyl radical $OH^\bullet$ is reported at a wavenumber that is very similar to that of hydroxyl anion; specifically, the radical absorbs at 3548.2 cm^{-1} (3). The proximity of the frequencies has been taken as evidence that the electron reducing the $OH^{\bullet}$ radical to an anion occupies a nonbonding orbital *(4)*. An important conclusion from this observation is that the O–H frequency should be largely insensitive to electron transfer away from the hydroxyl group and, as it will be shown below, rather depends on other factors.

The measured frequency of the hydroxyl anion differs from that expected for a harmonic oscillator, because molecular species behave like nonharmonic oscillators *(5–7)*. Briefly, the anharmonicity X_{12} (in units of cm^{-1}) of a molecular vibration can be calculated according to Equation (2.1) *(8)*:

$$X_{12} = 0.5 \times (2 \times \tilde{\nu}_{01} - \tilde{\nu}_{02}) \qquad (2.1)$$

where $\tilde{\nu}_{01}$ is the wavenumber of the fundamental frequency and $\tilde{\nu}_{02}$ that of the overtone frequency.

The harmonic frequency, ω_e (expressed in units of cm^{-1}), is easily obtained according to the following (Equation 2.2):

$$\omega_e = \tilde{\nu}_{01} + 2X_{12}. \qquad (2.2)$$

Calculations on the basis of these equations show that, for the gas–phase hydroxyl radical, the harmonic OH stretching vibration would be at $3738.4 \ cm^{-1}$ and, consequently, the O–H bond anharmonicity is $91.4 \ cm^{-1}$ *(8)*.

When the hydroxyl anion OH^{-} is bound to a positively charged center, M^{+}, two more vibrations appear in the IR spectra: a stretching mode $\nu(M–O)$ and a deformation mode $\delta(MOH)$. According to Nakamoto *(9)*, the $\nu(M–O)$ modes are observed in the region of $900–300 \ cm^{-1}$, whereas the $\delta(MOH)$ vibrations are observed between 1200 and $700 \ cm^{-1}$.

Within this model, the water molecule could be considered as a "hydrogen hydroxyl." However, in this case (i.e., with M=H), the $\nu(M–O)$ and $\nu(O–H)$ modes coincide and, as a result, are split into symmetric and antisymmetric vibrations. Therefore, from a spectroscopic point of view, water is a more complicated system than other hydroxyls.

Very often, the oxygen of a hydroxyl group is connected to more than one cation (in addition to being bound to hydrogen). This situation is realized in bridging hydroxyls $M_1–(OH)–M_2$. In this case, two different $\nu(M–O)$ modes will be observed and, because the M–O bond order in $M_1–(OH)–M_2$ hydroxyls is about half of the M–O bond order in M–OH hydroxyls, the $\nu(M–O)$ stretching vibration is expected at lower wavenumbers. The deformation modes also split into two components, called in-plane (δ) and out-of-plane (γ) modes. In addition, an M_1–O–M_2 deformation mode will appear. As a rule, the out-of-plane vibrations are situated at lower frequencies than the in-plane vibrations.

When the M (or M_1 and M_2) sites are located on a surface, the OH groups bound to them are commonly designated as surface OH groups. There is no consensus on the exact definition as to what constitutes a "surface OH group." In many cases, for example, in zeolites and metal-organic frameworks (MOFs), the hydroxyls are intrinsic to the structure. However, they are referred to as surface hydroxyls because of their accessibility. Hereafter, we will denote as surface hydroxyls all hydroxyls that are accessible to small molecules, because we believe this definition to be sensible in the context of catalysis.

The ν(OH) modes of surface hydroxyl groups are well distinguished in IR spectra, whereas in most cases, the M–O stretching vibrations and the deformation modes are masked by the strong absorbance of the bulk of the solid in the low-frequency region. However, in a few cases (predominantly concerning H-bonded hydroxyls), the direct detection of deformation modes is possible. This situation is encountered for metal oxides *(10)* and zeolites *(11)*.

Alternative to the direct detection of specific modes, indirect methods that take advantage of combination modes can be applied. Combination modes, which arise from addition or subtraction of two frequencies, appear in IR spectra, though generally with low intensity. By analyzing the combinations of masked vibrations with OH stretching vibrations, it is possible to deduce the positions of the masked vibrations (through knowledge of ν(OH)).

It is well established that the properties of a hydroxyl group and its spectral response depend on the nature of the cation or atom to which it is bound. Covalently bound, isolated hydroxyl groups (e.g., Si–OH, P–OH, or C–OH) have characteristic frequencies, and the angle M–O–H is smaller than 180°, that is, they are tilted.

According to the Stark effect, the stretching frequency of a dipole that is placed in a uniform electrostatic field should increase. Consequently, the frequency of the dipolar OH$^-$ group should be affected when the group is connected to a metal cation because of the cation's electrostatic field. To be precise, the entire field created by the cation and the surrounding surface should be taken into account. A detailed analysis of the influence of the bonding between OH groups and metal cations on the stretching frequency ν(OH) has been performed by Beckenkamp and Lutz *(12)*. The authors adopted the stretching frequency of the free OH^- ion (3556 cm^{-1}) as a reference and selected the metal–oxygen distance in the bulk of the material to establish correlations. For alkaline and alkaline-earth metal hydroxides, the authors found that, in the absence of any H-bonding, ν(OH) increases with decreasing M–O distance (r_{M-O}) in the first coordination sphere around the metal (see Table 2.1). For these hydroxides, the authors proposed the following empirical equation (Equation 2.3) to estimate the O–H stretching frequency (in cm^{-1}):

$$\tilde{\nu}(OH)[cm^{-1}] = 3541\,cm^{-1} + 1.156 \times 10^6\,cm^{-1}\,pm^2(r_{M-O})^{-2} \\ + 9.057 \times 10^8\,cm^{-1}\,pm^3(r_{M-O})^{-3} \qquad (2.3)$$

where r_{M-O} is the distance between metal and oxygen in the bulk of the material, in pm.

A plot of the OH stretching modes vs. the mean M–O distance in the first coordination sphere is presented in Figure 2.1. Data for metal hydroxides are represented by filled circles. The authors noted that no decrease of the bond order is expected as a result of the decrease of the total negative charge of the OH groups through electron transfer to the cations; this expectation arises from the very similar stretching frequencies of OH^- and $OH^{\cdot}$. Hence, the bond order concept cannot be used for an explanation of the variation in frequency.

The bond formed between alkaline and alkaline-earth metal cations and OH^- is essentially ionic. The trend in the frequencies is explained by the fact that the dipole moment and the strength of the O–H bond increase in the presence of a positive charge near the oxygen, probably proportionally to the cation field strength.

Table 2.1 OH stretching frequencies for the hydroxides of alkali and alkaline-earth metals, according to the data summarized in Refs. *(8,12)*

Compound	R_{M-O}, pm	$\tilde{\nu}$(OH), cm^{-1}	Compound	R_{M-O}, pm	$\tilde{\nu}$(OH), cm^{-1}
LiOH	196	3670.5	–		
NaOH	238	3635	Mg(OH)$_2$	210	3670
KOH	280	3597.5	Ca(OH)$_2$	237	3630
RbOH	292	3588	Sr(OH)$_2$	264	3611
CsOH	310	3578	Ba(OH)Cl	268	3593.5

Average values of ν(OH) are presented.

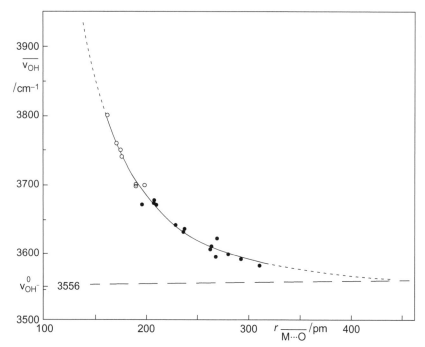

Figure 2.1 Mean OH stretching frequencies $\overline{\nu_{OH}}$ of hydroxide ions (not involved in OH$^-\cdots$X H-bonds) in alkali and alkaline-earth metal hydroxides (●), and on surfaces of CaO, Al$_2$O$_3$, and SiO$_2$, (○) vs. mean MO distances $\overline{r_{MO}}$ of the first coordination sphere. *Reproduced from Ref. (12).*

In addition to the alkaline and alkaline-earth metal hydroxides, Beckenkamp and Lutz *(12)* analyzed a series of other metal hydroxides and determined the O–H stretching frequencies. They concluded that an increase of the covalency of the M–O bond leads to a decrease of the OH stretching frequency. In this case, the concept of the bond order preservation could be applied, that is, with increasing M–O bond strength, the O–H bond strength and frequency decrease.

Ultimately, the frequency of an OH group attached to a cation is determined by the combination of two parameters: (i) the strength of the electrostatic field and (ii) the degree of covalency of the bond. The parameters have opposing effects on the frequency of the O–H stretching vibration: an increase in field strength implies an increase in frequency, whereas an increase in covalency of the M–O bond implies a decrease in frequency.

In the context of this review, the inclusion of data for surface OH groups into the correlation shown in Figure 2.1 is important. It is evident that for the selected surface hydroxyls (represented by the open symbols), the OH stretching frequencies are higher than those reported for hydroxides (filled symbols). This behavior can be explained, in the framework of the model, by the shorter M–O distance in oxides as compared to hydroxides.

The proposed model is based on a limited set of data and involves some simplifications. Specifically, silanol groups were considered together with alkaline and alkaline-earth hydroxides; linear and bridging hydroxyls were not distinguished; etc. Therefore, further and more accurate data are needed, as well as some refinement of the model. Nevertheless, the model clearly indicates that the OH stretching frequency, especially for ionic oxides, is largely independent of the chemical nature of the cation(s) to which the OH group is bound and that the determining factor is the electrostatic field.

Another important conclusion is that the ν(OH) value cannot be used to assess the properties (e.g., acidity or basicity) of the OH groups. The most basic hydroxyls (CsOH) are characterized by a frequency of 3578 cm^{-1}, which is the same as the OH frequency of a molecule of hypofluorous acid HOF in the gaseous state (13). The following data give a more general idea on the OH stretching frequencies of covalently bound hydroxyls: HOCl, 3581 cm^{-1} (14); HOBr, 3590 cm^{-1} (14); and matrix-isolated gaseous HNO$_3$, 3550 cm^{-1} (15).

It is emphasized that the Beckenkamp–Lutz model is not widely recognized in the field of surface science. The idea that complexation leads to a decrease of the O–H stretching frequency is prevailing in many papers. In his review from 1995 (16), Lutz underlined that "Because of the unknown or wrongly assumed OH stretching fundamental of hydroxide ions most research of hydroxide ions adsorbed on surfaces of solid oxides until very recently is questionable despite the enormous importance of such studies for the catalytic behaviour of oxide surfaces." We believe that consideration of this model could lead, in many cases, to the revision of current opinions.

Some results from computational chemistry support the general idea of the Beckenkamp–Lutz model. It was reported that interaction of an OH$^-$ group with a positive charge located near its oxygen atom strongly affects the OH stretching frequency (17–19). Calculations considering isolated Li$^+$OH$^-$, Mg^{2+}OH$^-$ and Al^{3+}OH$^-$ fragments indicated a substantial increase of ν(OH) as the M–O distance decreased from 0.18 to 0.14 nm; thus, the effect was most pronounced for Al^{3+}OH$^-$ (17). However, it was also found that with an increase of the electric field strength, ν(OH) initially increases, passes through a maximum, and then starts to decrease at very high field strengths. Consequently, it was pointed out that an increase of the charge and a decrease of the size of the metal cation (both of which enhance the field strength) should not always lead to an increase of the OH stretching frequency.

Evidently, to correctly predict the OH stretching frequency, one should take into account the ionic or covalent character of the M–O bond and the positions of the ions generating the electric field. Even for one of the simplest materials, MgO, the interpretation of the hydroxyl bands is still being debated.

Another question of interest is how the oxidation state of the cation or atom to which the hydroxyl group is bound affects the OH stretching frequency. Unfortunately, at least presently, there is no definite answer. An increase of the oxidation state leads to a decrease of the ionic radius and, consequently, should result in an increase of $\nu(OH)$. However, with increasing oxidation state, the covalency of the bond increases, which leads to a decrease in frequency. Therefore, the sum of the two effects (with different weighting factors in each case) should determine the frequency of the OH stretching modes. It should be noted that redox processes are often associated with a change in coordination. Therefore, it is difficult to find suitable data to extract the effect of the oxidation state; a comparison of the influence of cations in the same environment but differing in oxidation state would be needed. This problem can be illustrated by the OH frequencies of Fe^{2+}–OH and Fe^{3+}–OH groups in Fe-ZSM-5 zeolite (20). These groups are detected at very similar frequencies, Fe^{2+}–OH at 3665 cm^{-1} and Fe^{3+}–OH at 3670 cm^{-1}, indicating that there exists no simple dependence of the frequency on the oxidation state. Moreover, the groups strongly differ in their chemical reactivity and thermal stability. For example, the evacuation temperature needed for disappearance of 50% of the hydroxyls is 513 K for Fe^{2+}–OH, whereas it is 723 K for Fe^{3+}–OH.

There exist additional factors that can influence the $\nu(OH)$ of isolated (non-H-bonded) hydroxyls. One of them is the anharmonicity of the O–H bond, which is related to the shape of the Morse potential that is overlaid with the harmonic potential. The harmonic frequencies are higher than the measured ones; in general, an increase of the anharmonicity of a particular vibration lowers the O–H stretching frequency. For Morse potentials of equal shape, the anharmonicity decreases with decreasing wavenumber, because the zero-point energy of O–H bonds characterized by low frequencies lies closer to the bottom of the Morse potential curve, where the deviations from the harmonic oscillator model are smaller. However, the factors affecting the anharmonicity are far from being completely understood. For example, one would expect that SiOH groups are more anharmonic than the free hydroxyl anion because the SiOH stretching frequency is 3746 cm^{-1} and thus much higher than that of the free hydroxyl anion (3555.6 cm^{-1}). However, the OH anharmonicity of surface silanols was determined to be only 83.5 cm^{-1} *(21)*, whereas 92 cm^{-1} were reported for the free hydroxyl ion.

It is well established that ν(OH) bands of isolated hydroxyls shift to lower wavenumbers with increasing temperature *(22)*. However, the magnitude of this effect varies, and temperature shifts can be different even for hydroxyls of similar structure. For example, Niwa *et al.* *(23,24)* proposed the following empirical equation (Equation 2.4) for bridging hydroxyls in zeolites:

$$\Delta \widetilde{\nu}/\Delta T = k\, \mathrm{cm}^{-1}\mathrm{K}^{-1} \tag{2.4}$$

where k assumes values of -0.056 or -0.034, respectively, for the bridging hydroxyls in MOR and USY zeolites. Note that the opposite temperature effect is typical for H-bonded hydroxyls (see Section 2.4).

2.2. Why OH groups are encountered on surfaces

The reason for the existence of surface hydroxyl groups is generally a charge deficiency that needs to be, somehow, compensated. The details vary depending on the material. It is convenient to divide the surface hydroxyls into two major groups: stoichiometric OH groups that are represented in the chemical formula of a substance, and nonstoichiometric OH groups that exist as a result of the defect in the bulk structure that a surface presents. Note that hydroxyls of both groups can coexist on a particular sample.

Porous alumosilicates are a typical example of solids that possess stoichiometric OH groups. The substitution of Si^{4+} by Al^{3+} in zeolites leads to a charge imbalance, which is compensated by addition of a proton and formation of an OH group in the vicinity of the aluminum cations *(25)*. Thus, the general chemical formula of these compounds is $Si_nAl_mH_mO_{(2n+2m)}$, where m also indicates the number of structural hydroxyls.

Charge defects typically arise where a bulk crystal is terminated, as illustrated by the following simplified example. Consider a crystal lattice consisting of alternating layers of metal cations and oxygen anions. Splitting a crystal of such a material in a direction parallel to the layers will create one surface terminated by oxygen anions and one surface terminated by cations. Both new surfaces are characterized by charge defects in this case. However, charge stoichiometry can be reestablished if the cation-terminated plane is covered by OH groups and the oxide species on the oxygen-terminated plane are replaced by hydroxyls. Usually, such hydroxyls easily condense, thereby producing molecular water and O^{2-}. In many cases, however, the hydroxyls are associated with isolated charge defects (e.g., on crystal edges) and are characterized by significant thermal stability.

Hydroxyl groups may also be generated and stabilized when the oxidation state of a cation increases or decreases by one. The resulting charge defect is often compensated by an additional surface OH group (after oxidation of the cation) or replacement of O^{2-} by OH^- (after reduction).

It is worthwhile considering the origin of the surface hydroxyls. In most cases, the OH^- groups are biographic, that is, they originate from the preparation procedure. Indeed, during the synthesis of many catalysts, water is normally present. Aside from this possibility, surface hydroxyls are often formed as a result of the interaction of the solid with H-containing reagents.

An important route for the formation of surface hydroxyls is the dissociative adsorption of H_2 or protonic acids (HA) on surfaces of activated (i.e., thermally treated under vacuum or in H_2O-free gas flow) samples according to the following equation:

$$\begin{array}{ccc} & A & H \\ & | & | \\ Me^{n+} - O^{2-} + HA \rightleftharpoons & Me^{n+} & -O \end{array} \qquad (2.5)$$

The reaction described by Equation (2.5) is important for catalysis because in this way, hydroxyls can be produced in the course of catalytic reactions. If HA is water, two different kinds of OH group (located in the vicinity of each other) will be formed. During the reaction with water, a rearrangement of the bonding situation shown in Equation (2.5) can occur and two terminal OH groups may be formed. An example is the hydrolysis of siloxane bridges according to Equation (2.6):

$$Si - O - Si + H_2O \rightarrow 2\,SiOH \qquad (2.6)$$

Extraction of metal cations Me^{n+} from the solid can also lead to creation of surface hydroxyls, with n OH groups created per Me^{n+} species removed, in agreement with the rule of charge preservation. Typically, this chemistry can be formulated as an exchange reaction. However, more complicated situations are also encountered. For example, extraction of framework Al^{3+} ions from zeolites leads to the formation of so-called silanol nests (Figure 2.2). A perfect nest should consist of four Si–OH groups *(26)*.

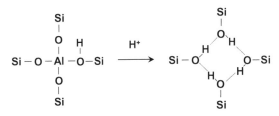

Figure 2.2 Scheme illustrating the formation of silanol nests.

Reduction of a solid by hydrogen can also produce surface hydroxyls. For instance, if Me^+ cations are present on the surface, their reduction will proceed according to Equation (2.7) and will lead to the formation surface hydroxyls.

$$n\left(-O^{2-} - Me^+\right) + n/2\,H_2 \rightarrow n\left(-O^{2-} - H^+\right) + Me^0{}_n. \qquad (2.7)$$

In some cases, the dissociative adsorption of H_2 on supported metals causes formation of nascent hydrogen. If the support is reducible (like TiO_2 or CeO_2), this hydrogen reduces the metal cations and is thereby converted to H^+, which react with lattice oxygen to form hydroxyl groups.

Finally, covalent hydroxyls (e.g., C–OH, S–OH, P–OH) can be formed on the surface if suitable anions are present such as carbonate, sulfate, or phosphate that can attract protons and give bicarbonate, etc.

2.3. Types of isolated OH groups

There are many ways to classify isolated surface hydroxyls; for example, they may be classified according to their acid–base properties as acidic, basic, or amphoteric. Depending on the coordination of the hydrogen atom, all types of surface hydroxyl groups can be divided into two main types, the isolated type and the H-bonded type. In this section, we will address the isolated hydroxyls, and the H-bonded hydroxyls will be discussed in the next section. We shall further classify the groups according to the number of surface cationic sites to which the hydroxyl oxygen is bound. It will be shown that this bonding is very important for the properties and spectral signature of the hydroxyls.

If one OH group is attached to one surface site, the stoichiometry of the complex is 1:1. These groups are commonly designated in the literature as type I hydroxyls (see Figure 2.3, left two images). When the bond is essentially ionic the OH groups are linear (type Ia), whereas covalent bonds lead to tilted (bent) structures (Ib). In the ionic model, the proton can be seen as

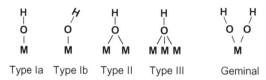

Figure 2.3 Types of surface OH groups. The hydroxyls shown in the scheme are "free," that is, not involved in any H-bonding. Depending on other factors, each type can be divided into subcategories. For example, depending on the M–O–H angle the type I hydroxyls are divided into linear (Ia) and bent (Ib) species. Other subcategories are introduced and explained in the text.

replacing a metal cation, and the atoms M–O–H are in line. In the case of covalent bonding, meaning there are directed bonds, the hybridization of the oxygen (typically sp^3) leads to the bent structure. It is important to realize type I is not synonymous to linear.

There are many examples of oxygen being simultaneously connected to two surface sites. These groups are considered to be of type II (Figure 2.3, middle image). Typical examples of such groups are the so-called acidic hydroxyls in zeolites that are connected to both an aluminum and a silicon atom. Depending on the surface structure, the OH groups can be attached to more than two surface atoms. The type signifies the number of connected surface sites; type n implies that the oxygen of the OH group is simultaneously connected to n atoms (Figure 2.3, left part).

The above classification was originally proposed by Tsyganenko and Filimonov *(27,28)* and has been adopted by researchers working in the field of catalysis. This classification evolved as a result of considering the surface structure of various oxides. It was assumed that the OH group occupies an oxygen vacancy on the surface and thus should be connected to fewer than n cations, where n is the coordination number of oxygen in the bulk.

A detailed analysis of the surface structures of oxides allowed the authors to formulate several important conclusions:

- Up to $n-1$ types of hydroxyls can be detected on the surface of a given oxide (where n is the oxygen coordination number in the bulk). Thus, only type I hydroxyls can be formed on silica, because the oxygen coordination number is 2 (see Table 2.2).
- Hydroxyls of type III cannot be proton acceptors, because the lone pairs of the oxygen atoms are engaged in bonding to metal atoms. This coordination implies that these hydroxyls should not possess basic properties.
- When the metal and oxygen ions on a given surface have two coordinative vacancies each, only type II hydroxyls will be stabilized.

Table 2.2 Vibrational frequencies of free surface hydroxyls on oxides with different crystal structures (27,28)

Oxygen coordination number	Oxide	OH frequency, cm^{-1}			
		Type I	Type II	Type III	Type V
2	SiO_2	3750			
3	TiO_2 (anatase)	3725	3670 3650		
4	γ-Al_2O_3	3800	3740	3700	
4	BeO	3735	–	3630	
4	ZnO	–	3675	3622	
4	ThO_2	3745	–	3655	
4	CeO_2	3710	–	3640	
4	HfO_2	3800	–	3690	
4	ZrO_2	3770	–	3670	
6	MgO	3750		–	3630
6	CaO	3700	–	–	3610
6	NiO	3735	–	3690	3630
6	CoO	–	–	3680	

- The O–H bond is weakened in the sequence from type I to type V hydroxyls. As a result, the hydroxyls with a higher type number participate more readily in H/D exchange and are characterized by higher proton mobility.

We note that another important conclusion can be drawn from the model introduced above: surface hydroxyls of type I are not allowed on univalent cations. The positive charge of a cation M^+ is, to a great extent, compensated by the bulk of the material, and thus such embedded M^+ site cannot compensate the single negative charge of an OH^- group. However, species similar to type I may be formed on metal surfaces during their initial stages of corrosion.

Tsyganenko and Filimonov (27,28) hypothesized that the formation of a coordinative bond should result in a lower stretching frequency of the hydroxyl, consistent with the idea of bond order preservation. The authors concluded that the increase in the number of metal cations to which an OH group is bound should lead to a decrease in the stretching frequency.

Although this relationship has been demonstrated for some particular cases, we note that the conclusion is based on a wrong premise (especially for ionic hydroxyls), and the dependence of the O–H stretching frequency on the type of hydroxyl should be carefully reconsidered.

It should also be noted that some heterogeneity can be observed within each type of hydroxyl depending on the coordination state of the cation(s) to which the OH group is bound, although this effect is considered to be smaller than the effect of the number of atoms bound to the hydroxyl oxygen.

There is also the possibility of two OH groups being simultaneously attached to one surface atom (Figure 2.3, right part). These groups are called geminal, and there are many reports on geminal $Si(OH)_2$ species (see, e.g., Refs. *(29–33)*). According to Ref. *(29)*, even three surface OH groups can be attached to one silicon atom. It is generally considered difficult to distinguish these multiplets of hydroxyls from isolated hydroxyls on the basis of IR spectra, and other techniques must be consulted for their detection. There are essentially no reports on other kinds of geminal surface hydroxyls (i.e., nonsilicon-bound), although some similar structures, for example, $V(OH)_2$, have been proposed *(34)*.

Many other structures, for example, surface sites connected simultaneously to isolated and bridging OH groups, are possible. If a particular structure has been established or suggested, it will be explicitly stated.

2.4. H-bonded OH groups

Hydrogen bonding has an enormous importance not only for catalysis but for the entire field of chemistry and biochemistry. As a result, thousands of articles, reviews, and books are devoted to this phenomenon. Here, we will briefly describe the main principles governing the behavior of surface hydroxyls. For more information, Refs. *(35–37)* are recommended to the reader.

Let us now consider a hydroxyl group engaged in H-bonding. There are three main possibilities for the hydroxyl group: (i) to donate its hydrogen atom, (ii) to attract a hydrogen atom from another molecule or ion to its oxygen (i.e., to donate its O-atom), or (iii) to be engaged in an H-bond with both of its atoms.

When the first possibility is realized, the H-bond can be divided into three main types *(38)*: linear (which is the preferred configuration), bent, or bifurcated (Figure 2.4). In all these cases, the hydrogen interacts with a

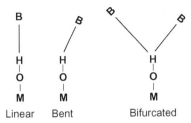

Figure 2.4 Three different types of H-bonds between a hydroxyl group and a species B, all arising from engagement of the hydrogen atom of the hydroxyl group.

negatively charged or negatively polarized and thus basic atom (B). It is usually assumed that, for an H-bond to form, the H$\cdots$B distance should be less than 350 pm *(16)*. The hydrogen bond is mainly electrostatic in nature, but a covalent contribution is possible in the case of very strong H-bonds.

The strength of H-bonds formed by OH groups is related to the O–H distance, which can vary by up to 10 pm. Because the O–H distance increases upon formation of an H-bond, ν(OH) decreases. This effect is very pronounced, and H-bonding is the most important factor affecting the position of the ν(OH) band. Strictly speaking, the OH modes are no longer isolated and should be described as O–H stretching vibrations of the O–H$\cdots$B complex.

Further consequences of the formation of an H-bond are the increase in intensity and the broadening of the ν(OH) bands. Figure 2.5A demonstrates that these effects increase with the H-bond strength. Adsorption of methane, a weak base, on SiO_2 leads to a red shift of the SiO–H modes by 34 cm^{-1}, and the shifted band has a higher integral intensity and is broader than the original band of the isolated OH group. With increasing basicity of the adsorbing molecule (CH_4, CO, CD_3CN in Figure 2.5A) and thus increasing strength of the H-bond, the OH band of the O–H$\cdots$B complex appears at lower frequency while its intensity and width increase. However, a decrease in intensity can occur in the case of very strong H-bonds; this phenomenon still requires explanation.

The increase of the absorption coefficient of the O–H stretching vibration is caused by the elongation of the OH bond and the resulting increase of the hydroxyl dipole moment and the associated enhanced variation of the dipole moment during vibration. In contrast to the IR intensities, the corresponding Raman intensities decrease upon formation of H-bonds.

It has been proposed that the H-bond-induced widening of the ν(OH) bands arises from coupling of the OH stretching vibration with the

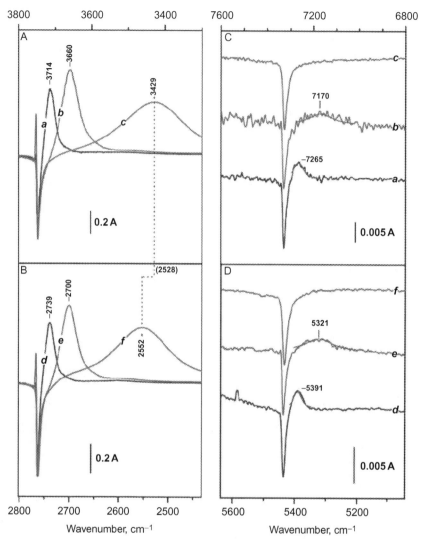

Figure 2.5 Difference FTIR spectra of silica (A, C) and deuteroxylated silica (B, D) after adsorption of CH_4 (a, d), CO (b, e), and CD_3CN (c, f). Equilibrium pressures $p(CO) =$ 500 Pa, $p(CH_4) = 1500$ Pa, and $p(CD_3CN) = 400$ Pa. Temperature of adsorption is 293 K for CD_3CN and 100 K for CH_4 and CO. (A) $\nu(OH)$ region, (B) $\nu(OD)$ region; (C) $2\nu(OH)$ region, and (D) $2\nu(OD)$ region. *Reproduced with permission from Ref. (21). Copyright 2013 American Chemical Society.* (See the color plate.)

hydrogen bond (H···B) stretching vibration (via both sum and difference combinations) *(39)*.

As a result of the decrease of the O–H bond order upon H-bond formation, the vibrations involving the metal atom (M–O stretching vibrations and the deformation modes) are shifted toward higher frequencies. The respective bands are also broadened, but to a smaller extent than the ν(OH) band.

As a rule, the strength of the H-bond decreases with increasing temperature, because of the increasing length of the H-bond. However, the opposite effect may occur with strongly bent H-bonds. It has been reported that, in contrast to the behavior of isolated hydroxyls, the band maxima of H-bonded OH groups are shifted to higher frequencies as the temperature increases (40). It has been noted that this trend, which is not expected on the basis of bond strength considerations, is only observed for symmetric H-bonds.

The overtones of the OH stretching vibrations, designated as 2ν(OH), are highly sensitive to the formation of H-bonds. It is well known that their band intensity fades with the strength of the H-bonds, and in the case of medium-to-strong H-bonds, the bands may be too weak to be observed (see Figure 2.5C and D) (21,41,42). The phenomenon is far from being well explained. It has been proposed that the formation of H-bonds leads to a cancellation effect between the mechanic and electric anharmonicity of the O–H oscillator (41). Unfortunately, this complication prohibits the calculation of the harmonic frequencies of strongly H-bonded hydroxyls.

The effect of H-bonding on the anharmonicity of the O–H oscillator has been extensively discussed in the literature (43–46), and analysis of the available data allows the conclusion that weak and medium H-bonds lead to a decrease of the O–H bond anharmonicity (43,44). The anharmonicity is increased when strong H-bonds are present (45,46), but this increase is not dramatic (45). To provide some values, the anharmonicity of the free O–H oscillator (OH radical) is about 90 cm^{-1}, whereas an anharmonicity of approximately 200 cm^{-1} was reported for very strongly H-bonded hydroxyls in a phenol–pyridine complex (47). If the hydrogen acceptor is an oxygen atom, the anharmonicity can reach values of up to 120 cm^{-1}.

Everything stated above is generally valid for all kinds of hydroxyls. Below, we shall consider in more detail the case of surface OH groups. The formation of H-bonds through interaction with various adsorptives will be discussed in Section 3. On surfaces, the H atom of a hydroxyl can often form a weak bond with another oxygen atom belonging to the solid or another OH group in the vicinity. The resulting complexes are called H-bonded surface hydroxyls. It can be difficult to make a clear distinction between free and H-bonded hydroxyls, because in some cases the H-bond is very weak.

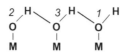

Figure 2.6 OH groups participating in the formation of H-bonds with their O atom (1), their H atom (2), and both their H and O atoms (3).

Figure 2.6 schematically presents the possibilities of H-bonding between surface hydroxyls. If only the oxygen atom of a hydroxyl is engaged in the H-bond (labeled example 1 in Figure 2.6), the O–H bond will be weakened and the O–H frequency should decrease. However, a much more pronounced effect results if the H atom participates in an H-bond (example 2 in Figure 2.6). If both the H and the O atom participate in the formation of an H-bond (example 3 in Figure 2.6), the largest shift of the O–H stretching modes will be obtained, but it will not be significantly larger than the shift for example 2.

In porous materials, the OH groups are often located inside the pores and thus can easily interact, through an H-bond, with lattice oxygen. Note that in this scenario, it is exclusively the H atom of the hydroxyl group that participates in the H-bond.

When an H-bond is formed between vicinal hydroxyls, the hydrogen atom or the oxygen atom may be involved in the bonding (see Figure 2.6). It is important to differentiate between the various cases, because the manner in which H-bonding occurs is decisive for the properties and spectral behavior of the OH groups. For example, the hydroxyls denoted as (1) in Figure 2.6 are only slightly affected by the H-bond. Sometimes, they are referred to as "H-bonded hydroxyls donating only oxygen" or "weakly interacting vicinal hydroxyls." A very appropriate term seems to be "H-acceptor hydroxyls." Figure 2.7 gives the systematization of silanol groups according to Ref. *(48)* and illustrates how vicinal silanol groups interact, with one group donating and one group accepting hydrogen (yielding the structure labeled "bridged").

Note that the H-bonded hydroxyls can be classified as types I, II, and so on, according to the same principle as the isolated hydroxyls, depending on the number of cations to which the oxygen is bound. However, in the case of H-bonded hydroxyls, it is nearly impossible to distinguish between the different types of OH groups, because the position of the bands is mainly determined by the strength of the H-bond.

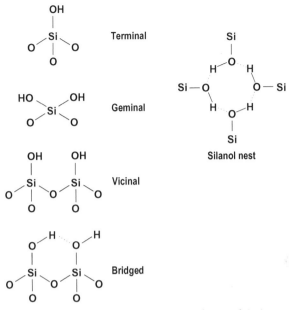

Figure 2.7 Different types of silanol groups. *Adapted from Ref. (48).*

2.5. Stability of surface hydroxyls

In this section, we shall provide a general picture of the stability of surface hydroxyl groups. For details on the stability of particular species, the reader is referred to Section 8. It is emphasized that there are serious contradictions in the reports on the stability of particular hydroxyls. These inconsistencies result from variations in the experimental conditions; specifically, the residual pressure is essential when seeking to decompose OH groups. Water evolved during a thermal treatment can adsorb on cold parts of the apparatus and may readsorb on the sample later, thus regenerating surface hydroxyls. The technical characteristics of equipment used for such measurements vary considerably, also because many groups use custom designs. Hence, observations regarding the stability show a significant spread. Panov *et al.* introduced a protocol to compare the stability of different hydroxyls that is less prone to error. The authors determined the temperature needed to reduce the OH band to half of its initial intensity by evacuation for a defined period (e.g., 1 h) *(20)*. This approach partly overcomes the above described problems because small amounts of readsorbed water have no significant impact on the spectra.

The thermal stability of surface hydroxyls varies over a wide range. Some hydroxyls are removed at a rather low temperature, and others (e.g., Si–OH) are observed on surfaces even after evacuation at temperatures above 1300 K *(29)*. One of the crucial factors determining the stability of a hydroxyl group is the presence (or absence) of another OH group in its vicinity. Two vicinal hydroxyls can interact according to the reverse reaction of Equation (2.5), thus producing water and a surface O^{2-} ion.

Vicinal hydroxyls formed after water dissociation are normally H-bonded. Consequently, many investigations demonstrated that H-bonded hydroxyls on various surfaces are characterized by low stability. Decomposition according to the reverse reaction of Equation (2.5) requires that one of the hydroxyl groups must have a suitable O^{2-} ion (e.g., not in a framework position) that can be eliminated as water. If such oxygen is lacking, the H-bonded hydroxyls can demonstrate a remarkable stability. This stability is exemplified by the silanol nests in SiBEA, which are characterized by a band at 3520 cm^{-1} (see Figure 2.8). Because there are no suitable oxide ions that can be eliminated as water during condensation of the silanols, the latter demonstrate a remarkable stability, up to a temperature of 723 K *(49–53)*. The recombination reaction to give water also does not occur with hydroxyls that are H-bonded to lattice oxygen.

Other important factors affecting the stability of hydroxyls are the nature and the oxidation state of the cation(s) to which the hydroxyl oxygen is bound. For example, the concentration of Fe^{2+}–OH groups in Fe-ZSM-5 (which absorb at 3665 cm^{-1}) decreases by about 50% after evacuation at a temperature of 500 K (Figure 2.9), whereas the respective temperature for a similar decline in Fe^{3+}–OH groups (detected at 3670 cm^{-1}) is 723 K (Figure 2.10) *(20)*. However, even in this case, the temperatures must be compared with caution, because the groups may decompose by different routes. The Fe^{2+}–OH groups most probably disappear according to the reverse reaction of Equation (2.5) since they were produced by dissociative adsorption of water. If the OH group is bound to a cation in a high oxidation state (here Fe^{3+}), the dehydroxylation can be accompanied by a redox reaction, as shown in Equation (2.8):

$$2\,Fe^{3+} - OH^- \rightarrow 2\,Fe^{2+} + H_2O + \tfrac{1}{2}O_2 \qquad (2.8)$$

A general trend that has been reported for many materials concerns the relative stability of hydroxyls bound to the same cation: the stability increases with increasing ν(OH) frequency. This trend is consistent with the fact that the less stable H-bonded hydroxyls vibrate at lower frequencies. As seen in

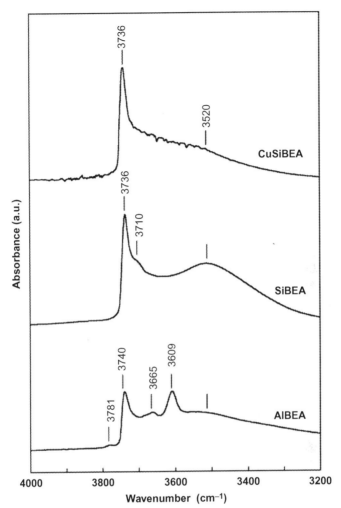

Figure 2.8 FTIR spectra of AlBEA, SiBEA, and CuSiBEA recorded at room temperature after calcination at a temperature of 773 K for 3 h in flowing air and subsequent outgassing at 10^{-3} Pa for 2 h at 573 K. *Reproduced from Ref. (49).*

Figure 2.11, a stepwise increase of the evacuation temperature applied to MgO leads to the successive disappearance of the hydroxyl bands located at 3615 and 3683 cm^{-1}, whereas the band at 3734 cm^{-1} persists even during evacuation at 973 K *(54)*. However, if only isolated hydroxyls are present on a surface, an inverse relationship between the wavenumber and stability is generally observed. This trend ensues because type I OH groups are less stable than type II, which are less stable than type III, and so on. This behavior is demonstrated in Figure 2.12, which shows that zirconium hydroxyls

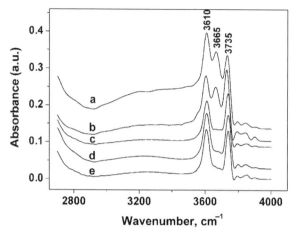

Figure 2.9 IR spectra recorded after H_2O adsorption on Fe-ZSM-5 at a temperature of 373 K and subsequent evacuation at various temperatures: 513 K (a), 623 K (b), 723 K (c), 773 K (d), and 823 K (e). *Reproduced from Ref. (20).*

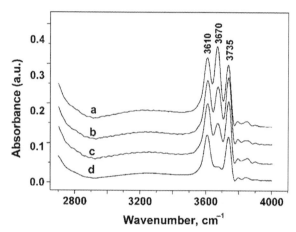

Figure 2.10 IR spectra recorded after H_2 interaction with O_α/Fe-ZSM-5 at a temperature of 373 K and subsequent evacuation at various temperatures: 513 K (a), 723 K (b), 773 K (c), and 823 K (d). *Reproduced from Ref. (20).*

characterized by an IR band at 3662 cm^{-1} are somewhat more stable than hydroxyls absorbing at 3680–3670 cm^{-1} and definitely more stable than hydroxyls vibrating at 3778 cm^{-1} *(55)*. Thus, for isolated hydroxyls of the same chemical composition, the stability increases with increasing coordination number of the hydroxyl oxygen.

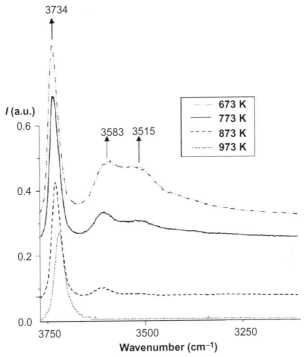

Figure 2.11 DRIFT spectra of MgO in the ν_{OH} vibration region after a series of dehydroxylation steps. Water was adsorbed at a temperature of 373 K, then the sample was heated under N_2 flow (70 cm³ min⁻¹) to the desired activation temperature and kept at this temperature for 15 min. DRIFT spectra were taken at the activation temperature at the end of the 15 min dwell time. *Reproduced from Ref.* (54).

Another important question is whether dehydroxylation is reversible or not. Often, the thermal treatment conducted to decompose OH groups causes changes of the surface structure of the solid, and rehydroxylation after dehydroxylation results only in partial recovery of the OH coverage.

For many applications, especially of zeolites, it is important to know the hydrothermal stability of the OH groups, as well as their stability during acid leaching or other treatments. These stabilities strongly depend on the properties of the solid itself.

2.6. Detection of surface hydroxyl groups and techniques for their characterization

As already stated, the most frequently used technique for detection and characterization of surface hydroxyl groups is infrared spectroscopy, which is the

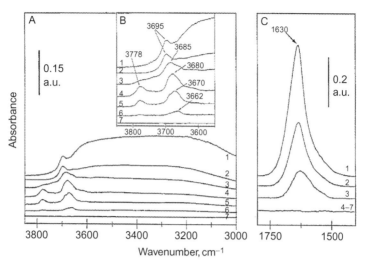

Figure 2.12 FTIR spectra recorded during dehydration of ZrO_2. (A) ν_{OH} region (3850-3000 cm^{-1}), (B) (inset) enlarged segments of the ν_{OH} region reported in (A), and (C) the δ_{HOH} region (1800-1400 cm^{-1}). Curves 1–3 refer to samples vacuum-activated at room temperature for 1 min, 5 min, and 1 h, respectively. Curves 4–7 refer to samples vacuum-activated for 1 h at 473, 673, 873, and 1073 K, respectively. *Reproduced from Ref. (55).*

focus of this review. It should be emphasized that samples exposed to air contain, in addition to OH groups, a large amount of molecularly adsorbed water, which forms H-bonds with the OH groups. As a result, a broad absorbance in the region of 3800–3200 cm^{-1} (extending sometimes to even lower wavenumbers) is detected in the IR spectra of as-prepared samples, and almost no information can be extracted. To investigate the individual OH groups, the materials are normally outgassed at various temperatures, which removes weakly adsorbed water. In the application of other characterization techniques, nonactivated samples or samples activated by other procedures may be used. Therefore, one should compare results obtained by different techniques with caution.

An important fact affecting the analysis of IR spectra is the strong variation of the OH absorption coefficient, which is influenced by many factors, above all by H-bonding. Hence, although the intensity of a particular OH band is proportional to the concentration of the respective hydroxyls, one cannot take the uncorrected intensities of bands as a measure of the relative populations of hydroxyl species. Rather, it is compulsory to know the absorption coefficients. This behavior is a disadvantage in some respects, but also provides opportunity for extracting more information from the spectra.

Raman spectroscopy, although having great potential for the investigation of surface hydroxyl groups and often being complementary to IR spectroscopy, has rarely been utilized for the characterization of surface hydroxyls. A peculiarity of this spectroscopy is that the intensity of the $\nu(OH)$ bands decreases upon formation of H-bonds. Some examples of Raman spectra of surface hydroxyls on oxide surfaces are summarized in Ref. *(56)*.

Another widely used technique for OH detection and characterization is solid-state NMR spectroscopy. The most frequently investigated nucleus is ^1H, and the magic angle spinning (MAS) technique allows the isotropic chemical shift to be measured. For surface OH groups, the shift is in the range between 0 and 15 ppm, with the higher values found for H-bonded hydroxyls (see Table 2.3 *(57)*). Another trend is that silanol groups (which are of type I) typically produce a signal at 1.2–2.2 ppm, whereas bridging (type II) hydroxyls in zeolites produce a signal at 3.6–5.2 ppm. These chemical shift ranges are illustrated by the NMR spectra of an HBEA zeolite sample (see Figure 2.13). It should be emphasized that neither IR nor NMR spectra of the OH groups as such provide complete information about their properties. To obtain additional information, various molecular probes may be applied that will coordinate with the OH group. Upon formation of a hydrogen bond with a molecular probe, the ^1H MAS NMR signal of the hydroxyl group signal is shifted to higher values.

Table 2.3 Chemical shift of various types of OH groups in zeolites *(57)*

Type of OH group	Detailed description	^1H NMR shift, δ_H, ppm
MOH	MOH groups on the outer surface or in large cavities	−0.5 to 0.5
SiOH	SiOH on the external surface or at defects	1.2–2.2
AlOH	Extraframework Al species, some H-bonding	2.4–3.6
MOH	H-bonded OH groups; M = Ca, Al	2.8–6.2
Si(OH)Al	Bridging OH groups in large cavities	3.6–4.3
Si(OH)Al	Bridging OH groups in small cages of channels	4.6–5.2
Si(OH)Al	Perturbed OH groups	5.2–8.0
SiOH	SiOH involved in strong H-bonding	~15

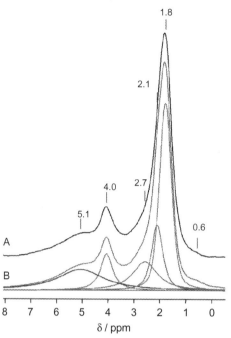

Figure 2.13 ¹H MAS NMR Hahn-echo spectrum of zeolite HBEA. (a) Experimental spectrum and (b) decomposition of the spectrum into six individual lines. The signals at 5.1 and 4.0 ppm are assigned to bridging Al(OH)Si hydroxyls, the signal at 2.7 ppm, to AlOH groups and the signals at 2.1 and 1.8 ppm, to silanol groups. *Reproduced from Ref. (58).* (See the color plate.)

More detailed information on the use of NMR spectroscopy for the characterization of surface hydroxyls can be found in several review and research papers *(33,53,57,59–68)*. In addition to ¹H MAS NMR, ²⁹Si and ²⁷Al MAS NMR techniques are widely used for the characterization of hydroxyls on silica and in zeolites. Appropriate nuclei have been selected for spectroscopy of other materials. Information can also been obtained from spectra of the probe molecule, which may be enriched in the isotope of interest, like, for example, ¹⁵N-pyridine *(66)* or ¹³CO. Combining the information from IR and NMR spectra can be extremely useful. On the basis of a comparison of ¹H NMR and IR spectra of various zeolites (HY, HMOR, and HZSM-5), it was concluded that the molar absorption coefficient of bridging hydroxyls is higher than that of free silanols *(68)*.

Inelastic neutron scattering (INS) is another technique that has been used for the characterization of surface hydroxyls, but the number of publications is limited *(69–72)*. Information on the hydroxyl coverage can also be obtained by X-ray photoelectron spectroscopy *(73–76)*. Various other methods that are not spectroscopic techniques are applied to characterize surface hydroxyls. A description of these methods is beyond the scope of this review, and the reader is referred to the according literature. The following methods are of particular importance: calorimetric measurements of the heats of adsorption of probe molecules *(60,77,78)*, temperature-programmed desorption of probe molecules *(60,79–82)*, neutron powder diffraction *(83)*, and catalytic test reactions *(59,84–87)*.

In the past two decades, the methods of theoretical chemistry have developed substantially, and now computational analysis is an important contribution to the investigation of surface OH groups. The attempts in this field started many years ago, and several valuable review papers have been published on this subject *(31,87–93)*. We will compare experimental observations with results from theoretical modeling throughout this review.

2.7. Role of hydroxyl groups in catalysis and hydroxyl coverage of catalytic materials

Discussions of OH groups in the context of catalysis normally focus on their role as active centers in a number of reactions. The work by Haag *et al. (94)* constitutes a classic example; the authors established a linear relationship between the concentration of aluminum in HZSM-5 (which implies an equal concentration of bridging hydroxyls) and the activity for cracking of *n*-hexane. It was concluded that all protonic acid sites in the zeolite are characterized by the same turnover frequency. Many other correlations between catalytic properties of materials and the strength and/or density of their Brønsted acid sites are well established. We will not discuss this aspect in detail and recommend instead a number of recently published reviews *(59,60,87)*. Two more points are worth mentioning. One point is that the cooperative action of Brønsted and Lewis acid sites has been demonstrated. The second is that, of course, OH groups must not necessarily be involved in a catalytic conversion; in fact, they can even block the catalytically active sites.

However, the role of OH groups in catalysis is not limited to their participation in elementary reaction steps. Surface hydroxyls widely participate in reactions during preparation of catalysts *(95)*. Hydroxyls are crucial for preparation techniques like ion exchange or grafting and play a role during a number of other procedures, such as impregnation. Surface reactions with hydroxyls may be used to deposit the catalytically active phase or for functionalization of surfaces. Finally, OH groups are also involved in processes like sintering, which are essential for the properties of some catalysts.

In the following paragraphs, we shall consider the main peculiarities of the hydroxyl groups present on the surface of various classes of catalytic materials. More details on selected materials will be provided in Section 8.

2.7.1 Zeolites and related materials

There are several groups of IR bands that are characteristic of the H-forms of zeolites. The charge defect brought about by the presence of an aluminum atom is compensated by a proton located at an oxygen atom connected to one silicon atom and one aluminum atom. The resulting bridging hydroxyl groups are highly acidic and are also called the Brønsted acid sites of the zeolite. The concentration of these hydroxyls correlates well with the aluminum content, as expected (96) (See Figure 2.14). Deviations from this relationship occur for low-silica zeolites that contain nonframework aluminum species.

The bridging hydroxyls are typical of microporous structures, and the OH stretching frequency is, to some extent, affected by the dimension of the cavity where the hydroxyl is located. In their classical work, Jacobs and Mortier (97) proposed a concept to explain the stretching frequencies of hydroxyl groups that are integrated into the 6- or 8-membered rings of zeolites. The authors postulated that the vibration of an unperturbed OH group should be at 3645 cm^{-1}. As a result of the electrostatic interaction of the hydroxyl proton with the nearest oxygen atoms, the ν(OH) of a hydroxyl in a ring will be red-shifted. The authors proposed the frequency shift, $\Delta\tilde{\nu}$ in cm^{-1}, to be inversely proportional to the average of the squared minimum distances between the proton and the framework oxygen, d_{min} in Å, according to the following (Equation 2.9):

$$\Delta\tilde{\nu} = 24.324 - 537.74\left(1/d_{min}^2\right). \tag{2.9}$$

The experimental data were convincingly described by this correlation (97) (see Figure 2.15). Recently, Equation (2.9) was used to identify the location of some hydroxyl groups in the MCM-58 material (98).

A new consideration is that the OH stretching frequency of the bridging hydroxyls may be affected by weak hydrogen bonding to framework oxygen atoms, and this effect is believed to take place even in 12-membered ring apertures. The stretching frequencies of the hydroxyls in several zeolites were recently summarized by Busca (59), and the data in Table 2.4 confirm

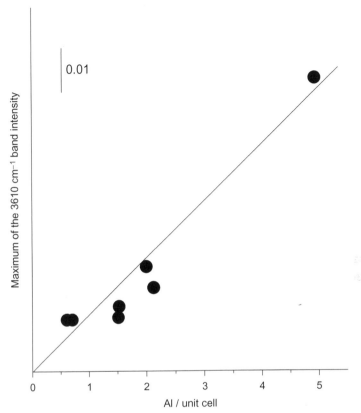

Figure 2.14 Maximum IR intensity at 3610 cm^{-1} (relative to sample weight) vs. framework Al content of HZSM-5 samples. *Reproduced from Ref.* (96).

the general trend of progressively lower OH stretching frequencies as the cavity dimension decreases.

Theoretical chemistry-based calculations (88) have indicated that the Al–O–Si angle also affects the vibrational frequency. Decrease of the angle strengthens the OH bond which leads to an increase of the O–H stretching frequency. Concomitantly, the acidity of the hydroxyls decreases. One additional factor that can influence the ν(OH) is the aluminum content. However, this effect is only noticeable with low-silica zeolites (Si/Al < 6) (97).

Isomorphous substitution of aluminum in zeolites also affects the stretching frequency of the bridging hydroxyls, as can be seen in Figure 2.16. For ZSM-5 zeolites containing various metal cations in their framework, ν(OH) increases in the sequence aluminum, germanium, and iron, whereas the acidity of the hydroxyls decreases in the same sequence

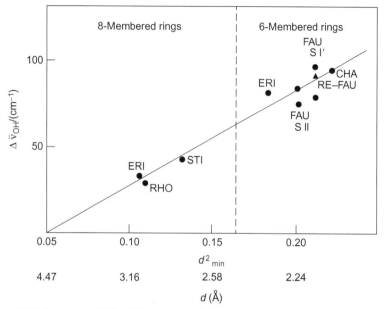

Figure 2.15 Frequency shifts of OH groups vibrating in six- and eight-membered rings with respect to their unperturbed vibration correlated with the minimum distances to the lattice oxygens in the same ring (d_{min}). *Reproduced from Ref. (97).*

(99). The conclusions are supported by density functional theory (DFT) calculations *(100)*.

Another type of hydroxyls that are typical of zeolites are isolated silanol groups, which are observed around 3745 cm^{-1} (see, e.g., Figure 2.16). These hydroxyls are located on the external surface of the zeolite and, consequently, are more abundant in samples characterized by a small crystallite size. A relationship between the external surface area and the ratio between the acidic and silanol groups can be seen in Figure 2.17 *(101)*.

In zeolites with defective structures, silanol groups can also be located at the internal surface. In this case, they are slightly H-bonded and are observed at somewhat lower frequencies (3740–3730 cm^{-1}) than free silanols.

If a fraction of the aluminum is, for whatever reason, not in framework positions, aluminum hydroxyls are observed in the spectra, typically in the range between 3670 and 3660 cm^{-1}. An according band can be discerned in Figure 2.16, although it has a low intensity. Another Al–OH band, at approximately 3782 cm^{-1}, was observed for some zeolites (mostly samples of BEA). This band has been attributed to hydroxyls bound to tetrahedrally coordinated aluminum, but a detailed interpretation is still lacking.

Table 2.4 Spectral characteristics of bridging OH groups located in different zeolite cages (59)

Zeolite	Channel/cage	Size, Å	ν(OH), cm^{-1}	^{1}H NMR shift, δ_H, ppm
FER	8-Ring channel [010]	3.5 × 4.8	3598	4.4
	10-Ring channel [001]	4.2 × 5.4	3591	
MFI	10-Ring channel [101] (sinusoidal)	5.1 × 5.5	3610, Channels	4.1 ÷ 4.3
	10-Ring channel [010] (straight)	5.3 × 5.6	3620, Intersection	
BEA	12-Ring channels [001]	5.6 × 5.6	3609	3.8 ÷ 5.6
	12-Ring channels [100]	6.6 × 6.7	3628, 3608, 3590	
MOR	8-Ring compressed channels [001]	2.6 × 5.7	3588, Side pockets	4.0 ÷ 4.2
	8-Ring side pockets [010]	3.4 × 5.8	3609, Intersection	4.6
	12-Ring main channels [001]	6.5 × 7.0	3605, Main channel	
FAU	Hexagonal prism accessed through 6-ring channels	2.7 × 2.7	3501	4.7 ÷ 4.8
	Sodalite cages accessed through 6-ring channels	2.7 × 2.7	3553	4.7 ÷ 4.8
	Supercages accessed through 12-ring main channels [111]	7.4 × 7.4	3625	3.7 ÷ 4.4

H-bonded hydroxyls are also often detected (although not always discussed) in zeolites of various kinds. It is ascertained that the already discussed silanol nests are characterized by a broad band centered around 3500 cm^{-1}. The nests constitute defects in the zeolite structure and are caused by extraction of a T-atom (a tetrahedrally coordinated metal cation of the framework) and filling of the gap with four protons. However, bridging hydroxyls in zeolites can participate in strong H-bonds with the framework oxygen atoms, and these hydroxyl structures can contribute to the broad absorption around 3500 cm^{-1}.

The picture is similar for porous alumophosphates. These materials can carry an additional type of hydroxyls, namely P–OH groups, which are

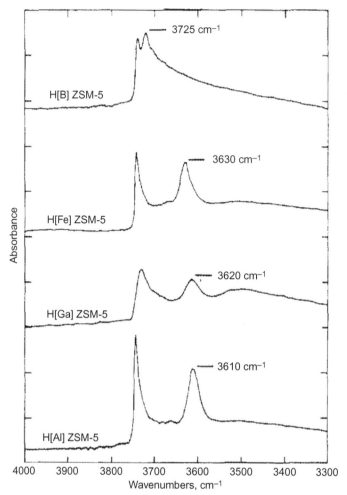

Figure 2.16 Hydroxyl stretching region of FTIR spectra of ZSM-5 zeolites isomorphously substituted with iron, gallium, or aluminum. The position of the bridging hydroxyls is labeled. *Reproduced with permission from Ref. (99). Copyright 1985 American Chemical Society.*

located at the external surface of the material or at defects. The corresponding bands are observed at around 3675 cm^{-1} *(59,102–104).*

Finally, OH groups attached to cations introduced into the zeolite by ion exchange have been reported.

2.7.2 Silica, mesoporous alumosilicates, and related materials

Mesoporous silicates, mesoporous alumosilicates, and silica are characterized by isolated silanol groups (with an absorption range of 3750–3740 cm^{-1}).

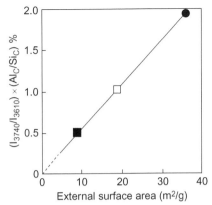

Figure 2.17 Correlation between the relative intensity of the 3740 cm^{-1} absorbance region corrected for the aluminum content, and the external surface area of HZSM-5 samples. Al$_C$/Si$_C$ stands for the Al/Si molar ratio in the crystallites. *Reproduced from Ref. (101).*

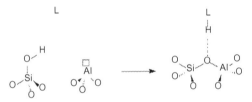

Figure 2.18 Structural rearrangement of silanol groups to bridging OH groups on alumosilicates. The rearrangement is induced by a base (L). *Reproduced from Ref. (106).*

Structural defects can cause the appearance of stable H-bonded hydroxyls. According to the model by Tsyganenko *(27,28)*, no type II hydroxyls can exist on pure siliceous materials. In principle, it would be possible for bridging hydroxyls to appear as a result of the presence of another cation, for example, Al^{3+}. Analysis of the literature data shows that bridging hydroxyls on silica–alumina and mesoporous alumosilicates are, if at all, detected in negligible concentrations. Seemingly inconsistent with this result, these materials demonstrate a remarkable acidity, much higher than that of pure silica. These observations have been rationalized by Busca and coworkers *(59,105,106)*. The authors proposed that bridging hydroxyls are formed only in zeolite pores and are not stable on open surfaces. However, upon reaction with bases, silanol groups can achieve a bridging configuration (see Figure 2.18). The fact that weak bands around 3610 cm^{-1} are occasionally registered with silica–alumina samples has been explained by the presence of a small amount of zeolite-like pores or the presence of impurities, for example, bicarbonates. Conclusions similar to those made by Busca and coworkers *(59,105,106)* regarding the presence of the so-called pseudo-bridging silanols have been made on the basis of DFT calculations *(107)*.

Geminal OH groups can also exist on silica *(108)*. It is believed that their bands are located a few wavenumbers below the bands arising from isolated silanols *(29)*. Positions at 3733–3732 cm^{-1} *(56,109)* or 3745–3742 cm^{-1} *(29,110)* have been suggested. Therefore, it is difficult to prove the presence of geminal silanols on the basis of IR spectra alone. Takei *et al.* proposed that $Si(OH)_3$ surface groups also exist and recorded their vibration at 3745 cm^{-1} *(29)*.

As for zeolites (see Section 2.7.1), hydrogen-bonded silanols and hydroxyls associated with extraframework aluminum (EFAL) sites have been found in spectra of the materials discussed in this section.

2.7.3 Metal oxides

With respect to hydroxyl groups, the situation on oxide surfaces is more complicated than that on the surface of zeolites. Some examples of the spectral signature of hydroxyl groups on oxide surfaces were already provided in Table 2.2. The M–O bond in metal oxides may possess an ionic, a covalent, or a mixed character, depending on the oxide. Consequently, the oxides demonstrate acidic, basic, or amphoteric properties.

Ionic (i.e., basic) oxides are characterized by well-defined OH stretching bands in their IR spectra. On the surfaces of these oxides, different types of hydroxyls exist (according to the classification by Tsyganenko). Because the microcrystals are terminated by various crystal planes that are characterized by different configurations of the ions on the surface, the hydroxyl distribution strongly depends on the sample morphology and the crystal habitus. Thus, different families of hydroxyl groups (that may be associated with particular planes) are observed for many oxides including ZnO *(111)*, γ-Al_2O_3 *(112)*, α-Fe_2O_3 *(113)*, α-Cr_2O_3 *(114)*, TiO_2 *(115)*, ZrO_2 *(116)*, and CeO_2 *(117)*. It must also be considered that hydroxyls are often located at surface defects, such as steps, corners, or edges. The concept of the ionic bond is valid for the oxides of the alkaline-earth metals (MgO, CaO, SrO, BaO) *(118)*. A reduction of the ionicity is reported for Al_2O_3, but even in this case, the departure from the ionic model is not exceptionally large.

Covalent oxides (e.g., silica, see Section 2.7.2) have OH groups that, when isolated, exhibit characteristic OH stretching frequencies. The same is valid for other covalently bound hydroxyls, such as POH. Bonds with a high degree of covalency are formed in oxides containing metals in a high oxidation state, for example, in V_2O_5 or MoO_3. Supported forms of such oxides are of practical interest because of the low specific surface area and the low melting points of the pure substances. While the OH stretching frequencies are again characteristic of the chemical composition of the hydroxyl *(119)*, these materials are characterized by ill-defined hydroxyl bands. This spectral behavior has been associated with delocalization of the hydroxyl protons. In some cases, geminal hydroxyls have been suggested. In summary, a complete interpretation of the spectra of such compounds is pending.

As can be expected, the hydroxyl spectra of metal oxides strongly depend on a possible redox pretreatment *(117)*. Generally, an increase of the oxidation state of the metal cation leads to a more covalent character of the metal–oxygen bond.

2.7.4 Other materials

As mentioned in Section 1, most of the materials in our surroundings have surface hydroxyl groups. Similar to oxides, other ionic crystals (e.g., MgF_2 *(120)*) also have hydroxyl groups present on their surfaces.

Hydroxyls can also be attached to metal particles and can be associated with the initial stages of metal corrosion. For example, the reaction of oxygen with hydrogen-covered supported metal nanoparticles of the elements iron, cobalt, nickel, rhodium, iridium, and platinum resulted in the appearance of hydroxyls. The O–H stretching modes of these species were reported in the region of $3685-3500 \text{ cm}^{-1}$ *(121)*. Hydroxyl groups on metals can also be formed through dissociative adsorption of water *(122)*. In fact, these OH groups are connected to metal cations (produced by oxidation of the metal) that could be partially surrounded by other cations.

Recently, MOFs have attracted considerable interest. Hydroxyl groups on such samples are generally characterized by highly covalent bonds. They can be connected to the inorganic or the organic part of the structure. A review on the use of IR spectroscopy to characterize MOF samples has recently been published *(123)*.

2.8. Acid–base properties

The first complete theory of acids and bases was formulated by Arrhenius in 1897 *(124)*: in water, acids dissociate to produce H^+, whereas bases produce OH^-. The more advanced and general Brønsted–Lowry theory was proposed in 1923 *(125,126)*. According to this theory, acids are substances able to donate protons, whereas bases are able to accept protons. Even more general is the definition by Lewis *(127)*: acids are substances able to accept a lone electron pair, whereas bases are substances able to donate such pairs. Walling *(128)* defined surface acidity of solids as the ability to convert an adsorbed neutral base to its conjugate acid.

Without doubt, surface acidity and basicity are decisive for the properties and the catalytic performance of a solid. Consequently, measuring the acidity and basicity of a given family of OH groups is of great importance. Below, we shall briefly consider the main principles of such measurements, thereby focusing on vibrational spectroscopy. Several monographs and review papers addressing this problem have recently been published *(56,60,88,129–139)*.

There are various acidic and basic sites on surfaces. Coordinatively unsaturated cations act as electron pair acceptors and are called Lewis acid sites. The Brønsted acid sites on catalyst surfaces are hydroxyl groups able to donate protons. The basic sites are, in most cases, oxygen anions, including oxygen from hydroxyl groups.

Both the acidity and the basicity of surface OH groups extend over a wide range, depending on the nature, type, and exact location of the group. There are two main factors determining the acid–base properties of surface hydroxyls: (i) the nature of the site(s) to which they are bound and (ii) the type of the hydroxyl group (taking also into account the coordination of the metal cations to which the OH is bound). Hydroxyls with a largely ionic character, such as Ba–OH groups, will be characterized by strong basicity, whereas hydroxyls connected by covalent bonds will typically, but not necessarily, exhibit strong acidity. The type of the hydroxyl is also important. The OH groups of type I demonstrate, as a rule, weak or no acidity. In comparison with that of type I groups, the acidity of bridging hydroxyls is enhanced (provided the bond is nearly covalent), because of the stronger bond between oxygen and surface and the resulting weakening of the O–H bond. A more detailed analysis shows that other factors can also affect the acid–base properties of surface hydroxyls. For example, proton transfer has been promoted through solvation, that is, the transfer is the result of the combined action of several basic molecules *(11,135)*. Such processes can be impeded by steric hindrances (that do not allow the simultaneous interaction of the acidic site with two or more basic molecules).

Because the proton in acidic hydroxyls is very "loose," it can be easily replaced by another cation. This reaction is widely used to modify surface properties. A large variety of cations can be introduced into many types of zeolites, thus providing the possibility to tune their adsorption and catalytic properties.

The exchange is not restricted to metal cations. Acidic OH groups can easily protonate basic molecules, and the resulting cation (e.g., NH_4^+) replaces H^+. The proton can be exchanged during catalytic reactions by a surface intermediate. Also, cationic products of surface reactions, for example NO^+, can replace H^+ *(140)*.

There are many methods for measuring acidity, and they were often originally conceived for solutions. To describe solutions, the most frequently used parameter is the pH. However, the pH (which depends on the acid concentration) is characteristic of a given system and does not account for the intrinsic properties of the acid. To estimate the properties of a given acid, usually the negative logarithm of the acid dissociation constant, the pK_a value, is used. Note that this value depends on the measurement conditions, such as temperature or ionic strength. Another peculiarity of the pK_a value is that it was originally defined for aqueous solutions. Broadening the definition for other solvents is possible and has been done, but this approach results in different sets of pK_a values, because the solvent participates in the chemical equilibrium. Therefore, the pK_a value does not represent an intrinsic property of an isolated species.

Another quantitative measure of acidity is the proton affinity (PA), which was originally introduced to describe the basicity of molecules in their gaseous state. The PA is an intrinsic property of a molecule, defined in the absence of a solvent. Therefore, the PA seems to be a more appropriate parameter for the characterization of acidity or basicity (i.e., the acid or base strength) of surface OH groups than the pK_a value.

The PA is defined as the negative enthalpy of a proton addition reaction, and thus the term "affinity" is used. The PA of an M–OH group, PA_{OH}, is the enthalpy (ΔH) of the reaction in Equation (2.10) *(129)*.

$$M - OH \rightarrow MO^- + H^+ \tag{2.10}$$

and the PA of a base B is the enthalpy (ΔH) of the reaction:

$$BH^+ \rightarrow B + H^+. \tag{2.11}$$

The total enthalpy of the two reactions combined will depend on the PAs of the hydroxyl group and of the interacting base.

The PA is used to interpret spectroscopic measurements of hydroxyls brought in contact with a base. For this type of reaction, it was reported that the enthalpy of formation of the hydrogen bond is proportional to the square root of the shift of the O–H stretching modes toward lower wavenumbers, $\Delta\nu(OH)$. If one and the same base is used to probe different OH groups,

$\Delta\nu(OH)$ correlates with the PA of the hydroxyls. The relationship was originally proposed for solutions *(37,141)*. The measurement of the acidity of hydroxyl groups by the shift of their OH stretching modes to lower frequencies upon complexation with bases is known as the H-bond method.

The estimation of the acidity of surface hydroxyls with the help of the shift of $\nu(OH)$ started soon after the development of the IR technique for surface studies *(142)* and is now widely used. Details on how to measure the OH acidity according to the H-bond method are provided in Section 3. It should be noted that, because the proton is not actually transferred, the results obtained by the H-bond method may somewhat deviate from other acidity measurements.

Another widely used method for measuring surface acidity is the so-called ion-pair method *(59,129)*. This method is based on the detection of the protonated species, BH^+, after protonation of a basic molecule according to Equation (2.12):

$$M - OH + B \rightarrow MO^- + BH^+ \qquad (2.12)$$

By utilizing molecules with varying basicity, one may rank the acidity of different types of surface hydroxyls. Another approach is to investigate the thermal stability of the complexes formed. The ion-pair method provides direct information on the proton transfer but its use is restricted to the characterization of surface hydroxyls with relatively high acid strength.

Basic sites on surfaces are generally associated with oxygen anions, also as part of an OH group. Presently, there is no generally accepted method for measuring the basicity of surface hydroxyls *(131)*. In general, acidic molecules (e.g., CO_2, CH_3OH) interact more readily or preferentially with more basic hydroxyls. Also, basic hydroxyls are easily replaced by anions such as Cl^- and F^-. In this way, the most basic OH groups can be selectively eliminated from a surface.

3. INTERACTION OF OH GROUPS WITH PROBE MOLECULES: FORMATION OF H-BONDS

3.1. Probe molecules

The interaction of molecules with OH groups is investigated for two reasons: (i) to characterize the surface hydroxyls and (ii) to obtain information on the activation (and conversion) of particular molecules. In the first case, the molecule has to act as a probe. To be a good probe, a molecule should fulfill a number of requirements, as summarized in several review articles

(56,129–131,143–145). For characterization of surface hydroxyls via the hydrogen-bond method, the specific requirements are reduced, because the OH groups themselves can take over some of the functions that otherwise the probe would have to provide. The requirements in this case can be summarized as follows:

- The probe molecule should be basic enough to form stable H-bonded complexes with the OH groups. At the same time, to avoid protonation, the basicity should not be very high. To provide some guidance, the PA of the probe molecule should typically be lower than 800 kJ mol^{-1}. However, more basic molecules are appropriate when the hydroxyls are weakly acidic.
- Although the probe molecule may interact with other sites (e.g., Lewis acid sites), an H-bonded probe molecule should not interact simultaneously with other surface sites.
- The reactivity of the probe molecule at the experimental conditions should be low to avoid any reactions other than simple H-bonding. The formation of surface compounds via chemical transformations would modify the intrinsic acid–base properties of the original material.
- Because the main spectral parameter analyzed is the shift of the OH stretching vibrations, the molecular probe should not exhibit vibrational modes in the $\nu(OH)$ region. Therefore, OH-containing molecules, such as water, are not suitable probes for this purpose.
- The probe molecule should be as small as possible to permit access even to hydroxyls located in narrow pores and to avoid steric hindrance of its interaction with a surface site. However, the use of molecular probes with varying molecular dimensions is very helpful in the determination of the location of hydroxyls.
- It is advantageous if the spectral response of the probe is sensitive to the bonding with the surface. A quantitative determination of the number of interacting hydroxyls becomes possible under these circumstances.

The most commonly used molecular probe for measuring the acidity of surface hydroxyls by the hydrogen bond method is definitely CO. Other widely applied probes are benzene and CD_3CN, followed by N_2 and alkenes.

3.2. General principles of the hydrogen-bond method

The basic spectral properties of the H-bonded surface hydroxyls were already described in Section 2.4. Here, the interaction of hydroxyls with molecular probes via formation of an H-bond will be considered, and we shall focus on the spectral features of the hydroxyl groups and of the probe molecules, as well as on the kind of information that can be extracted from the spectral parameters.

In principle, when a probe molecule forms an H-bond with only the hydrogen atom of an OH group, the interaction yields information on the acidity of the hydroxyl. On the contrary, if the molecule (hypothetically) interacts only with the oxygen atom, the information concerns the basicity. When the interaction is with both, the oxygen atom and the hydrogen atom, the situation is more complicated and needs a careful analysis. A complicated situation also arises when the probe molecule interacts not only with an OH group, but with one or more other surface sites.

The formation of H-bonds between OH groups and various probe molecules is widely used to assess the acidity of surface hydroxyls. A potential problem is that a given probe may be protonated by one kind of surface hydroxyls while it forms only an H-bond with another (less acidic) kind of OH. Another important phenomenon is that a probe molecule may form an H-bond with a hydroxyl group at low surface coverage but may accept a proton at high surface coverage when the OH group interacts simultaneously with two probe molecules. In this section, we will exclusively consider H-bond formation. However, whenever data for protonation exist, it will be explicitly mentioned.

As a consequence of the H-bond formation, the O–H bond order decreases and the OH stretching modes are shifted to lower frequencies. Concomitantly, the integral intensity and the full width at half maximum (FWHM) of the OH stretching band increase. It is well established that for relatively weak H-bonds (i.e., when the $\Delta\tilde{\nu}(OH)$ shift is below 400 cm^{-1}) the following trends apply: the stronger the H-bond, the larger the shift and the broadening of the band, and the higher the intensity.

Makarova et $al.$ (146) investigated the interaction of a series of weak bases (Ar, H_2, O_2, N_2, CH_4, C_2H_6, C_3H_8, CO, and C_2H_4) with the bridging hydroxyls in various porous materials. They reported good correlations between the shift of the OH modes ($\Delta\tilde{\nu}_{OH}$ in cm^{-1}) and the broadening and increase in intensity of the OH band:

$$\Delta A/A_0 = 0.018 \, \Delta\tilde{\nu}(OH) \qquad (2.13)$$
$$\Delta a/a_0 = 0.010 \, \Delta\tilde{\nu}(OH) \qquad (2.14)$$

where A_0 is the integral intensity of the $\nu(OH)$ band before the interaction and ΔA is the change in the integral intensity caused by the formation of the

H-bonded complex, and (using the same notation) Δa and a_0 are the change in line width and the original line width.

Unfortunately, in most cases only the shift of $\tilde{\nu}(OH)$ is reported, and no details on the FWHM or integral absorbance are provided.

Datka and Gil *(147)* established a correlation between the final half-width ($\Delta\tilde{\nu}_{1/2}$) and the frequency shift ($\Delta\tilde{\nu}$) of stretching vibrations by letting a series of compounds with uniform hydroxyls, specifically, phenol, acetic acid, trifluoroacetic acid, NaHX zeolite (Si/Al = 1.0), and SiO_2 interact with various electron donor molecules. The deviations from this correlation were used as a measure of the heterogeneity of OH groups in several zeolites.

A useful review on the application of infrared spectroscopy for the measurement of surface acidity was published by Paukshtis and Yurchenko about 30 years ago *(129)*. The authors analyzed the literature data that were available at the time and extracted that, for weak H-bonds, the shift of the OH stretching modes is proportional to the enthalpy of the following reaction:

$$OH + B \rightarrow OH \cdots B \qquad (2.15)$$

and is therefore a function of the PAs of the hydroxyl and the base. Because the PAs of the bases are known from the literature (see Table 2.5), the PA of the OH group can be easily calculated. To find PAs not reported in Table 2.5 or more precise values, the reader can consult the updated literature, for example, Ref. *(148)*.

It has been established that, for a specific hydroxyl group, $\Delta\tilde{\nu}(OH)$ correlates well with the PA of the base used as probe *(141)*. On the contrary, using a specific base, the shift of the stretching modes of different hydroxyls depends on the PA of the hydroxyls. This dependency is the foundation of the so-called Bellamy–Hallam–Williams (BHW) plots *(149)*. When considering the interaction of two different OH groups with a family of basic molecules, a plot of the $\Delta\tilde{\nu}(OH)$ values of one OH group versus the respective values of the other (with the PA of the base as parameter) should yield a straight line, and the slope will be a measure of the relative acid strength of one of the OH groups (relative to the acid strength of the other). In surface chemistry, silanol groups are usually used as a standard. The BHW plot has been verified for hydrogen bonds of small to medium strength. It was established that this correlation is valid for shifts that do not exceed 1000 cm^{-1} *(135)*.

Another possibility to draw the BHW plot is to consider the interaction of different hydroxyls with two basic molecules. Although $\Delta\tilde{\nu}(OH)$ is

Table 2.5 pK_a, proton affinity, and shift of the SiOH stretching modes for various bases (129)

No.	Base formula	Base name(s)	pK_a	PA (kJ mol^{-1})	$\Delta\tilde{\nu}$(OH) for SiOH (cm^{-1})
1	Ar	Argon	–	377	8
2	O_2	Dioxygen (oxygen)	–	423	12
3	Kr	Krypton	–	435	16
4	Xe	Xenon	–	477	19
5	N_2	Dinitrogen (nitrogen)	–	477	24
6	C_6H_{12}	Cyclohexane	–	586	45
7	CO	Carbon monoxide	–	598	90
8	CH_3Cl	Chloromethane (methyl chloride)	–	670	110
9	CH_3NO_2	Nitromethane	11.93	753	160
10	CH_3CHO	Ethanal (acetaldehyde)	–	774	290
11	$CH_3C(O)CH_2CH_3$	Butan-2-one (methyl ethyl ketone)	−7.2	774	330
12	$C_4H_8O_2$	1,4-Dioxane (dioxane)	−3.22	799	390
13	$(CH_3)_2CO$	Propanone (acetone)	−7.2	816	340
14	$(CH_2)_4O$	Oxolane (tetrahydrofuran, THF)	−2.02	820	470
15	$(C_2H_5)_2O$	Ethoxyethane (diethyl ether, ether)	−2.42	824	460
16	$C_6H_5C(O)CH_3$	1-Phenylethanone (acetophenone)	−6.15	833	320
17	CH_3COOCH_3	Methyl acetate	–	816	270
18	$(CH_3)_2SO$	Dimethyl sulfoxide	0	870	550
19	C_6H_5CN	Benzonitrile	−10.43	–	300
20	CH_3CN	Acetonitrile	−10.13	783	300

Table 2.5 pK_a, proton affinity, and shift of the SiOH stretching modes for various bases (129)—cont'd

No.	Base formula	Base name(s)	pK_a	PA (kJ mol^{-1})	$\Delta\tilde{\nu}$(OH) for SiOH (cm^{-1})
21	NH_3	Azane (ammonia)	9.25	846	750
22	$(C_3H_7)_3N$	Triisopropylamine (tripropylamine)	11.0	940	990
23	$C_6H_5NH_2$	Phenylamine (aniline)	4.81	874	500
24	$CH_3C_3H_3N_2$.	1-Methylimidazole	6.95	–	830
25	$CH_3CH_2NHCH_2CH_3$	Diethylamine	10.7	929	930
26	$N(CH_3)_3$	Trimethylamine	9.67	929	990
27	$(CH_2)_5NH$	Piperidine (hexahydropyridine)	11.22	933	980
28	$N(CH_2CH_3)_3$	Triethylamine	11.02	958	990
29	C_5Cl_5N	2,3,4,5,6-Pentachloropyridine (perchloropyridine)	−8	–	270
30	$C_5H_3Br_2N$	2,5-Dibrompyridine	−3	–	450
31	C_5H_4BrN	5-Bromopyridine	0.8	891	580
32	$C_5H_2Cl_3N$	3-Chloropyridine	3.6	891	630
33	C_5H_5N	Pyridine	5.21	912	750
34	C_6H_7NO	2-Methoxypyridine	3.28	912	550
35	C_6H_7N	2-Methylpyridine (2-picoline)	5.97	925	750
36	H_2S	Hydrogen sulfide	–	714	150
37	$(CH_3)_2S$	Dimethyl sulfide	−5.25	824	370

widely used, the ratio of $\Delta\tilde{\nu}(OH)/\tilde{\nu}_0$ (where $\tilde{\nu}_0$ is the wavenumber of the unperturbed hydroxyl) is often preferred.

Paukshtis and Yurchenko (129) proposed the following correlation for weak H-bonds ($\Delta\tilde{\nu}(OH) < 400\,\text{cm}^{-1}$) formed after adsorption of different bases:

$$PA^{OH} = PA^{SiOH} - 442.5\log(\Delta\tilde{\nu}(OH)/\Delta\tilde{\nu}(SiOH)). \qquad (2.16)$$

where PA^{OH} and PA^{SiOH} are the proton affinities of the OH groups under investigation and of silanol groups, each in kJ/mol.

In this correlation, the shift measured for surface silanols is used as a reference. The PA^{SiOH} is assumed to be 1390 kJ mol^{-1}. To be able to use Equation (2.16) in conjunction with a given basic molecule, one needs to know the shift of the OH modes of silanol groups that is induced by this base. Data for the shift of silanol modes induced by a series of bases can be found in Table 2.5. Note that the $\Delta\widetilde{\nu}(OH)/\Delta\widetilde{\nu}(SiOH)$ ratio does not depend on the base used. However, the equation is not applicable for strong H-bonds because then the relationship between PA and $\Delta\nu(OH)$ is no longer logarithmic. For the situation of $\Delta\widetilde{\nu}(OH)$ exceeding -400 cm^{-1}, another equation has been proposed (the so-called method of the strong H-bond)

$$\Delta\widetilde{\nu}(OH)/\widetilde{\nu}^0(OH) = 0.216 - 0.0096\Delta pK_a, \qquad (2.17)$$

where $\widetilde{\nu}^0$ is the OH stretching frequency before formation of the H-bond, and ΔpK_a is the difference between the pK_a of the OH group and that of the base participating in the formation of the H-bond.

The pK_a values of bases that are often employed in IR spectroscopy are given in Table 2.5. However, as will be shown below, the spectra of strongly H-bonded hydroxyls are complicated because of the occurrence of Fermi resonance, and the maxima of the shifted bands cannot be accurately measured. For this reason, the acidity of hydroxyl groups is preferably measured through formation of weak H-bonds.

It is important to know that the formation of H-bonds does not only affect the OH stretching but also the OH bending modes. The latter are shifted to higher frequencies. In addition, an important phenomenon encountered in the case of medium or strong H-bonds is the split of the $\nu(OH)$ modes into two or three components as a result of Fermi resonance. These phenomena, originally reported by Hadži *(150)* for solutions, have been summarized and transferred to surface chemistry by Paze *et al. (11)*. The authors distinguished three cases of interaction between the bridging OH groups in zeolites and various bases: formation of weak, medium, or strong H-bonds.

When a weak H-bond is formed, the $\nu(OH)$ modes are shifted to lower frequencies by up to 400 cm^{-1}. Concomitantly, the δ and γ bending modes shift to higher frequencies (see Figure 2.19B and C). The changes in $\nu(OH)$ position, FWHM, and integral intensity are proportional to the enthalpy of reaction. In Ref. *(11)*, the authors have assumed that the O–H bond will

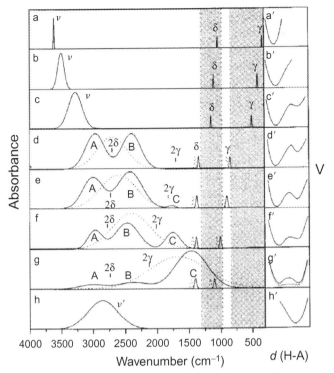

Figure 2.19 Left part of panel, (a–h): Qualitative representation of the IR spectroscopic features of weak, medium, and strong A–H···B or A⁻···H–B⁺ H-bonded complexes. The half-width of the bending modes is (somewhat arbitrarily) assumed to slightly increase upon increasing the hydrogen bond strength. The shaded areas correspond to regions obscured by the skeletal modes of the zeolite framework. Right part of panel, (a′–h′): Schematic representation of the correlated evolution of the proton potential as function of the A–H distance in A–H···B or A⁻···H–B⁺. The separation barrier in e′–f′ can be very low, and a potential curve characterized by an asymmetric single flat minimum may be used alternatively. *Reproduced with permission from Ref. (11). Copyright 1997 American Chemical Society.*

become more anharmonic upon H-bonding. However, as already noted, weak H-bonds cause an increase in the harmonicity of the O–H oscillators *(43,44)*.

When medium-strength H-bonds are formed ($\Delta\tilde{\nu}_{OH}$ between -400 and -1000 cm^{-1}), the same phenomena are observed. However, when the position of the maximum of the shifted $\nu(OH)$ band coincides with that of the first overtone of the δ modes (2δ), a Fermi resonance occurs. As a result, the $\nu(OH)$ band shape is no longer represented by the dotted line in Figure 2.19D, and two separate bands appear, represented by the solid line, that are the so-called A and B components of the band.

Usually, Fermi resonance leads to the appearance of two bands with equal intensities (as in Figure 2.19D). The situation is different if the OH band is shifted to frequencies slightly lower than the value of 2δ. Normally, the Fermi resonance should disappear. However, the shifted OH stretching modes are better described as consisting of a continuum of bands with slightly different positions (which is the explanation of the band broadening). Therefore, in this case, the so-called Fano-type Fermi resonance will occur *(151)*. This effect implies that if the 2δ mode falls within the range spanned by the baseline width of the shifted ν(OH), splitting will always occur; however, the two bands may exhibit different intensities. This situation is illustrated in Figure 2.19E, which shows a hypothetical spectrum characterized by a B component that is more intense than the A component.

Another phenomenon that is illustrated in the same figure is the appearance of a third component (the so-called C-component). The third band is also due to Fermi resonance, but in this case the 2γ modes are involved. When the red shift of the OH stretching mode is very large, the C-component is more pronounced (Figure 2.19F), and it may become the principal component of the shifted ν(OH) band if a very strong H-bond is formed (Figure 2.19G).

The ABC structure is typical of H-bonded hydroxyls but usually does not appear if protonation occurs (see Figure 2.19H). However, we note that even after proton transfer the spectra may show an ABC structure, namely if the hydrogen from the BH^+ product forms H-bonds.

Although the Fermi resonance is typical of hydroxyls with medium or strong H-bonds, it was recently reported that the resonance can also occur when only weak H-bonds are present *(152–155)*. A detailed analysis of the picture presented in Figure 2.19C helps with the assessment of the spectral behavior. It can be seen that the 3δ (second overtone) mode falls into the region of the shifted ν(OH) when a weak H-bond is formed. The intensities of second overtones are typically by one to two orders of magnitude lower than those of first overtones. Although the fundamental δ mode strongly decreases in intensity with the formation of an H-bond, it remains relatively intense as long as the H-bond is weak. This behavior is a factor favoring a relatively high intensity of the second overtone, in contrast to cases with formation of stronger H-bonds. In any case, because of the relatively low intensity of the 3δ mode, the Fermi resonance is less pronounced than the resonances involving 2δ or 2γ modes.

It has been reported *(152, 153)* that the Fermi resonance occurs when the OH stretching vibrations are shifted to approximately 3300 cm^{-1}.

Consequently, the effect is usually observed after interaction of bridging zeolite hydroxyls with CO *(152–154)* or molecules of similar basicity, as, for example, C_6H_6 *(155)*. In contrast, weaker bases (e.g., N_2) cause smaller shifts, and no resonance occurs. No resonance is detected when CO is coordinated to the 3640 cm^{-1} hydroxyl band of HY zeolite, because the shifted band is located at ca. 3350 cm^{-1} *(102)*.

An important conclusion from these findings is that a split of ν(OH) modes of OH$\cdots$CO complexes is not necessarily indicative of heterogeneity of the hydroxyl groups, as often reported *(156–160)*.

The Fermi resonance of the ν(OH) mode with the 2δ and 2γ modes renders the spectra more complicated and hampers the exact determination of the ν(OH) band shift. It was proposed that the shift can be calculated on the basis of the frequencies and the intensities of the A, B, and C bands *(102,161)*. However, this calculation can give only approximate values and an error of roughly 10% has been estimated *(11)*.

Another possibility of H-bonding is the interaction of a probe molecule with the oxygen atom of the hydroxyl. In the case of hydroxyls that are covalently bound to the surface, this bonding configuration should result in a decrease of the O–H bond order and, accordingly, of the O–H stretching frequency. However, the effect should be smaller as compared to the "normal" H-bond configuration (via the H atom of the hydroxyl). The IR absorption coefficient should increase because of the increased polarity of the hydroxyl. Note that, as a rule, the hydroxyl oxygen of covalent oxides is weakly basic. Surface adsorption complexes with H-bonding exclusively via the oxygen of the hydroxyl are not known. However, it is possible for one molecule to interact simultaneously with the hydrogen and oxygen atoms of a hydroxyl group. Such a bonding situation was recently proposed for methane interacting with silanol groups *(162)*.

Many theory-based investigations of H-bonds with surface hydroxyls have been made, and yet complete agreement of computational and experimental results has not been reached. It should be emphasized that the shift of the OH modes is temperature-dependent *(31)*, whereas calculated frequencies are valid for absolute zero temperature. Hence, the temperature difference can account for some of the disagreement between theoretical and experimental results. Another source of discrepancy is the fact that the computed frequencies are usually harmonic. Thus, anharmonicity corrections are required to match experimental data.

In the following sections, we will discuss the interaction of hydroxyls with various probe molecules. The CO molecule will be considered first, because some of the conclusions also pertain to other weakly basic molecular probes.

3.3. Carbon monoxide

The use of CO as a probe for the characterization of oxide surfaces and zeolites has been reviewed about 10 years ago *(163)*. The CO stretching frequency depends on the coordination (through C- or O-atom) and the adsorption geometry (linear, bridged). Coordination of CO through the π-electron cloud was found not to be possible *(164)*. For linearly bound CO, the CO stretching frequency depends on the balance of the contributions from electrostatic, σ electron, and π electron interactions that together make up the bond to the surface site. However, the interaction between CO and OH groups is essentially electrostatic. The PA of CO at the carbon atom is 594 kJ mol^{-1} and at the oxygen atom 426.3 kJ mol^{-1} *(148)*. Consequently, CO prefers a linear bonding to hydroxyl groups via its carbon end, and the CO stretching frequency is determined by the strength of the electrostatic field. The stronger the bond, the higher the CO stretching frequency. Depending on the acidity of the surface hydroxyls, the carbonyl bands are detected in the region of 2183–2145 cm^{-1} (usually between 2176 and 2155 cm^{-1}). In principle, the position of the carbonyl vibration could be used for estimation of the acidity of the OH groups, but it is much less sensitive than the shift of the ν(OH) bands. Note that, because the interaction is relatively weak, the experiments are performed at low temperature, typically around 100 K, and often in the presence of gas-phase CO. Recording the spectra at various coverages can help distinguishing between the different adsorption forms, because the most acidic sides react first (i.e., at low coverage) with CO, and then the weaker sites are successively occupied. A similar approach is to use difference spectra reflecting the changes occurring with small variations of the CO coverage. This approach is demonstrated in Figure 2.20, in which spectrum "a" (high coverage) demonstrates a complex interaction while spectrum "b" (low coverage) indicates almost exclusive interaction of CO with the 3615 cm^{-1} OH hydroxyl groups.

There exist numerous reports with IR investigations on the interaction of CO with different hydroxyls. The results of most of them are summarized in Tables 2.6–2.9. For the sake of brevity, data that differ only slightly have been merged.

We will start the discussion with the bridging hydroxyls in zeolites, which are by far the most extensively investigated OH groups of catalysts. According to Kubelková *et al.* *(102)*, the PAs of these OH groups vary within the interval of 1204–1139 kJ mol^{-1}. The authors used Equation (2.16) and assumed a value of 90 cm^{-1} for $\Delta\tilde{\nu}$(SiO − H). Analysis of

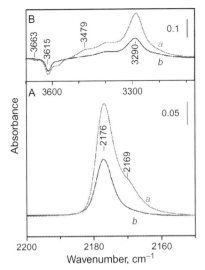

Figure 2.20 Changes in the FTIR spectra of HZSM-12 induced by adsorption of CO at a temperature of 100 K. Spectrum (a) is the difference between two spectra registered at slightly different medium coverages, and spectrum (b), the difference between two spectra registered at slightly different low coverages. (A) CO stretching region and (B) OH stretching region. *Reproduced from Ref.* (165). (See the color plate.)

the data presented in Table 2.7 indicates that the value of 90 cm^{-1} is not exact, but it has generally been adopted as a standard.

The CO-induced shift of $\widetilde{\nu}(OH)$ of the bridging hydroxyls is typically around -300 cm^{-1} (Table 2.6), corresponding to a PA of 1160 kJ mol^{-1}. A similar shift was found by DFT calculations *(285)*. The highest acidity of zeolite hydroxyls was reported for dealuminated faujasites *(171–173)*; in this case, the PA reaches a value of about 1080 kJ mol^{-1}.

Although the picture seems to be clear, there are some peculiarities that deserve attention:

1. *Temperature effect.* The position of the hydroxyl band maxima is weakly temperature-dependent. Introduction of CO to the system usually leads to a decrease of the pellet temperature and a small blue shift of $\nu(OH)$. The effect may be minimized by recording difference spectra reflecting small coverage changes and then determining the shift from the minimum of the negative and the maximum of the positive IR band (see, for example, Figure 2.20).

2. *Static effect.* Adsorbed CO modifies the surface by donating electrons, and at high CO coverages, the electrophilicity of the surface sites is

Table 2.6 Shift of $\tilde{\nu}$(OH) of bridging hydroxyls in zeolites and in porous materials induced by low-temperature CO adsorption

Sample	$\tilde{\nu}$(OH), cm^{-1}	$|\Delta\tilde{\nu}_{OH}|$, cm^{-1a}	$\tilde{\nu}$(CO), cm^{-1}	Note	Ref.
HY	(a) 3636	301–328	2178–2176	(a), (b) different samples; coverage-dependent shift; minor fraction of the LF OH groups affected	(166)
	3552	227–262	2167–2165		
	(b) 3641	269–295	2176–2173		
	3548	192–222	2162–2160		
HY	3637–3632	272–324	2178–2174	Different samples	(167)
HY	3638	305	2176		(167)
HY	3640	300	2172		(168)
HY	3629	288	2173		(136)
HY	3641	271–287		Coverage-dependent	(169)
HY	3642	283	2174		(170)
DAY (USY)^b	3623	443	2183		(171)
DAY (USY)	3632–3627	434–440	2180–2179	Heterogeneity	(172,173)
		348–350			
DAY (USY)	3604	416	2180		(174)
	3630	360	2170		
DAY (USY)	3608	382			(175)
	3631	327			

DAY	3605	380	2182		(176)
DAY	3600	380			(177)
	3525	245			
DAY	3627–3601	374–344	2180–2177		(170)
DAY	3633	353	2179		(178)
LiHY	3642	302	–		(179)
KHY	3650	220			
RbHY	3653	168			
CsHY	3658	160			
HNaY	3650–3635	237–296	–	Depends on Na content, Si/Al ratio, and CO coverage	(102,131,180)
REY[c]	3635	305	2176		(172)
HEMT	3637	291–320	2178–2175	Coverage-dependent; minor fraction of the LF OH groups affected	(166)
	3550	222–250	2165–2163		
HX	3665	160	–		(181)
SBA-15[d]	3610	390	2175	Weak intensity	(182)
HMOR	3619	359–389	2175		(183)
HMOR	3670	370			(183)

Continued

Table 2.6 Shift of $\nu(OH)$ of bridging hydroxyls in zeolites and in porous materials induced by low-temperature CO adsorption—cont'd

Sample	$\tilde{\nu}(OH)$, cm^{-1}	$|\Delta\tilde{\nu}_{OH}|$, cm^{-1}	$\tilde{\nu}(CO)$, cm^{-1}	Note	Ref.
HMOR	3606	330–344	2176	Heterogeneity	(184)
		192–248	2171		
HMOR	3617–3612	332–338	2177–2176		(167,184)
HMOR	3620–3618	326	2175	Dealuminated sample	(185)
HMOR	3625	311	2174		(185)
HMOR	3614	310	2176–2172		(186)
HMOR	3610	300		Heterogeneity	(158)
		206			
HMOR	3609	296	2172		(131)
HMOR	3585	260	2169	Pyridine-precovered sample	(184)
HMAZ	3626	379	2180		(187)
	3606	276	2165		
2D zeolite	3584	379	2184		(188)
SBA-16[d]	Not resolved	(365)	–	Assuming unperturbed OHs vibrate at 3600 cm^{-1}	(189)
HZSM-5	3625–3610	335–350	2175		(190)
HZSM-5	3622	330	2180		(105)

HZSM-5	3640	320		2175	(191)
HZSM-5	3620–3615	310–320		2175	(170,192–194)
HZSM-5	3610	319		–	(195)
HZSM-5	3618	313 / ca. 220	Heterogeneity	2175	(157)
HZSM-5	3617–3615	310–313		2175–2173	(102,103,105,152,153)
HZSM-5	3610	308–310		2174–2175	(196–198)
HZSM-5	3601	310		–	(146,199,200)
HZSM-5	3616	305–310	Different Samples	2180–2170	(201)
HZSM-5	3621	307		2177	(202)
HZSM-5	3610	306 / 130	Heterogeneity	2175	(202)
HZSM-5	3618	303		2173	(203)
HZSM-5	3618–3614	301–303		2173–2174	(167,204)
HZSM-5	3616	294		2173	(205)
HZSM-5	3619	290		2175	(206)
HNaZSM-5	3610	290		–	(167)

Continued

Table 2.6 Shift of $\nu(OH)$ of bridging hydroxyls in zeolites and in porous materials induced by low-temperature CO adsorption—cont'd

Sample	$\tilde{\nu}(OH)$, cm^{-1}	$|\Delta\tilde{\nu}_{OH}|$, cm^{-1}	$\tilde{\nu}(CO)$, cm^{-1}	Note	Ref.
H[Ge]ZSM-5[d]	3614	300–315	2175–2172		(207)
H[Fe]ZSM-5	3637	289	2172	Fe(OH)Si	(131,200)
H[Fe]ZSM-5	3630	270	2172	Fe(OH)Si	(208)
H[Ga]ZSM-5	3622	292	2172	Ga(OH)Si	(200)
H[Ga]ZSM-5	3622–3621	279–282	2172	Ga(OH)Si	(157,209)
HSAPO-34	3625–3614	335–349	2170	Different samples; the 3630 cm^{-1} band of low intensity	(210,211)
	3630	281–287	2170		
	3601–3600	190–191	2164		
HSAPO-34	3627	270	2173		(212)
	3603	263			
HBEA	3610	312–343	2178	Different samples	(213)
HBEA	3610	340	–		(58)
HBEA	3606	327–331	2176–2175	Heterogeneity	(156,159)
		226–231	2167		
HBEA	3615	322	2175		(98)
HBEA	3615–3614	319–320	2177–2176	$\Delta\nu(OH)$ does not depend on the Si/Al ratio (214)	(11,214)

HBEA	3610	300–320	2175		(215)
HBEA	3616–3612	312–315	2175		(216–218)
HBEA	3612	307	2175		(170)
HBEA	3610–3608	298–305	2175–2173		(219–221)
HBEA	3615	305	2174		(222)
Fe[Si]BEA	3632	280	2174	Si(OH)Fe	(223)
[Ti]BEA	3620	300	2180		(224)
H-saponite[d]	3590	340	2178		(225)
MCM-58	3628	330	2175		(98)
	3556	166	–		
MCM-68	3617	328	2175		(98)
	3574	148	–		
ITQ-2	3628	328	2176		(226)
MCM-22	3627	326	2174		(109)
MCM-22	3626	320–321	2176–2174		(227,228)
MCM-49	3627	326	2174		(109)
HZSM-12	3611	325	2176		(165)
	3575	294	2175		

Continued

Table 2.6 Shift of $\nu(OH)$ of bridging hydroxyls in zeolites and in porous materials induced by low-temperature CO adsorption—cont'd

Sample	$\tilde{\nu}(OH)$, cm^{-1}	$|\Delta\tilde{\nu}_{OH}|$, cm^{-1}	$\tilde{\nu}(CO)$, cm^{-1}	Note	Ref.
HZSM-12	3615	322	2175		(98)
	3577	165			
HZSM-12	3510	306	2175		(195)
SSZ-24 (AFI)	3612	317	2177		(104)
Theta (TON)	3598	316	2173	Not all sites accessible	(229)
HSSZ-13 (CHA)	3613	316 (284)	2177–2174	Equal acidity considered because the two bands are shifted to 3300 cm^{-1}	(212)
	3584				
SiO$_2$–Al$_2$O$_3$	(3635)	315	2177	$\nu(OH)$ deduced	(174)
SiO$_2$–Al$_2$O$_3$	3607	315	–	Weaker sites not present after 800 K evacuation	(181)
		165			
SiO$_2$–Al$_2$O$_3$	3609	156	2174	Very low intensity	(230)
HCHA	3620	300	2174	Ill-defined maximum at 3620 cm^{-1}	(231)
		150	2163		
HFER	3603	286–299	–	Acidity increases with dealumination	(232)
HFER	3607–3605	296–297	2173–2172		(160,167,203)
HFER	3610–3606	290–292	2172–2171		(154,233,234)

Material					
CoHFER	3606	295	2172		(235)
HMCM-41[d]	(3610)	290	2175	Low intensity	(236)
HMCM-41[d]	3609	154	2174	Very low intensity	(230)
HMCM-41[d]	3605	202 / 156	–	Weaker sites not present after 800 K evacuation	(181)
SAPO-40	3643–3630	280–290	2176		(237)
SAPO-5	3626	268	2173		(104)
HMCM-48[d]	3607	205 / 162	–	Weaker sites not present after 800 K evacuation	(181)
FAPO-36	3678	185	2165		(238)

[a] The absolute values of the shifts are presented.
[b] DAY—dealuminated Y zeolite; USY—ultrastable Y zeolite.
[c] REY—Y zeolite containing rare-earth elements.
[d] Contains aluminum.
The data are arranged according to the materials, starting with FAU-type zeolites hosting the most acidic hydroxyls. For a particular material, the data are arranged according to the decrease of the shift.

Table 2.7 Shift of ν(OH) of silanol groups on various materials induced by low-temperature CO adsorption

Sample	$\tilde{\nu}$(OH), cm^{-1}	$|\Delta\tilde{\nu}_{OH}|$, cm^{-1}	$\tilde{\nu}$(CO), cm^{-1}	Note	Ref.
SiO$_2$	3745	155	2157	Heterogeneity	(172)
		75	2155		
SiO$_2$	3743	98	2158		(239)
SiO$_2$	3750	95	2160–2156		(240)
SiO$_2$	3748–3746	87–88	2157		(21,241)
SiO$_2$	3751	78	2157		(242)
Silicalite	3742	100	2156		(243)
	3704	120			
Silicalite	3749	96	2169	O-bonded CO also proposed	(244)
Silicalite	3750	90	2162		(224)
SBA–15	3747	94	2158		(245)
SBA–15	3745	85	2156		(182)
[Si]BEA	3745	90	2155		(219)
1,4-Phenylen–silica	3730	60	2153	Ordered periodic mesoporous organosilica material	(246)
SiO$_2$–Al$_2$O$_3$	3747	356–367	2178		(247)
		293	2174		
		189	–		
		93	2157		

Sample					Ref.
SiO$_2$–Al$_2$O$_3$	3745	75	2169–2164		(174)
	3740	335, 190			
SiO$_2$–Al$_2$O$_3$	3748–3728	129–228	2169–2164	Acidity increases with increase of silica content up to 40%	(173)
SiO$_2$–Al$_2$O$_3$	3745	250–300	2173	Heterogeneity	(172)
		165	2156		
		78	2156		
SiO$_2$–Al$_2$O$_3$	3747	87	2158		(248)
HZSM-5	3749	98	2156		(152)
Na-ZSM-5	3747	90	2156		(249)
HBEA	3740	300	–		(58)
HBEA	3749–3737	82–95	2158–2156		(11,156,214,219,222)
HMOR	3740	102	2156		(183)
HMOR	3747–3745	85–90	2157–2156		(184,186)
NaMCM-41[a]	3748	83	–		(250)
LiMCM-41[a]	3740	80	–		(251)
Saponite	3742	110	2160		(225)
HMCM-41[a]	3744	170	2159		(236)
		86	2157		

Continued

Table 2.7 Shift of $v(OH)$ of silanol groups on various materials induced by low-temperature CO adsorption—cont'd

Sample	$\tilde{v}(OH)$, cm^{-1}	$|\Delta\tilde{v}_{OH}|$, cm^{-1}	$\tilde{v}(CO)$, cm^{-1}	Note	Ref.
SBA-16[a]	3745–3733	83	2155		(189)
V/SBA-15	3742	99	2157	Shoulders corresponding to $\Delta\tilde{v}(OH)$ of -124 and -72 cm^{-1}, respectively	(252)
[Ti]BEA	3750	90	2160		(224)
TS-1	3750	70–90	2162		(224)
SiO$_2$–TiO$_2$	3745–3740	78–79	2156		(110,253)
H[Ga]ZSM-5	3750	96	–		(209)
H[Fe]ZSM-5	3747	94	–		(208)
	3720	140			
Fe[Si]BEA	3739	150	2157		(223)
		84			
SO$_4$/ZrO$_2$/SBA-15	3745	144	2160		(245)
		106	2158		

[a]Contains aluminum.
Table starts with data for pure siliceous materials.

Table 2.8 Shift of $\nu(OH)$ of various aluminum hydroxyls induced by low-temperature CO adsorption

Sample	$\widetilde{\nu}(OH),$ cm^{-1}	$\|\Delta\widetilde{\nu}_{OH}\|,$ cm^{-1}	$\widetilde{\nu}(CO),$ cm^{-1}	Note	Ref.
$\alpha\text{-Al}_2O_3$	3746	110		Average value	(254)
	3668				
$\delta/\gamma\text{-Al}_2O_3$	3728	128	2152		(255)
	3680	140			
$\gamma\text{-Al}_2O_3$	3695	95	2157		(241)
$\gamma\text{-Al}_2O_3$	3785, 3775	0	–		(114)
	3725	70	2152		
	3715	80	2152		
	3685	95	2152		
$\gamma\text{-Al}_2O_3$	3790, 3775	0	–		(256)
	3735	125	2157		
	3700	190	2167		
$\gamma\text{-Al}_2O_3$	3732	115	2157	–	(257)
	3680	120	2163	More stable	
$\gamma\text{-Al}_2O_3$	3730	100–150	2154	Coverage dependent	(258)
$\gamma\text{-Al}_2O_3$	3730	145	2158–2156		(259)
	3787	(200)			
K/Al_2O_3	3730	100–150	2154	Not acidic at high concentration of K	(260)
	3675	0			
SO_4/Al_2O_3	3640	205			(261)
	3695	170			
Imogolite	3743	233	2166	$(OH)_3Al_2O_3$ SiOH, nanotubes	(262)
	3725	245			
	3660	210			
HZSM-5	3670	260	2175	Assigned to bridging OH	(192)

Continued

Table 2.8 Shift of ν(OH) of various aluminum hydroxyls induced by low-temperature CO adsorption—cont'd

Sample	$\tilde{\nu}$(OH), cm^{-1}	$|\Delta\tilde{\nu}_{OH}|$, cm^{-1}	$\tilde{\nu}$(CO), cm^{-1}	Note	Ref.
HZSM-5	3670–3665	220–215			(190,194)
HZSM-5	3668–3665	202–197	2168		(103,205)
Na-ZSM-5	3673	138	–	–	(249)
HZSM-5	3786–3780	Unknown		Not acidic	(190,205)
HBEA CsHBEA	3670	215–175	2174	The shift decreases with Cs concentration	(215)
HBEA	3660	270			(58)
HBEA	3665	195	2175		(219)
HBEA CsHBEA	3670	220	2174		(221)
	3780	130	2156		
	3673	203	2174		
HY	3670	189		Steamed sample	(177)
ITQ-2	3670	210	–		(226)
MAZ	3660	210	2160		(187)
MCM-22	3670	190	2165		(227)
SiO$_2$–Al$_2$O$_3$	3730–3720	150–180	–		(248)
	3660	220	2163		
SiO$_2$–Al$_2$O$_3$	3677	190	2171		(174)
SBA-15	3700–3650	200–250	ca. 2172		(182)

noticeably weakened. This phenomenon is called static effect and is typical of cations on oxide surfaces (where the Lewis acidity is affected by CO coverage) and was reported also for cations in zeolites (286). For zeolites with a low Si-to-Al ratio (e.g., faujasites), the measured CO-induced shifts of the O–H modes are coverage-dependent and decrease with increasing coverage (see the band at 3179-3268 cm^{-1} in Figure 2.21). This phenomenon has been discussed and related to the static effect rather than to a heterogeneity of the hydroxyls (166). The

Table 2.9 Shift of ν(OH) of OH groups on various oxides and related materials induced by low-temperature CO adsorption

Sample	M[a]	$\widetilde{\nu}$(OH), cm^{-1}	$\|\Delta\widetilde{\nu}_{OH}\|$, cm^{-1}	$\widetilde{\nu}$(CO), cm^{-1}	Note	Ref.
MgO	Mg	3746, 3640	0	–		[114]
CeO$_2$	Ce	3664, 3640	0	–		[114]
ZnO	Zn	3670	40	2148		[263]
		3625	45			
		3445	25			
α-Cr$_2$O$_3$	Cr	3678, 3661	0			[114]
		3638	63			
		3618	64			
		3606	74			
Cr/ZrO$_2$	Cr^{n+}	3633	158	2162	Oxidized sample	[264]
		3645	65	(2147)	Reduced sample	
MIL-100[b]	Cr^{n+}	3585	90	2160		[265,266]
La$_2$O$_3$	La	3647	34	2157	$\Delta\widetilde{\nu}$(OH) of -58 and -107 cm^{-1}, respectively, for LaOCl and LaCl$_3$	[267]
SnO$_2$	Sn	3740–3730	90–100	2155		[268]
		3660	90	2155		
		3625	155	2164		
TiO$_2$(a)[c]	Ti	3724	119			[114]
		3676	144			
TiO$_2$(a)[c]	Ti	3675	ca. 115	2155		[269]
NH$_3$/TiO$_2$	Ti	ca. 3675	ca. 65	2152		[269]
TiO$_2$(r)[c]	Ti	3688	144			[114]
		3660	148			
TiO$_2$(r)[c]	Ti	3690–3670	130–150	2141		[270]
TiO$_2$(r)[c]	Ti	ca. 3675	ca. 130	2160–2155		[271]
TiO$_2$(a+r)[c]	Ti	3720 / 3675	128	2153	One shifted band at 3570 cm^{-1}	[272]

Continued

Table 2.9 Shift of $\nu(OH)$ of OH groups on various oxides and related materials induced by low-temperature CO adsorption—cont'd

Sample	M	$\tilde{\nu}(OH)$, cm^{-1}	$|\Delta\tilde{\nu}_{OH}|$, cm^{-1}	$\tilde{\nu}(CO)$, cm^{-1}	Note	Ref.
$TiO_2(a+r)^c$	Ti	3721	100	2158–2152	Average value	(273)
		3678–3636	150			
F^-/TiO_2	Ti	3675	120	2158		(274)
$TiO–SiO_2$	Ti	3720	200	2165		(110)
ZrO_2	Zr	3770	0	2158–2156		(275)
		3660	70			
$m-ZrO_2$	Zr	3780	0	2168		(276,277)
		3680	70			
SO_4/ZrO_2	Zr	3639	129	2166		(278)
SO_4/ZrO_2	Zr	3655	100	2162	Washed sample	(278)
SO_4/ZrO_2	Zr	3640	160	2177–2167	Weak intensity	(276,277)
			220			
PO_4/ZrO_2	P	3664	131–175	2170–2169		(277)
SAPO-5	P	3677	195–175			(102,104)
SAPO-34	P	3678	178	2174		(104)
Nb_2O_5	Nb	3714	200–224			(279)
WO_3/ZrO_2	W	3650	160	2165		(280)
WO_x/TiO_2	W	3800–3600	ca. 200	2168–2158		(281)
V-MCF	V	3665	175	2158–2157	Different samples	(282)
		3687	180			
		3642	250			
V/SiBEA	V	3647–3646	142–160	2165–2160		(216,283)
FeBEA	Fe^{3+}	3686–3684	100	2158		(217)
FeFER	Fe^{3+}	3674	145			(284)

[a]Atom(s) to which the OH group is bound.
[b]Chromium carboxylate MOF sample.
[c]Letters a and r stand for anatase and rutile, respectively.

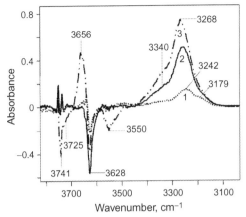

Figure 2.21 Changes in the IR spectrum of "fresh" USY 2 zeolite after adsorption of 0.06 (1), 0.25 (2), and 1.67 (3) μmol mg^{-1} of CO at a temperature of 77 K. A spectrum of the sample itself before CO addition was subtracted. *Reproduced from Ref. (171).*

presence of Lewis acid sites with adsorbed CO molecules may also affect the measured shift of the OH modes in the same direction. If the shift is found to be coverage-dependent, there are two correct methods, to extrapolate to zero coverage, or to specify the shift for a particular coverage. Fortunately, the static effect is negligible for high-silica zeolites.

3. *Solvent effect.* As already noted, when CO interacts only with a fraction of the OH groups present in high-silica zeolites (low and medium coverages) the maxima of the carbonyl band and the shifted OH bands are coverage-independent. However, at high coverages (usually in the presence of CO in the gas phase), both bands, $\nu(OH)$ and $\nu(CO)$, are shifted to lower frequencies (see Figure 2.22). These shifts are explained by "solvation" of the OH groups by CO. Therefore, to correctly measure the CO-induced shift, it is necessary to conduct the experiment such that only 1:1 complexes are formed. It appears that many discrepancies in the data for the same kind of hydroxyls reported in the literature are the result of different experimental conditions, which sometimes favor the solvation of OH groups. Note that the solvent effect is opposite to the static effect, and combination of both effects may lead to the sequential occurrence of blue and red shifts of the $\nu(OH)$ of H-bonded hydroxyls.

4. *Fermi resonance.* As already noted, Fermi resonance occurs when the shifted OH band is around 3300 cm^{-1}. The split associated with the

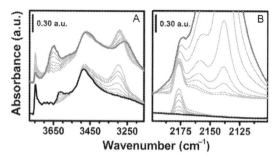

Figure 2.22 FTIR spectra recorded after increasing dosages of CO on HSSZ-24 at a temperature of 77 K. (A) The OH stretching region, and (B) the CO stretching region (background subtracted). The black bold curve corresponds to the spectrum of the activated sample, and the gray bold curve corresponds to the spectrum taken at the highest CO loading. *Reproduced from Ref.* (104).

resonance leads to the appearance of a shoulder on the shifted band and may lead to wrong conclusions about the heterogeneity of the hydroxyl groups. To unambiguously demonstrate or exclude heterogeneity, comparison with another probe (e.g., N_2) is necessary. Analysis of heterogeneously distributed hydroxyls can become very complicated if Fermi resonance occurs. For more details on the effect, see Section 6 on H/D exchange.

5. *Bonding of CO through its oxygen end.* Isocarbonyls are well known for CO adsorbed on alkali and alkaline-earth cations in ion-exchanged zeolites *(163,287).* The fraction of O-bonded species (among the adsorbed CO molecules) is larger at higher temperatures and larger cationic radius of the adsorption site. The C–O stretching frequency is red-shifted with respect to gas-phase CO and appears in the region of 2123–2089 cm^{-1} *(163).* About 10 years ago, it was reported that both C- and O-bonded CO species are formed with the silanol groups in silicalite *(244).* The authors made the conclusion on the basis of the appearance of a carbonyl band located at 2131 cm^{-1}. Later, O-bonded CO on HZSM-5 was reported, also based upon analysis of the carbonyl stretching region, where a band at 2112 cm^{-1} was detected *(287).* If linkage isomerization of CO occurred during measurements, it could seriously complicate the determination of surface acidity and would limit the usefulness of CO. However, we are of the opinion that the fraction of carbon monoxide that is eventually O-bonded to hydroxyl groups is negligible. Based on the PA of the O-end of CO, it can be estimated that this adsorption form should induce a shift of the OH modes that is only about 13% of the shift

caused by the C-bonded form. In other words, if a shift of -300 cm^{-1} is measured (as a result of formation of OH–CO complexes), an additional shifted band with $\Delta\tilde{\nu}(OH)$ amounting to about -40 cm^{-1} should correspond to isocarbonyls (OH–OC complexes). Analysis of the literature data indicates the lack of such a band, which suggests no measurable amounts of OH$\cdots$OC complexes are produced. Possibly, the 2131 cm^{-1} band reported in Ref. *(244)* is associated with another adsorption form, for example, CO O-bonded to oxygen anions *(288)*.

6. *Dual sites.* The existence of so-called dual sites in zeolites has been discussed in the past decade *(289,290)*. It is established that, if a suitable geometric fit exists, CO attached to a cationic site in a zeolite can interact with another cation via its oxygen atom. Because CO is normally more strongly bound to cationic sites than to OH groups, a CO molecule coordinated to a metal cation could affect the O–H group via its oxygen atom. However, no experimental evidence for such an effect has been reported.

7. *Measuring acidity of hydroxyls that are already H-bonded.* The theoretical considerations of the acidity of hydroxyls concern isolated groups. However, the O–H stretching modes of the bridging hydroxyls in zeolites are *a priori* affected by H-bonding with zeolite framework oxygen atoms. Formation of an H-bond with CO may result in the breaking of existing H-bonds. Consequently, as has already been noted *(98,165)*, the measured acidity (that is deduced from the shift) will not correspond to the true intrinsic acidity of the hydroxyl. The problem could be resolved by estimating the stretching frequency of the hydroxyls in the absence of any H-bonding (as proposed by Jacobs and Mortier *(97)*) and deserves the attention of the theoretical chemists.

Notwithstanding these complications, CO is a powerful probe. In the following, we report important findings made with the help of CO and will discuss the surface hydroxyls of various materials.

CO clearly reveals the effect of isomorphous substitution on the acidity of the bridging hydroxyls. For example, for MFI zeolites, the acidity decreases in the following sequence: H[Al]ZSM-5 > H[Ga]ZSM-5 > H[Fe]ZSM-5 *(208,209)*. Comparison of isostructural samples also indicates that the acidity of the bridging hydroxyls is always higher in zeolites than in the corresponding alumophosphates *(212)*.

By analyzing difference spectra after CO adsorption on mazzite, Shigeishi *et al. (187)* established important details that were not distinguished in the original spectra. Bare Lewis acid sites, Al^{3+}, interact with the bridging

OH groups in mazzite, thus shifting the OH modes from 3626 and 3606 cm^{-1} to 3590 and 3567 cm^{-1}, respectively. When CO is adsorbed on these Lewis acid sites, the interaction between Al^{3+} and OH groups subsides, and the bands are shifted to their intrinsic positions.

It has been reported that the absorption coefficient of the carbonyl band around 2175 cm^{-1} is 2.6–2.7 cm µmol^{-1} (174,178), and this value can be used to quantify the number Brønsted acid sites.

We start the discussion of silanol groups with those of silica (Table 2.7). An absolute value of the shift of the silanol OH stretching vibrations of 90 cm^{-1} is generally accepted as a standard for calculating the PAs of other hydroxyls. Inspection of the first rows of Table 2.7 indicates small (and sometimes more serious) deviations from this value. This variation introduces some uncertainty into the calculation of PAs based on spectral data.

Pure siliceous materials show, in general, very similar shifts of their silanol bands. It is well established that the presence of a second ion (most frequently Al^{3+}, but also others, such as Fe^{3+} or sulfates) enhances the acidity of some of the silanols (226,241). This behavior is demonstrated in Figure 2.23 where two shifted bands appear as a result of CO interaction with Si–OH groups on a silica–alumina sample: it is evident that the presence of aluminum has led to an increase in acidity of a fraction of the silanol groups.

The attribution of the observed acidity to silanols is sometimes questionable. As just discussed, the presence of aluminum is thought to enhance the acidity of nearby silanols; however, upon complexation with CO the SiOH groups are supposedly converted into bridging hydroxyls (see Figure 2.18). Thus, it appears that in this case, the hydrogen bond method is associated with some uncertainty. In principle, the wavenumber of the shifted OH band should be referenced to the wavenumber of the bridging hydroxyls. Therefore, in some instances (when high acidity of the silanols was established), the measured shifts may have been misleading, and the acid strength might have been overestimated.

Interestingly, analysis of published spectra (174) shows that when the silanol band around 3740 cm^{-1} is perturbed by incremental addition of CO first a band at 3550 cm^{-1} develops, followed by a band at 3405 cm^{-1}. This unexplained behavior still needs a thorough analysis.

Alumina is an oxide characterized by a highly ionic (56%) metal–oxygen bond. Most authors are of the opinion that the Al–OH groups absorbing above 3770 cm^{-1} (type I) are not affected by CO and are therefore not acidic (Table 2.8). Similar observations and interpretations have been made

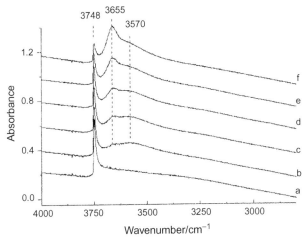

Figure 2.23 FTIR spectra (ν_{OH} region) of a silica–alumina (SIRAL) sample (30% SiO$_2$) after pretreatment (at a temperature of 823 K) under vacuum (a), and adsorption of 10 (b), 40 (c), 100 (d), 200 (e), and 500 Pa CO (f) at 85 K. *Reproduced from Ref. (241).*

regarding the 3780 cm^{-1} band of zeolites. Other hydroxyl groups on alumina have been reported to shift in frequency as a result of the formation of carbonyl complexes. The change in $\widetilde{\nu}(OH)$ is -70 to -200 cm^{-1}, indicating that the OH groups exhibit weak to moderate acidity. It is also evident that the presence of another cation strongly affects the acidity of aluminum hydroxyls.

Aluminum hydroxyls connected to EFAL species in zeolites absorb at 3665 cm^{-1}. The CO-induced shift of $\widetilde{\nu}(OH)$ of EFAL species present in H–zeolites was reported to span a wide interval, -190 to -260 cm^{-1}. The presence of alkali metal cations in zeolites decreases the acidity of the residual hydroxyls *(215,221,249)*. Similarly, potassium cations on alumina decrease the hydroxyl acidity *(260)*. On the contrary, sulfation results in an increase of acidity.

Because of the high basicity of MgO, magnesium hydroxyls are not affected by CO *(114)* (Table 2.9). The $\Delta\widetilde{\nu}(OH)$ measured for the hydroxyls on ZnO is very small *(263)*. Some hydroxyls on chromia are not acidic; others demonstrate weak acidity *(114)*. Evidently, the oxidation state is very important, as demonstrated by analysis of Cr/ZrO$_2$ *(264)*. Higher acidity is measured for OH groups on covalent oxides (bulk and supported) including vanadia *(216,282,283)*, tungsten oxide *(280,281)*, and niobia *(279)*. Like alumina, other oxides demonstrate enhanced acidity after sulfation

(276–278). Modification with chloride also enhances the acidity of the hydroxyls. For example, it was shown that the acid strength of La–OH groups increases in the order $La_2O_3 < LaOCl < LaCl_3$ *(267)*. Covering the surface of TiO_2 with ammonia, although hardly affecting $\nu(OH)$, leads to a drastic decrease of the hydroxyl acidity *(269)*. This behavior is evidence of the capability of titania to transmit electrons through the solid. Moreover, the results indicate that surface modification, for example, in the course of catalytic reactions, may seriously modify hydroxyl acidity. The acidity of Fe^{3+}–OH groups attached to iron cations in zeolites (with the iron located in the exchangeable ion positions) strongly depends on the nature of the zeolite *(217,284)*. In any case, the acidity of hydroxyls on oxides is weaker than that of bridging hydroxyls in zeolites.

Catalytic reactions often occur in the presence of water that is adsorbed on the catalyst surface. Several research groups have measured the acidity of the hydroxyl groups of the adsorbed water. Paze *et al.* *(11)* reported that H_2O was bonded to the acidic hydroxyls of BEA zeolite in such a way that one of the hydroxyl groups formed an H-bond with the framework oxygen. The other "free" OH group was characterized by a band at 3710 cm^{-1}. This free OH group shifted to 3606 cm^{-1} ($\Delta\tilde{\nu} = -102$cm^{-1}) upon low-temperature CO adsorption, whereby the corresponding CO stretching frequency was detected at 2162 cm^{-1}. However, Maache *et al.* *(184)* reported, for water in a MOR zeolite, a hydroxyl band at 3660 cm^{-1} and measured a significantly higher acidity ($\Delta\tilde{\nu}(OH) = -210$cm^{-1}, $\tilde{\nu}(CO)$ at 2158 cm^{-1}). According to Garrone *et al.* *(250)*, molecular water adsorbed on Al^{3+} sites in alumosilicates absorbs at 3698 and 3611 cm^{-1} and exhibits significant acidity, as evidenced by a CO-induced shift of $\tilde{\nu}(OH)$ of -200 to -250 cm^{-1}. Vimont *et al.* *(265,266)* adsorbed water on coordinatively unsaturated chromium ions in a chromium carboxylate MOF. Their measurements indicated moderate acidity of the water OH groups; the CO-induced shift of $\tilde{\nu}(OH)$ was found to be -160 cm^{-1} and $\tilde{\nu}(CO)$ was 2163 cm^{-1}. A second water molecule bound to the complex exhibited weaker acidity ($\Delta\tilde{\nu}(OH) = -75$cm^{-1}). The authors pointed out that the higher the polarizing power of the cation, the stronger the created Brønsted acidity. Moreover, hydroxyl groups of adsorbed alcohols have also demonstrated high acidity: $\Delta\tilde{\nu}(OH)$ of -180 and -235 cm^{-1}, respectively, for CF_3CH_2OH and $(CF_3)_2CHOH$ *(266)*.

Following investigations of zeolites, a correlation between the shift of the OH stretching frequency and $\tilde{\nu}(CO)$ was proposed *(178,247)*. This correlation is theoretically expected because the more acidic the OH groups, the higher the shift of both, $\widetilde{\nu}(CO)$ and $\widetilde{\nu}(OH)$. Figure 2.24 shows the relationship between $\widetilde{\nu}(CO)$ and $\Delta\widetilde{\nu}(OH)$ according to the data presented in Tables 2.6–2.9. It should be taken into account that the relative uncertainty along the Y-axis is higher than along the X-axis because the perturbation of the carbonyl stretching frequency is much weaker than that of the OH stretching frequency. In any case, the general trend of an increase of $\widetilde{\nu}(CO)$ with increasing $\Delta\widetilde{\nu}(OH)$ is evident. However, even considering the error, there are some deviations from this correlation. Recent DFT calculations indicate that the acidity of the hydroxyls does not directly correlate with the $\widetilde{\nu}(CO)$ shift because the CO stretching frequency is affected by the electrostatic shift induced by the whole surface *(291)*.

As a conclusion, it can be stated that CO is the most frequently used probe for measuring acidity of surface hydroxyl groups according to the H-bond method. The CO-induced shift of $\widetilde{\nu}(OH)$ is, in most cases, large enough to permit correct measurement. There is a substantial collection of data available in the literature that can be used for comparison purposes.

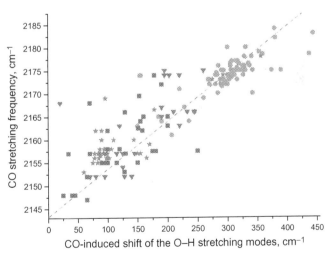

Figure 2.24 Correlation between the CO stretching frequency and the CO-induced shift of the O–H stretching modes of surface hydroxyl groups. Data from Table 2.6, circles; data from Table 2.7, stars; data from Table 2.8, triangles; and data from Table 2.9, squares. (See the color plate.)

Some phenomena (e.g., static shift, Fermi resonance) affect the correctness of the measurements but the uncertainty is only a few cm^{-1} and hence not relevant.

3.4. Dinitrogen (N_2)

Dinitrogen has been widely applied for measuring the acidity of hydroxyl groups, though there are far fewer investigations using N_2 than there are using CO. The PA of N_2 is 493.8 kJ mol^{-1} *(148)* and, consequently, the N_2-induced shift of $\tilde{v}(OH)$ is around 2.5 times smaller than the CO-induced shift. Hence, there is considerable uncertainty in the measurements when probing weakly acidic OH groups, not only because of the low absolute value of the shift but also because of the partial superposition of the original and the shifted band. However, as it will be shown, N_2 has some advantages over CO when highly acidic hydroxyls are characterized.

Dinitrogen is an IR-silent molecule. The symmetry is lowered after adsorption and the N–N stretching modes become IR-active, but the intensity of the N–N band is low. Dinitrogen is polarized by the OH groups and, as a result of the Stark effect, the N–N stretching frequency increases. Because there is no value for the IR absorption of gaseous N_2, it was proposed to use the frequency of N_2 adsorbed on silicalite as the reference *(136)*. As with CO, the stronger the bond with the surface, the higher the N–N frequency. However, the N–N stretching vibrations are even less sensitive to the strength of the bond with the surface sites than the CO stretching vibrations. Therefore, the position of the N–N band is used to identify the adsorption form, and not to make detailed conclusions on the acid strength of the hydroxyls. Typically, when attached to strongly acidic hydroxyls, N_2 vibrates around 2333 cm^{-1}.

Because free N_2 is IR-silent, there is no interference in the spectra from gas-phase absorption. However, the $^{14}N-^{14}N$ modes are in the region of absorption of CO_2, which is present in the atmosphere and hence in the laboratory environment. Even small changes in the concentration of CO_2 in the gas phase can strongly affect the appearance of the low-intensity $^{14}N-^{14}N$ stretching band. Hence, when seeking to analyze the N–N modes, the use of the $^{15}N_2$ isotope is recommended. The $^{15}N-^{15}N$ stretching vibration is red-shifted by about 80 cm^{-1} with respect to the $^{14}N-^{14}N$ vibration and is thus not disturbed by atmospheric CO_2. It has been demonstrated that the perturbation of the OH stretching modes is essentially the same with $^{14}N_2$ and $^{15}N_2$ *(199)*. However, using $^{15}N_2$ as a probe, one should consider the possibility of formation of some $^{14}N^{15}N$ molecules, which absorb at a frequency roughly 40 cm^{-1} above that of $^{15}N_2$.

Typically, the N_2-induced shift of the OH modes of bridging hydroxyls ranges from -100 to -120 cm^{-1} (see Figure 2.25). An average shift of about -40 cm^{-1} is reported for silanol groups. Tables 2.10 and 2.11 summarize the data obtained after interaction of dinitrogen with various surface hydroxyl groups. A BHW plot correlating the CO- and N_2-induced shifts of OH modes using the data presented in Tables 2.6–2.11 is shown in Figure 2.26. It is evident from the figure that although the data were collected in different laboratories, the N_2-induced shift correlates well with the CO-induced shift. As with CO, the highest shift is detected for dealuminated Y (DAY) zeolites (170).

The relatively small N_2-induced shift ensures the absence of Fermi resonance when highly acidic hydroxyls are investigated. For this reason, it has recently been concluded that N_2 is a more appropriate probe for the characterization of hydroxyls than CO, especially if the OH groups are highly acidic (152). Because N_2 is a homonuclear molecule, complications due to linkage isomerization must not be feared.

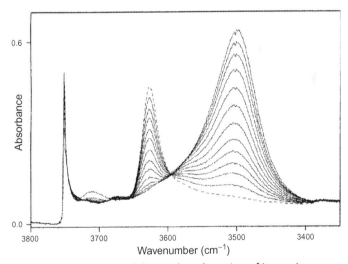

Figure 2.25 FTIR spectra recorded during the adsorption of increasing amounts of N_2 on MCM-22 at a nominal temperature of 77 K (OH stretching region). A shift of the band at 3626 cm^{-1} (bridging OH groups) to 3508 cm^{-1} is observed. In addition, a smaller shift of some of the silanol groups from 3747 to 3710 cm^{-1} can be seen. Dashed curve: activated sample without adsorbate. *Reproduced from Ref. (227).*

Table 2.10 Shift of ν(OH) of bridging hydroxyls of various materials induced by low-temperature N_2 adsorption

Sample	$\tilde{\nu}$(OH), cm^{-1}	$|\Delta\tilde{\nu}_{OH}|$, cm^{-1}	$\tilde{\nu}$(N$_2$), cm^{-1}	Note	Ref.
HY	3642	120			(170)
HY	3650	108	2326	Solvation	(136)
HY	3645	98	2334		(203)
DAY	3627	139			(170)
HBEA	3615	131	2333		(222)
HBEA	3614	126	2334		(11)
HBEA	3612	119			(170)
MCM-22	3626	123	2332		(228)
MCM-22	3626	120	2335–2330		(227)
HZSM-5	3619–2618	120–121	2334		(292)
HZSM-5	3616	120	–	$^{15}N_2$	(152)
HZSM-5	3621	112–114	2334	Higher shift in the presence of Lewis acid sites	(202)
HZSM-5	3615	109	2332		(199)
[Fe]ZSM-5	3630	100	2331		(208)
[Ga]ZSM-5	3617	100	2331		(208)
SAPO-40	3640	115	2334		(237)
SAPO-40	3628	120			(226)
HTON	3598	117	2333	Only partially perturbed	(229)
HMOR	3616	116	2335		(293)
HFER	3605	110	2331		(160)
HFER	3606	106	–	$^{15}N_2$	(154)
HFER	3605	103	2331		(233)

Table 2.11 Shift of ν(OH) of nonbridging hydroxyls in porous materials induced by low-temperature N_2 adsorption

Sample	$\tilde{\nu}$(OH), cm^{-1}	$|\Delta\tilde{\nu}_{OH}|$, cm^{-1}	$\tilde{\nu}(N_2)$, cm^{-1}	Note	Ref.
I. Silanol groups					
HY	3749	42	2326	Solvation	*(136)*
HZSM-5	3749	40	–	$^{15}N_2$	*(152)*
SAPO-40	3747	40			*(226)*
MCM22	3747	37	2326–2324		*(227)*
HBEA	3748	37	2325		*(11)*
HBEA	3737	32	–		*(222)*
II. Aluminum hydroxyls					
HZSM-5	3666	76	–	$^{15}N_2$	*(152)*
SAPO-40	3670	70			*(226)*
III. Adsorbed water					
HBEA	3710	30	2329		*(11)*

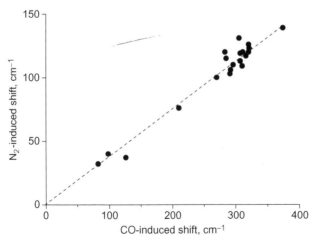

Figure 2.26 Correlation between the CO- and N_2-induced shifts of ν(OH) of various materials. *Data from Tables 2.6–2.11.*

3.5. Other diatomic molecules

Other homonuclear diatomic molecules are IR-silent like N_2, but become IR-active after adsorption. Dihydrogen has been introduced as a very convenient probe for measuring surface acidity. It has been pointed out that the H–H stretching modes are highly sensitive to the bonding, and their shift (with respect to the Raman frequency at 4162 cm^{-1} (2)) is much larger than that observed for the corresponding modes of CO or N_2 (136). However, because of its low PA (422.3 kJ mol^{-1} (148)), dihydrogen interacts only weakly with OH groups and the experiments require very low temperatures. Moreover, whereas CO and N_2 are linearly coordinated, theoretical calculations (294,295) and experimental results (136) indicate a side-on configuration of adsorbed H_2 with one H-atom interacting with a neighboring lattice oxygen atom. Hydrogen is also highly reactive and could interact, even at low temperature, with catalysts of complex composition. Because of these reasons, H_2 has not found wide application, and its main application is the characterization of very acidic hydroxyls.

Table 2.12 summarizes data on the H_2-induced shift of $\tilde{\nu}(OH)$ for bridging hydroxyls. Analysis of the data indicates that the results do not agree with those obtained using CO or N_2 as molecular probes. For example, H_2 suggests a very high acidity of the bridging hydroxyls in HSSZ-13, which was not confirmed by CO adsorption. Current understanding is that the geometry of adsorption is the reason for these inconsistencies.

Although silanols have weak acidity, their bands are shifted by H_2. However, differing values were reported: from −5 (183) to −20 cm^{-1} (296).

Interestingly, D_2 induces noticeably larger (by 6–7 cm^{-1}) shifts of the OH modes of zeolite acidic hydroxyls than H_2, which is attributed to the higher mass of deuterium (200).

Dioxygen is occasionally used as a molecular probe for measuring the acidity of hydroxyls (136,292). The PA of O_2 is almost the same as that of H_2 (421 kJ mol^{-1} (148)). A shift of −32 cm^{-1} was reported for the bridging OH groups of HZSM-5 and CuHZSM-5 zeolites (292,298). A larger shift (−48 cm^{-1}) was reported for the 3650 cm^{-1} band that characterizes the bridging hydroxyls in HY (136). The O–O modes appear around 1553–1549 cm^{-1} and are not very sensitive to the coordination. Slightly smaller shifts (by 5–6 cm^{-1}) were measured for the bands characterizing aluminol and silanol groups (298).

Nitrogen monoxide is a widely used molecular probe for the characterization of coordinatively unsaturated cationic sites (299). The PA of NO is

Table 2.12 Shift of $\nu(OH)$ of bridging hydroxyls induced by low-temperature H_2 adsorption

Sample	$\tilde{\nu}(OH)$, cm^{-1}	$|\Delta\tilde{\nu}_{OH}|$, cm^{-1}	$\tilde{\nu}(H_2)$, cm^{-1}	Note	Ref.
HSSZ-13 (CHA)	3612	64	4109	20 K	(231,296)
HCHA	3620	63	–	15 K	(231)
HMOR	3616	61	4097		(297)
HDAY	3627	56	–		(170)
HSAPO-34	3636	(58)	4104	15 K; not well resolved band	(231)
HMOR	3619	53–63	4103	Smallest shift with a high-silica sample	(183)
HNaY	3650	51	4115		(136)
HZSM-5	3621–3618	49	4103		(170,297)
HZSM-5	3609	49	4109	15 K	(194)
HZSM-5	3618	45 (52)	4106	D_2-induced shift in parentheses	(200)
H[Ga]ZSM-5	3622	35 (41)	4112	D_2-induced shift in parentheses	(200)
H[Fe]ZSM-5	3637	37 (44)	4112	D_2-induced shift in parentheses	(200)
HBEA	3612	49	–		(170)
HEMT	3637	46	4108		(297)
HY	3641	45	–		(170)
HY	3644	43	4109		(297)
SAPO-37	3646	41	4113–4111		(297)
HFER	3695	30	4103		(233)

531.8 kJ mol^{-1} (148), which suggests that it could also be a suitable probe for measuring the acidity of hydroxyls. However, at low temperature, NO tends to dimerize. It was reported that NO shifts the OH modes of the 3610 cm^{-1} band of HZSM-5 by -237 cm^{-1}. At higher coverages, the OH–NO complexes are converted into $OH(N_2O_2)$ species, and the associated red-shift of the OH band amounts to 447 cm^{-1} (300).

3.6. Noble gases

A common property of noble gas probes is that they are IR-silent both, in the gaseous and in the adsorbed state. Therefore, the only observable effect of their adsorption is the change induced in the OH region. An important advantage is the lack of any reactivity. The PA increases with atomic mass: helium, 177.8 kJ mol^{-1}; neon, 198.8 kJ mol^{-1}; argon, 369.2 kJ mol^{-1}; crypton, 424.6 kJ mol^{-1}; and xenon, 499.6 kJ mol^{-1} (148). It is reported that helium is not basic enough to induce any change in the OH spectra, not even if the hydroxyls are highly acidic (136,308). The shifts induced by argon are similar to those observed with O_2, although the basicity of the two molecules noticeably differs. The Xe-induced shifts are similar but slightly smaller than those observed after adsorption of N_2 (136,292,302,308).

3.7. Methane and alkanes

As a weak acid, methane is a probe molecule used for evaluating surface basicity (131). However, as a result of its amphoteric character, CH_4 also coordinates to OH groups. The PA of methane is 543.5 kJ mol^{-1} (148), thus between those of CO and N_2. Data on the methane-induced shift of $\tilde{\nu}(OH)$ for various surface hydroxyls are presented in Table 2.13. A general tendency can be seen of the shifts to be slightly higher than the corresponding N_2-induced shifts. This finding is not consistent with the difference between the PAs of CH_4 and N_2. It was also reported that methane is much more strongly adsorbed on OH groups than N_2 (162). These observations have been attributed to a multicenter adsorption of methane, with coordination to the H-atom of the hydroxyl group and to a basic oxygen center. Because the balance between the two bonds depends on the acid–base properties of the adsorbent, the methane-induced shift of the OH modes cannot be a reliable measure of the hydroxyl acidity. Evidently, the same conclusion can be drawn for other alkanes.

Interestingly, the CH_4-induced shift of the OH stretching mode seems to be strongly temperature-dependent. Most authors have reported the HZSM-5 bridging hydroxyls to be red-shifted by 106–120 cm^{-1} after low-temperature CH_4 adsorption (146,162,303–305), but two other groups (306,307) reported a much smaller red shift (45–57 cm^{-1}) at ambient temperature.

Other alkanes have also been used for probing the acidity of surface hydroxyls. With an increasing number of carbon atoms in the chain, the PA increases (596.3 kJ mol^{-1} for C_2H_6, 625.7 kJ mol^{-1} for C_3H_8, and 677.8 kJ mol^{-1} for i-C_4H_{10} (148)), which leads to a slight increase of the value

Table 2.13 Shift of ν(OH) of hydroxyl groups induced by adsorption of methane

Sample	$\widetilde{\nu}$(OH), cm^{-1}	$|\Delta\widetilde{\nu}_{OH}|$, cm^{-1}	Note	Ref.
I. Bridging hydroxyls				
HZSM–5	3601	120		*(146)*
HZSM–5	3622	112		*(303)*
HZSM–5	3601	ca. 110		*(304)*
HZSM–5	3622	107	128 cm^{-1} at high coverage	*(162)*
HZSM–5	3616	106		*(305)*
HZSM–5	3607	57	$T = 298$ K	*(306)*
HZSM–5	3605	45	$T = 288$ K	*(307)*
HY	3635	116	127 cm^{-1} at high coverage	*(162)*
	3640	113		
HMOR	3607	101		*(157)*
II. Silanol groups				
SiO$_2$	3750	36		*(162)*
SiO$_2$	3748	34		*(21)*
SiO$_2$	3750–3749	32		*(305,308)*
HZSM–5	3748	47		*(305)*
HZSM–5	3740	37		*(305)*
MCM–41	3740	20	Shift appears larger in figure	*(309)*
III. OH groups on oxides				
Al$_2$O$_3$	3750	43		*(304)*
	3665	25		
Al$_2$O$_3$	3760	53		*(305)*
ZnO	3499	7		*(310)*
MgO	3746	0		*(301)*

The experiments were performed at low temperature (around 100 K) unless otherwise noted.

of $\Delta\tilde{\nu}(OH)$. However, the size of the molecule also increases, preventing them eventually from entering small pores. Therefore, the C_{2+} alkanes are dis-advantageous as compared to methane and suffer from the same problem of multicenter adsorption. Consequently, although there are many reports on the adsorption of alkanes, the number of investigations in which they have been used for measuring the acidity of hydroxyls is not large. The results are generally consistent with expectations. The shift of $\tilde{\nu}(OH)$ of acidic hydroxyls in zeolites induced by various alkanes is relatively small and ranges from -110 to -150 cm^{-1} (311–316). In detail, the propane-induced shift was reported to be -110 cm^{-1} (314) or to be ill-defined and between -60 and -130 cm^{-1} (313). The shift caused by n-butane is -130 cm^{-1} (311). However, i-butane is characterized by two different interactions and two $\Delta\tilde{\nu}(OH)$ values: -50 and -150 cm^{-1}, respectively (311). The reported shift induced by n-heptane is -124 to -135 (315,316) or -150 cm^{-1} (311), and that induced by n-hexane is -120 (314) or -145 cm^{-1} (312).

3.8. Unsaturated aliphatic hydrocarbons

Of the class of unsaturated aliphatic hydrocarbons, alkenes have been widely applied for measuring hydroxyl acidity. Alkenes are bases of intermediate strength (PA of $C_2H_4 = 680.5$ kJ mol^{-1} and PA of $C_3H_6 = 751.6$ kJ mol^{-1} (148)), and the measurements could in principle be performed at ambient temperature. However, in most cases low-temperature experiments are per-formed to avoid oligomerization. The high reactivity of olefins, together with the already discussed possibility of multicenter adsorption (see meth-ane), is the main disadvantage of these probes.

Alkenes are hydrogen-bonded to OH groups via their π-electron cloud (56). The C=C stretching modes (which are IR-silent in the gas phase) are activated because of symmetry reduction and are shifted from 1625 cm^{-1} (the Raman gas-phase frequency) to 1618–1610 cm^{-1}. This small shift of the C=C vibration is insufficient to make definite conclusions on the strength of the interaction. However, the shift of the O–H stretching vibra-tions is relatively large but small enough to avoid strong Fermi resonance phenomena. Table 2.14 summarizes the C_2H_4- and C_3H_6-induced shifts of the O–H stretching modes for various hydroxyls in porous materials. Notwithstanding the above-mentioned disadvantages of the alkenes as molecular probes, a relatively good correspondence to the results obtained with CO as a probe is often observed.

It is obvious that, as in the case of CH_4, the measured shift is temperature-dependent, which often hinders the immediate comparison

Table 2.14 Shift of the OH stretching modes of hydroxyls in porous materials caused by adsorption of C_2H_4 and C_3H_6

Sample	$\tilde{\nu}(OH)$, cm^{-1}	$|\Delta\tilde{\nu}_{OH}|$, cm^{-1}, induced by		Ref.
		C_2H_4	C_3H_6	
I. Bridging hydroxyls				
2D zeolite	3584	487	–	*(188)*
HY	3660	375	–	*(317)*
HSAPO-4	3630	374	–	*(210)*
	3614	449		
	3600	250		
ITQ-2	3628	426; 390[a]	526; 480[a]	*(226)*
HZSM-5	3601	420	–	*(146)*
HZSM-5		400[a]	–	*(318)*
HZSM-5	3610	390[a]		*(147)*
HZSM-5	3609	389[a]	539[a]	*(315)*
HZSM-5	3610	360[a]		*(319)*
HZSM-5	3610	–	110[a]	*(314)*
HBEA	3614	410	514	*(11)*
SAPO-5	3625	330[a]	–	*(319)*
HY	3640	300[a]		*(319)*
NaHX	3668	265[a]	344[a]	*(320)*
NaHA	3620	240[a]	–	*(147,320)*
II. SiOH groups				
SiO_2	3748	140[a]	183[a]	*(147)*
ITQ-2	3747	140; 104[a]	190; 153[a]	*(226)*
HBEA	3749	126	155	*(11)*
III. AlOH groups				
HZSM-5	3665	290	–	*(315)*
ITQ-2	3670	310	410	*(226)*

[a]Ambient temperature experiments; all other experiments were performed at low temperature.

of results obtained in different laboratories. However, this dependence is not large enough to justify the small shift of only -110 cm^{-1} for the bridging hydroxyls in HZSM-5 induced by propene *(314)*. Evidently, in this case, a chemical transformation must have taken place. It has been reported that the oligomerization rate increases with the hydroxyl acidity *(319)*.

In this context, interesting results were reported by Bjørgen *et al.* *(321)*. After low-temperature adsorption of 1-butene on HBEA, these authors observed a red shift of the 3610 cm^{-1} band (bridging hydroxyls) by -500 cm^{-1} and assigned this phenomenon to formation of a hydrogen bond between the OH group and 1-butene. Simultaneously, a shift by -178 cm^{-1} was observed for the 3746 cm^{-1} silanol band. With time, butene oligomerized and the shifted OH bands disappeared while new shifts were then measured, amounting only to -120 cm^{-1} for the bridging hydroxyls and to -48 cm^{-1} for the silanol groups. These two shifts are rather typical of H-bonding with alkanes. Indeed, these shifts were immediately observed when 1-butene was adsorbed at RT, since the oligomerization was very fast. The described phenomena can be seen in Figure 2.27. In agreement with these results, shifts of OH stretching vibrations by -110 and -40 cm^{-1}, respectively, for the bands due to bridging hydroxyls and silanols *(322)* after adsorption of *cis*-2-butene have been attributed to H-bonding with oligomerization products.

The shift of the bands of bridging hydroxyls induced by different butenes (at 170 K, to avoid polymerization) is around -600 cm^{-1} *(311)*. The interaction of 1-butene with HFER was investigated in detail *(323)*. It was established that the band characterizing Brønsted OH groups was shifted and broadened to a band between 3450 and 2600 cm^{-1}. In addition, a weak feature appeared at 3500 cm^{-1} and was assigned to the interaction of the OH groups with the aliphatic part of the molecule. A careful investigation of the interaction of 1-butene with HZSM-5 *(324)* and DZSM-5 was performed *(325)*. At a temperature of 210 K, the OD modes were shifted by -369 cm^{-1} (which corresponds to a shift of the OH modes by -510 cm^{-1}). The IR bands of the H-bonded complex disappeared with time, and the shift of the acidic OH band approached that observed during adsorption of saturated hydrocarbons, thus indicating that polymerization occurred.

An interesting phenomenon was reported in Ref. *(316)*. In addition to the usual interaction of the propene oligomers with the bridging hydroxyls in HMOR zeolite (leading to a shift of the OH modes by about -120 cm^{-1}), a new type of interaction was noticed. This interaction was characterized by a shift of $+60$ cm^{-1}, that is, a shift in the opposite direction.

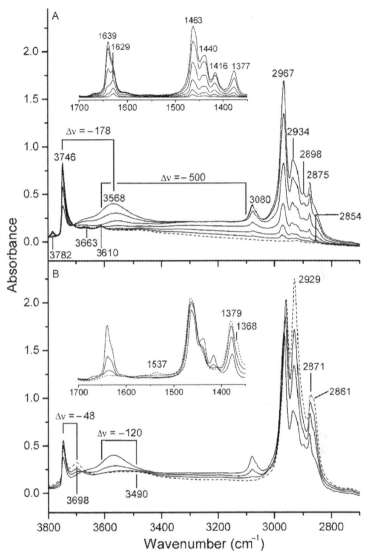

Figure 2.27 FTIR spectra of HBEA after adsorption of 1-butene at a temperature of 77 K and subsequent heating to 300 K. (A) Low-temperature range with adduct formation and shifts of the silanol and bridging hydroxyl groups by -178 and -500 cm^{-1}, respectively. The background spectrum of the zeolite is presented by a dotted line. (b) High temperature range with oligomerization of the 1-butene and hydroxyl shifts of -48 and -120 cm^{-1} that are typical of the interaction of hydroxyls with saturated hydrocarbons. The dotted line represents the last collected spectrum. The spectra in the region between 1700 and 1350 cm^{-1} are background-corrected and shown as insets. *Reproduced with permission from Ref. (321). Copyright 2004 American Chemical Society.*

The authors explained the phenomenon by repulsive interaction of the OH groups with short, positively charged oligomers and supported the assignments by quantum chemical calculations.

3.9. Benzene and aromatic hydrocarbons

Benzene is frequently used to probe the surface acidity of hydroxyl groups. It has been widely applied in the past, but numerous recent investigations are also available. A key advantage of benzene is the simplicity of the adsorption experiment: although benzene is a relatively weak base (PA $= 750.4$ kJ mol^{-1} (148)), its adsorption is performed at ambient temperature, and there is no need for a low-temperature IR cell. Benzene is adsorbed on both, Lewis acid sites and OH groups via π-bond donation of electrons from the aromatic ring. Upon complexation, the CH stretching modes and the ring vibrational bands are shifted, but these shifts are not exploited for the assessment of the acidity of hydroxyls. The benzene-induced shift of the OH stretching modes is slightly larger than that induced by CO.

Benzene does not meet some of the requirements for a good probe. The molecule is too large to penetrate small pores; for example, it can enter only the large channels in MOR (158). Trombetta et al. (326) noted that compared with olefins, benzene perturbs the internal hydroxyls in zeolites less strongly and the external ones more strongly. This contrasting behavior suggests that the C_6H_6–OH interaction is partly hindered inside the zeolite cavities. In the ideal case, the geometry of a complex consisting of benzene and a hydroxyl group should be as shown in Figure 2.28B. As noted by Lercher et al. (130), metal cations can interact uniformly with all delocalized electrons in the aromatic ring, thus forming a symmetric complex, whereas the smaller proton forms nonuniform bonds, which leads to a nonsymmetric complex and weaker interaction. In analogy to the behavior of other hydrocarbons, the hydrogen atoms of benzene can interact with basic surface centers. It was reported that at

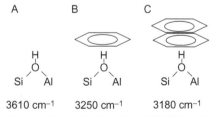

Figure 2.28 Scheme of the bridging hydroxyls of HZSM-5 before (A) and after (B, C) interaction of benzene. Structure B shows a complex with orientation of the aromatic ring perpendicular to the OH group, and structure C shows the cooperative action of two benzene molecules. *Reproduced from Ref.* (155).

low coverages on SAPO-5, benzene is indeed oriented with the aromatic ring perpendicular to the O–H bond as shown in Figure 2.28B *(327)*. However, according to Ref. *(319)* (concerning again SAPO-5), the C_6H_6 molecule may be oriented with the edge of the aromatic ring toward the OH group. Steric constraints can thus cause different interaction modes that prevent measurement of the intrinsic acidity of the hydroxyls with the help of the frequency shift. At high benzene coverages, ordered structures can be formed *(328)*. It was proposed that, as a result of the cooperative action of two or more benzene molecules, the shift of the OH band can increase, as depicted in Figure 2.28C. Finally, as with the use of CO as a probe molecule, Fermi resonance can occur, leading to a split of the shifted hydroxyl band *(155)*.

Notwithstanding these shortcomings of benzene, a large amount of data is available on the interaction of benzene with various hydroxyls (see Table 2.15). Several observations require detailed discussion. A gradual increase of the benzene coverage allows the discrimination of different C_6H_6–OH interactions. Such an experiment is illustrated in Figure 2.29. It is obvious that benzene interacts preferentially with the bridging hydroxyls in SAPO-5 and SAPO-11 (3628 cm^{-1}), shifting the OH bands by -280 (SAPO-5) and -310 cm^{-1} (SAPO-11). Only at higher coverages, benzene affects the POH groups (3676 cm^{-1}) and shifts their stretching modes by -210 cm^{-1}.

Taking into account the above discussion, the substantial heterogeneity of zeolite bridging hydroxyls that is inferred in many articles could be, at least in part, an artifact related to the probe, which produces multiple bands for steric reasons and/or Fermi resonance. However, benzene unambiguously confirms the heterogeneity of the silanols in silica–alumina materials.

Another observation that deserves attention is the detectable, though small (-30 cm^{-1} *(351)*), shift of the OH modes of magnesia hydroxyls upon benzene adsorption. These hydroxyl bands are not shifted upon CO adsorption. The lack of interaction with CO and the small shift seen with benzene could be due to the low acidity of the OH groups, or the small benzene-induced shift could be caused by the interaction of hydrogen from benzene with the oxygen atom of the hydroxyl groups.

Some authors have used aromatic compounds other than benzene, and the behavior is generally similar to that described in this section.

3.10. Acetonitrile (CH_3CN and CD_3CN) and other nitriles

A probe molecule for measurements of surface acidity that is also suitable for porous materials is acetonitrile *(31,56,59,60,130,134,138,361–364)*. As

Table 2.15 Benzene-induced shift of the OH stretching modes of different surface hydroxyls

Sample	$\tilde{\nu}(OH)$, cm^{-1}	$|\Delta\tilde{\nu}_{OH}|$, cm^{-1}	Note	Ref.
I. Bridging hydroxyls				
HZSM-5	3608	228, 358, 438	Different orientation	(329)
HZSM-5	3610	245, 309, 370, 435	Heterogeneity	(330)
HZSM-5	3610	360, 410	Steric heterogeneity	(331,332)
HZSM-5	3610	360		(314)
HZSM-5	3612	260, 360	Fermi resonance	(155)
HZSM-5	3605	355		(333)
HZSM-5	3615–3610	350		(192,311, 312,319, 334,335)
HZSM-5	3610	340	Heterogeneity suggested	(147)
HZSM-5	3610	310		(336)
NaHZSM-5	3609	158–170, 256–285, 361–363, 414–415	Heterogeneity	(337,338)
Ferrisilicate	3625	219, 270, 336, 405	Heterogeneity	(330)
HBEA	3605	352–365	Different samples	(339)
HBEA	3610	346		(334)
HBEA	3608	340	⋅ Dealuminated sample	(147)

HBEA	3602	337		(340)
HBEA	3605	205, 335	Heterogeneity	(341)
HBEA	3608	290		(147)
HNaBEA	3612	340		(342)
HMAZ	3610	270–376	$\Delta\nu(OH)$ increases with dealumination	(343)
HMOR	3610	348		(334)
HMOR	3607	357		(344)
HMOR	3610	345		(312)
HMOR	3611	310		(147)
HEMT	3631	345		(345)
HY	3640	280–320		(319,335)
CuHY	3670	220		(346)
HX	3660	240–260		(335)
NaHX	3668	210		(147)
ITQ-2	3628	310		(226)
SAPO-37	3640	320–330		(345,347)
SAPO-5	3625	325		(319)

Continued

Table 2.15 Benzene-induced shift of the OH stretching modes of different surface hydroxyls—cont'd

Sample	$\tilde{\nu}(OH)$, cm^{-1}	$|\Delta\tilde{\nu}_{OH}|$, cm^{-1}	Note	Ref.
SAPO-5	3628	280		(348)
SAPO-5	3625	275	After evacuation at 373 K; not all hydroxyls interact	(327)
	3520	170		
CoAPO-5	3628	315		(348)
SAPO-11	3628	310		(348)
CoAPO-11	3628	345		(348)
Borate	3720–3700	100–115	Si(OH)B	(349)
II. POH groups				
Cloverite[a]	3675	250		(350)
AlPO-5	3680	240		(319)
SAPO-5	3676	210		(348)
CoAPO-5	3676	210		(348)
SAPO-11	3676	210		(348)
CoAPO-11	3676	205		(348)

III. Silanol groups

SiO$_2$	3748		139	(147)
SiO$_2$	3745		140	(312)
SiO$_2$	3748–3747		120	(311,351)
SiO$_2$	3749		110	(308)
SiO$_2$–Al$_2$O$_3$	3750	Heterogeneity	110, 220	(352)
SiO$_2$–Al$_2$O$_3$	3750	Heterogeneity	120, 250	(351)
MCM-41	3745		118	(353)
	3715		135	
HZSM-5	3749		235	(311)
HZSM-5	3745	Heterogeneity	120, 160	(155)
HZSM-5	3746	Heterogeneity	140, 241	(333)
HZSM-5	3745	Heterogeneity	140	(331)
HMOR	3610	Heterogeneity	240, 360	(158)
ITQ-2	3747		120	(226)
HBEA	3740		100	(340)

Continued

Table 2.15 Benzene-induced shift of the OH stretching modes of different surface hydroxyls—cont'd

Sample	$\tilde{\nu}(OH)$, cm^{-1}	$|\Delta\tilde{\nu}_{OH}|$, cm^{-1}	Note	Ref.
HBEA	3730	31–40	Internals silanols; different samples	(339)
HNaBEA	3745	140		(342)
IV. Aluminol groups				
η-Al$_2$O$_3$	3740	65		(351)
	3690	80		
γ-Al$_2$O$_3$	3790, 3771, 3731	110–170	One shifted band	(354)
γ-Al$_2$O$_3$	3740	65		(351,355)
	3680	80		
F/Al$_2$O$_3$	3720–3700	110–140	Average shift	(351,355)
Mg/Al$_2$O$_3$	3740	65		(351)
Na/Al$_2$O$_3$	3735	35		(351)
K/Al$_2$O$_3$	3720	20		(351)
HZSM-5	3665	215		(314)
HZSM-5	3665	280		(192)
HZSM-5	3663	225		(155)

HZSM-5	3737	146, 242		(332)
HBEA	3680–3660	233–251	Different samples	(339)
	3782	66–82		
HNaBEA	3789	95		(342)
ITQ-2	3670	250		(226)
AlPO-5	3800	180		(319)
Cloverite[a]	3702	220		(350)

v. Other hydroxyls

TiO$_2$	3735	260	The band at 3690 cm^{-1} due to presence of sulfates; the bands at 3735–3725 cm^{-1} associated with Si impurities	(356)
	3725	160		
	3715	100		
	3690	200		
	3670	30		
TiO$_2$	3720–3675	165	Average shift	(357)
NH$_3$/TiO$_2$	3715–3670	80	Average shift	(357)
K/TiO2	3712	62		(358)
V/TiO$_2$	3680	210	V–OH groups	(119)

Continued

Table 2.15 Benzene-induced shift of the OH stretching modes of different surface hydroxyls—cont'd

Sample	$\tilde{\nu}$(OH), cm^{-1}	$|\Delta\tilde{\nu}_{OH}|$, cm^{-1}	Note	Ref.
ZrO$_2$	3670	100		(335)
ZrO$_2$	ca. 3665	70		(359)
SO$_4$/ZrO$_2$	3640	200		(335)
SO$_4$/ZrO$_2$	ca. 3640	70, 185		(359)
α-Fe$_2$O$_3$	3670	ca. 100		(113)
α-Fe$_2$O$_3$	3670–3640	65	Average shift	(360)
MgO	3745	30		(351)

[a]Gallophosphate.

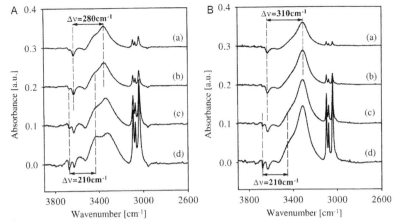

Figure 2.29 FTIR difference spectra of the OH stretching region during adsorption of benzene on SAPO-5 (A) and SAPO-11 (B) ($T = 313$ K): (a) benzene partial pressure 10^{-3} mbar, (b) 10^{-2} mbar, (c) 10^{-1} mbar, and (d) 1 mbar. At low benzene coverage, the bridging hydroxyls are perturbed (shifts of -280 and -310 cm^{-1}, respectively). At high benzene coverage, also the POH groups are perturbed (shift of -210 cm^{-1}). *Reproduced from Ref. (348).*

other nitriles, CH$_3$CN is coordinated to acid sites (both Lewis and Brønsted) via its nitrogen atom. The PA of CH$_3$CN is 779.2 kJ mol^{-1} *(148)*, thus between those of CO and NH$_3$. Therefore, the surface complexes are stable enough to be formed and detected at ambient temperature. Among the commonly applied probe molecules, acetonitrile is the strongest base that is usually not protonated, not even by very acidic hydroxyls; rather, it forms H-bonds with the hydroxyls. These conclusions are supported by theoretical considerations *(365–368)*. In a few cases, concerning some very acidic hydroxyls, the possibility of proton transfer to mono- *(369)* or dimeric species *(134,370)* has been reported, but some of these conclusions have been questioned *(371)*. However, proton transfer does take place during interaction of acetonitrile with the bridging hydroxyls of zeolites at elevated temperatures *(167)*.

Important advantages of acetonitrile are its small size (allowing interaction even with hydroxyls located in the small eight-membered ring pores of FER zeolite *(372)*) and the access to the lone pair electrons on the nitrogen atom with less steric hindrance as compared to pyridine. It is also established that acetonitrile is adsorbed in an "on-top" configuration on hydroxyl groups *(327)*. These properties, together with the large shift of the OH bands, make acetonitrile a preferred molecular probe in many cases.

The C–N stretching modes of acetonitrile are highly sensitive to hydrogen bonding, coordination, and protonation and thus could, in principle, be used to determine the nature and the strength of the interaction. However, as a result of Fermi resonance with the combination mode of $\nu(C–C) + \delta(CH_3)$ (appearing at 2287 cm^{-1}) *(373)*, $\nu(C–N)$ of liquid acetonitrile is observed as a doublet with one intense band at 2254 cm^{-1} and a less intense band at 2294 cm^{-1}. This split makes the analysis of the spectra very difficult. Therefore, when analysis of the C–N region is necessary, deuterated acetonitrile is preferably used. The C–N stretching modes of the free CD_3CN molecule are observed at 2263.1 cm^{-1} *(374)*. Since deuteration shifts the frequency of the $\nu(C–C) + \delta(CD_3)$ combination vibration to much lower frequencies, there is no split of the C–N modes.

Interaction of CD_3CN with weakly acidic hydroxyls shifts the OH stretching modes and produces a well-defined symmetric band. An example is given in Figure 2.5, which shows that $\Delta\tilde{\nu}(OH)$ for the silanol groups on silica is about −320 cm^{-1}. If the hydroxyls are strongly acidic, the shift is much larger and the shifted OH band will split into two (AB) or three components (ABC). Such a split can be seen in the spectra in Figure 2.30, which were registered after CD_3CN adsorption on HZSM-5 *(375)*. The ABC maxima are detected at 2780, 2415, and 1695 cm^{-1}, respectively. Consequently, the value of $\Delta\tilde{\nu}(OH)$ cannot be directly measured, but it may be estimated by the position of the center of gravity of the A, B, and C components. An additional band at 2110 cm^{-1} can be assigned to $\nu_s(CD_3)$; the corresponding asymmetric vibrations are responsible for the appearance of a weak band at around 2250 cm^{-1}. These two modes are largely insensitive to the coordination.

Upon coordination of the nitrile group, the C–N modes of CD_3CN shift to higher frequencies. When an H-bond is formed, the $\nu(C–N)$ is typically observed in the region of 2331–2270 cm^{-1}, and the higher the frequency, the stronger the H-bond. A relationship between the $\nu(CN)$ of CD_3CN involved in H-bonding in solutions and the Hammett acidity function (H_0) of the solution was reported by Anquetil *et al.* *(369)*. The large variation of the shift of the C–N modes permits a fine distinction of slight differences in hydroxyl acidity. The analysis of the C–N modes is particularly important when the O–H stretching region is too noisy or, because of an AB(C) split, $\Delta\tilde{\nu}(OH)$ cannot be exactly determined. For example, the small differences between the acidity of OH and OD isotopologues could be measured *(376)*. In this respect, CD_3CN is a unique molecular probe.

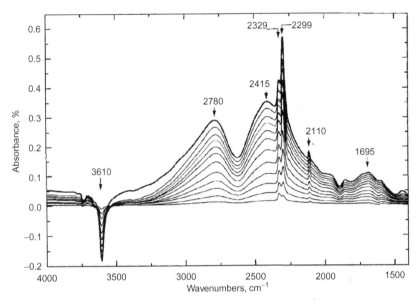

Figure 2.30 Difference FTIR spectra of CD$_3$CN adsorbed at ambient temperature on HZSM-5 with increasing coverage. Spectrum of the pretreated catalyst is subtracted. Hydroxyls absorbing at 3610 cm^{-1} are shifted and split to an ABC pattern with maxima at 2780, 2415, and 1695 cm^{-1}. The bands at 2329 and 2299 cm^{-1} are assigned to C–N modes of CD$_3$CN bound to Al^{3+} Lewis acid sites and bridging hydroxyls, respectively. *Reproduced from Ref. (375).*

When CD$_3$CN is protonated, the C–N modes shift to 2324–2315 cm^{-1} *(369)* and for protonated dimers, to 2230 cm^{-1} *(370)*, that is, H–bonded and protonated forms can easily be distinguished. When acetonitrile is coordinated to Lewis acid sites, the shift is much larger, and the band position may reach a value of 2335 cm^{-1}. In some circumstances, ν(CN) of the acetonitrile coordinated to a Lewis site may coincide with that of protonated acetonitrile and even with that of the H–bonded form, which makes the distinction between these forms difficult. In the presence of olefins, CD$_3$CN can form *N*-alkylacetonitrilium ions, which are detected above 2375 cm^{-1} *(377)*. It should be emphasized that the interaction of acetonitrile with OH groups does not stop at the formation of 1:1 complexes, and dimers can be formed at high coverage *(311)*. Therefore, removal of gas–phase acetonitrile from the apparatus is recommended if spectra representing solely strong interactions are desired.

Figure 2.31 shows the C–N stretching region of CD$_3$CN adsorbed on HZSM-5. The band at 2297 cm^{-1} characterizes CD$_3$CN H-bonded to

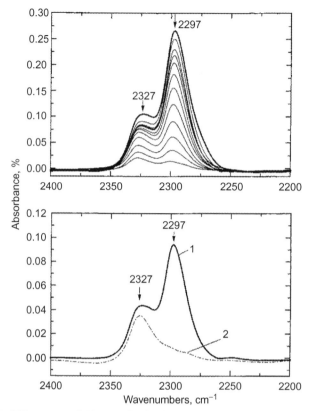

Figure 2.31 FTIR spectra of CD$_3$CN adsorbed on HZSM-5 in the C–N stretching region (top) with increasing coverage; (bottom) after evacuation at (1) 373 K and (2) 423 K. Complexes of CD$_3$CN with Al^{3+} Lewis acid sites (2327 cm^{-1}) are thermally more stable than CD$_3$CN H-bonded to bridging hydroxyls (2297 cm^{-1}). *Reproduced from Ref. (375).*

bridging hydroxyls, and the band at 2327 cm^{-1} is due to CD$_3$CN coordinated to Al^{3+} Lewis acid sites. It can be seen that the H-bonded complexes decompose during evacuation at a temperature of 423 K. For comparison, the complexes of CD$_3$CN with silanol groups are unstable already at ambient temperature. A table with the thermal stability of the complexes of various hydroxyls with acetonitrile was published in Ref. (378).

Unfortunately, there are several serious drawbacks of using acetonitrile, which arise from its high reactivity (379). For example, CH$_3$CN can dissociate on oxides with pronounced acid–base pairs, producing a hydroxyl group and CH$_2$CN$^-$ (380). The anion of the acetonitrile dimer can be formed as a result of oligomerization (56). Finally, CH$_3$CN may react with hydroxyl groups producing acetamide (CH_3CONH^-) species *(379,381–383)*, acetic acid, and ammonia *(56)*. A possibility that needs to be considered when analyzing the spectra of CD_3CN adsorbed on oxides is that some OH groups may be affected by acetonitrile adsorbed on cation sites in the proximity (if coordinatively unsaturated cations are available) *(379,384)*. An additional complication of using CD_3CN is the possibility of H/D exchange with the surface *(385)*, but this process occurs only at elevated temperatures *(167,376)*.

It should be mentioned that an acidity scale of zeolites was proposed on the basis of the temperature of CD_3CN protonation *(167)*. This scale correlates well with the *n*-hexane cracking activity.

Quantitative determination of the hydroxyl groups can be made with the knowledge of the $\nu(CN)$ absorption coefficient, which has been reported to be, for the 2300–2295 cm^{-1} band, 2.05 ± 0.1 cm µmol^{-1} *(372,386)*. This value is smaller than the absorption coefficient for acetonitrile adsorbed on Lewis acid sites (which is 3.6 ± 0.1 cm µmol^{-1}).

Table 2.16 summarizes data from acidity measurements of surface hydroxyls by adsorption of CD_3CN (respectively CH_3CN) as a probe. Because the AB or ABC structure of the shifted OH band often prevents exact determination of $\Delta\tilde{\nu}(OH)$, and conclusions must be made on the basis of an analysis of the $\nu(CN)$ region, additional data for the $\nu(CN)$ mode are summarized in Table 2.17.

Analysis of the data presented in Tables 2.16 and 2.17 shows that CD_3CN is a preferred probe for porous materials but is rarely used for measuring acidity of OH groups on oxides. This selective use results from the chemical transformations often occurring with acetonitrile on oxide surfaces. Some particular findings that should be mentioned are:

- The enhanced acidity of some of the silanols in silica–alumina samples, detected by CO, is also observed with CD_3CN as a probe *(106)*.
- The decrease in acidity of hydroxyls in the presence of metal cations after partial exchange of hydroxyl groups is demonstrated not only for alkali metals *(394)* but also for copper *(395)*.
- The variation in the Brønsted acidity depending on the nature of the cation in isomorphously substituted materials is clearly seen *(390,394)*.
- CD_3CN is widely used to characterize POH groups, and a characteristic C–N vibration around 2284 cm^{-1} is observed *(389,392)*. Highly acidic POH were also detected on phosphated silica ($\tilde{\nu}(CN)$ at 2290 cm^{-1}) *(387)*.
- By using deuterated TiO_2 samples (to reduce the noise), different acidity of the hydroxyls was observed for samples of different origin; the C–N

Table 2.16 CD_3CN-induced shift of $\nu(OH)$ of various hydroxyls and the respective C–N modes

Sample	$\tilde{\nu}(OH)$, cm^{-1}	$\|\Delta\tilde{\nu}_{OH}\|$, cm^{-1}	$\tilde{\nu}(CN)$, cm^{-1}	Note	Ref.
I. Silanol groups					
SiO_2	3746–3745	300–330	–	CH_3CN	*(106,378)*
SiO_2	3745	325	2274		*(387)*
SiO_2	3748	318	2275		*(21)*
SiO_2–Al_2O_3	3740	ca. 300	–	CH_3CN	*(106,378)*
	3742	ca. 800			
Fe/SiO_2	3682	216–260	2273		*(388)*
HY	3738	320	2278		*(389)*
HZSM-5	3745	335			*(390)*
HZSM-5	3745	305	2278		*(361)*
HBEA	3745	145	2275		*(391)*
SAPO-34		340	2278		*(392)*
SAPO-5	3743	310	2278		*(389)*
Bentonite	3740	310	2278		*(393)*
II. POH groups					
P/SiO_2	3675	600	2290		*(387)*
$AlPO_4$-5	3677	610	2280		*(389)*
$AlPO_4$-11	3677	610	2281		*(389)*
SAPO-5	3678	620	2284		*(389)*
SAPO-34	–	575	2284–2280		*(392)*
SAPO-11	3678	620	2284		*(389)*
CoAPO-5	3677	590–610	2283		*(389)*
CoAPO-11	3677	60	2283		*(389)*
CoAPO-44	3677	555	2286		*(389)*
III. Bridging hydroxyls					
HX	3655	703[a]	–		*(366)*
HY	3630	~1100[a]	2300	Band at	*(361)*
	3550	~500[a,b]	2284	$3550\ cm^{-1}$ only slightly affected	

Table 2.16 CD$_3$CN-induced shift of ν(OH) of various hydroxyls and the respective C–N modes—cont'd

Sample	$\widetilde{\nu}$(OH), cm^{-1}	$|\Delta\widetilde{\nu}_{OH}|$, cm^{-1}	$\widetilde{\nu}$(CN), cm^{-1}	Note	Ref.
HY	3629	1039[a]		Si/Al = 18	(394)
	3630	950[a]		Si/Al = 5	
HY	3649–3645	837–933[a]	–	Depends on Si/Al	(366)
HY	3629	1030[a]	2290		(389)
NaHY	3647	747			(394)
CsHY	3554	424			(394)
HZSM-5	3612	1068[a]	–		(366)
HZSM-5	3610	1130–1150[a]	2298		(390)
HZSM-5	3610	1150[a]	2297		(375,395)
HZSM-5	3610	1110[a]	–		(394)
CuHZSM-5	3610	1020–975[a]	2297	$|\Delta\widetilde{\nu}$(OH)$|$ decreases with increasing Cu concentration	(375,395)
H[Fe]ZSM-5	3630	1020[a]	–	Si(OH)Fe	(394)
H[Fe]ZSM-5	3635	945[a]	2298	Si(OH)Fe	(390)
HMOR	3608	1038[a]	–		(366)
HMOR	3609	1069[a]	–	Si/Al = 10	(394)
	3610	1050[a]		Si/Al = 6.7	
SAPO-5	3626, 3600	1000[a]	2290		(389)
SAPO-11	3631	1000[a]	2290		(389)
SAPO-34	–	ca. 1000[a]	2300		(392)
SAPO-44	3625, 3600	1000[a]	2286		(389)
IV. Other hydroxyls					
Al$_2$O$_3$	3790, 3738	200–250	–	One shifted band	(378)
MgO	3750	210	–		(382)

[a]Calculated by center of gravity.
[b]The authors suggested that the shift does not characterize the OH acidity.

Table 2.17 C–N modes of CD_3CN H-bonded to different hydroxyl groups

Sample	$\tilde{\nu}(OH)$, cm^{-1}	$\tilde{\nu}(CN)$, cm^{-1}	Note	Ref.
I. Bridging hydroxyls				
HY	3638	2298		*(167)*
DAY	3600	2295		*(396)*
	3622	2287		
DAY (USY)	3632	2302		*(173)*
HZSM-5	3613	2301.2–2301.6	Slightly coverage dependent	*(376)*
HZSM-5	3610	2300		*(361,397,398)*
HZSM-5	3614	2298		*(167)*
HZSM-5	3617–3605	2297		*(399)*
[Fe]ZSM-5	3639–3626	2298	Si(OH)Fe	*(400)*
[Ga]ZSM-5	3630–3560	2297	Si(OH)Ga	*(399)*
[B]ZSM-5	3616	2276–2273	Si(OH)B	*(399)*
HMOR	3617	2298		*(167)*
HMOR	3610	2297	Band at 2314 cm^{-1} considered as due to protonated CD_3CN	*(369)*
HMOR	3610	2295	–	*(371)*
	3590	2314, 2275	Two CD_3CN molecules	
HFER	3607	2298		*(167)*
HFER	3603	2298–2296		*(401)*
HFER	3604	2297		*(402)*
HFER	3601	2296	Solvation	*(362)*
HFER	3580	2292		*(403)*
HBEA	3614–3612	2298–2297		*(391,404)*
MCM-58	3485	2297		*(386)*
MCM-22	3620	2296		*(386)*

Table 2.17 C–N modes of CD_3CN H-bonded to different hydroxyl groups—cont'd

Sample	$\tilde{\nu}(OH)$, cm^{-1}	$\tilde{\nu}(CN)$, cm^{-1}	Note	Ref.
SAPO-18	3626, 3600	2291, 2284		(397)
Bentonite	3670	2302	Al(OH)Fe	(393)
II. Other hydroxyls				
H–Nafion[a]	3000	2303	SOH	(135,398)
LaHZSM-5	–	2286	LaOH, tentative assignment	(375)
AlPO-18	3678	2282	P–OH	(397)
SAPO-18	3678	2282	P–OH	(397)
Al_2O_3	–	2261	Al–OH, chemical transformation	(405)

[a]A perfluorosulfonic polymer.

modes of adsorbed CD_3CN were detected at 2265 or 2270 cm^{-1} (406,407).

- Adsorption of CD_3CN on zeolites allows detection of the γ(OH) modes of the zeolitic OH groups because they are shifted to higher frequencies and move out of the region of strong framework absorption (366). The intensity of these vibrations decreases with the increase of the strength of the H-bond formed.
- CD_3CN is also used as a probe molecule in NMR spectroscopy, and combination of the two techniques can be very useful (391).

Several other nitriles have been employed for the characterization of hydroxyl groups. Their use will not be discussed in detail here; instead, we mention only two important points. Nitriles that are less reactive than CH_3CN may be advantageous for the characterization of oxide surfaces (130). Some larger nitriles (e.g., pivalonitrile) are widely used to explore the accessibility of hydroxyl groups (see Section 5).

3.11. Water

Water is not suitable as a probe for measuring surface acidity because the bands from the water hydroxyls superimpose those of the free OH groups. Nevertheless, since the water–OH interaction is very important, we shall briefly consider the associated spectral behavior.

The PA of water is 691 kJ mol^{-1} (148), a value that is considerably smaller than the PA of CD_3CN (779.2 kJ mol^{-1}). Consequently, one would expect no protonation of water, not even by highly acidic hydroxyls. This expectation is fulfilled in most cases. For example, adsorption of water on oxides produces a band around 1640 cm^{-1}, which is indicative of molecular H_2O *(408,409)*. However, for most oxides, the situation is very complicated because of the presence of Lewis acid–base pairs where water can dissociate. Moreover, H_2O is preferentially adsorbed on coordinatively unsaturated metal cations, and the respective IR bands superimpose those of water interacting with OH groups. Thus, broad bands are usually observed in the 3700–2600 cm^{-1} region resulting from overlapping contributions of the O–H stretching vibrations, the $2\delta(H_2O)$ modes of H-bonded water molecules, and the stretching modes of H-bonded surface hydroxyls *(409)*. Therefore, detailed IR spectroscopic investigations have mostly been performed with materials lacking Lewis acidity. For silica, a shift of the silanol band by -250 to -300 cm^{-1} was observed *(11,29,409,410)*. Even in the absence of Lewis acidity, the superposition of the OH bands complicates the detailed analysis.

A large number of experimental and theoretical investigations have been devoted to the interaction of acidic hydroxyls in zeolites with water, to clarify the central question of whether the H_2O molecule is protonated or not.

We shall first consider the situation of 1:1 complexes, which are present when the number of adsorbed water molecules is lower than the number of available bridging hydroxyls. The published spectra *(11,140,186,411–421)*

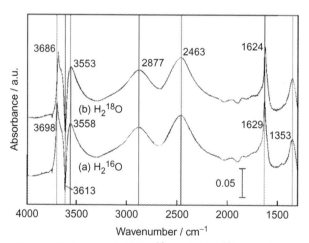

Figure 2.32 Difference FTIR spectra of (a) $H_2^{16}O$ and (b) $H_2^{18}O$ adsorbed on HZSM-5 at a temperature of 303 K. The water equilibrium pressure was 67 Pa. Vertical lines indicate the peak positions observed for $H_2^{16}O$. *Reproduced with permission from Ref. (419). Copyright 1996 American Chemical Society.*

are similar and are characterized by the following main features (see Figure 2.32, spectrum a):

- Appearance of a negative peak (in difference spectra, otherwise partial disappearance) of the bands of acidic hydroxyls at 3640–3605 cm^{-1}.
- Parallel development of broad bands around 3000–2820 cm^{-1} and 2540–2440 cm^{-1}, as well as of a weak feature at about 1700 cm^{-1}.
- Development of a band at 1370–1350 cm^{-1}. In some cases, an additional band at 880–875 cm^{-1} was detected that formed concomitantly.
- Parallel (or almost parallel) rise in intensity of bands at 1633–1620, 3710–3660, and 3625–3540 cm^{-1}.

These spectral features have received controversial assignments. In early reports *(414)*, the new $\nu(OH)$ bands formed upon water adsorption were assigned to new types of silanol or aluminol groups. In more recent publications, the bands around 2900 and 2500 cm^{-1} were attributed to two different shifted bands and were taken as evidence of heterogeneity of the surface hydroxyls *(412)*. Another interpretation *(415–417)* assumed the formation of a hydroxonium ion:

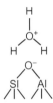

The band at about 3700 cm^{-1} was attributed to the free OH vibration of this structure, and the bands at about 2900 and 2500 cm^{-1}, to the antisymmetric and symmetric vibrations of the hydroxonium ion, respectively.

More recent theoretical investigations *(368)* indicated that the water molecule is hydrogen-bonded to the bridging OH groups of the zeolite. An experimental attempt to resolve this discrepancy was made by Parker *et al.* *(421)*; their approach was to investigate the adsorption of H$_2$O and D$_2$O on HZSM-5 and DZSM-5. The authors found that the bands around 2900 and 2500 cm^{-1} appeared after adsorption of H$_2$O or D$_2$O on HZSM-5, but not after their adsorption on DZSM-5. These results allowed concluding that the two bands (at 2900 and 2500 cm^{-1}) are associated with the proton from the bridging OH group and are not related to hydroxonium ions. Unfortunately, some D/H exchange occurred during the experiments which introduced some uncertainty concerning predominantly bands of low intensity. Indeed, the authors concluded that a band around 3700 cm^{-1} is also associated with the OH group, which seems not to be the case (see below).

Key experiments were performed by Wakabayashi *et al. (419)*, who compared the IR spectra produced by adsorption of H_2O and $H_2{}^{18}O$ on HZSM-5 (Figure 2.31). No isotopic exchange occurred at the experimental conditions. On the basis of the results, the authors conclusively assigned the bands at 2877, 2463, and 1353 cm^{-1} (not sensitive to isotopic substitution) to perturbed bridging hydroxyls, and the bands at 3698, 3558, and 1629 cm^{-1} (sensitive to isotopic substitution) to adsorbed water. Moreover, the authors assigned the bands at 2877 and 2463 cm^{-1} to the AB components of the shifted $\tilde{\nu}(OH)$ modes and the band at 1353 cm^{-1} to shifted $\delta(OH)$ vibrations.

After the hydrogen-bond model of water adsorption was accepted, additional details have been established. Theoretical modeling *(422)* indicated that water is adsorbed via two hydrogen bonds, one bond to the proton from the acidic OH groups and one to framework oxygen. These conclusions were confirmed by combined DFT and INS investigations *(423)*. Paze *et al. (11)* identified two OH stretching modes of water adsorbed on HBEA: one narrow band (FWHM = 15 cm^{-1}) at 3710 cm^{-1} that was ascribed to "free" (i.e., not H-bonded) water hydroxyls, and a band at 3525 cm^{-1} (FWHM = 60 cm^{-1}) that was assigned to water hydroxyls experiencing a weak H-bond with framework oxygen.

We will now consider the situation of an excess of water in zeolites, that is, the presence of more than one water molecule per acidic hydroxyl. At high water concentrations, the ABC bands disappear, as do the bands of the $\delta(OH)$ and $\gamma(OH)$ modes *(11,186,413,416,420)*. The bands of coordinated water also fade. These spectral changes point toward a conversion of the H-bonded forms into dimeric and/or polymeric structures. Theoretical investigations *(424,425)* indicated that the dimeric structures are protonated. It was also noted *(424)* that the protonation process depends not only on the acidity of the zeolite hydroxyls but also on the topology of the zeolite framework, which determines the geometry of adsorption. It was concluded that dimers formed in FAU zeolites are protonated, whereas in ZSM-5, protonated and neutral dimers coexist. A band around 1700 cm^{-1} was reported to be indicative of protonation *(186,420)*. As a conclusion, it can be said that the cooperative action of two H_2O molecules can lead to a proton transfer. Therefore, protonated species are expected when zeolites are suspended in water, which should be taken into account when considering reactions in aqueous media, such as the ion-exchange process.

Finally, it should be mentioned that the adsorption of water on the superacidic Nafion® material is followed by protonation even at a low coverages, that is, before the appearance of 2:1 adducts *(426)*.

3.12. Concluding remarks

In addition to the molecules discussed in this section, many other weak bases have been adsorbed and have been found to shift OH stretching modes. While the data are important for a particular scientific problem, these molecules have no practical relevance as probes for measuring protonic acidity. In our opinion, CO and N_2 have the greatest potential for future utilization among the weak probes, and xenon is a possible third candidate. Among the medium and strong bases, CD_3CN seems to be the most useful. The right choice of probe depends on the material and the expected acidity of the hydroxyl group.

4. CHEMICAL TRANSFORMATIONS OF SURFACE OH GROUPS

The outcome of many chemical reactions is that the surface hydroxyl groups lose their identity. A typical example is the protonation of strong bases, whereby the O–H bond of the hydroxyl is broken. However, many other interactions also lead to the disappearance of the hydroxyl groups as such, because the M–O or O–H bonds, or both, are broken. In all cases, the proton is transferred to another species.

4.1. Proton transfer to strong bases

Strong bases are protonated by acidic surface hydroxyls; the products of this reaction are conjugated acids and bases. The stabilization of the conjugated acid on the surface plays an important role for the progress of the reaction. In the following sections, we shall consider the proton transfer from OH groups to ammonia, pyridine, and some substituted molecules. It is important to bear in mind that sometimes proton transfer occurs as a result of the cooperative action of two or more basic molecules.

4.1.1 NH_3 and alkylamines

Ammonia is a strong base with a PA of 853.6 kJ mol^{-1} *(148)* and a pK_a of 9.2 *(134)* and can be protonated even by surface hydroxyl groups of moderate acidity. The NH_3 molecule is very small with a kinetic diameter of 0.26 nm *(334)* and can thus penetrate even small cavities in porous materials. Upon adsorption on metal oxides or zeolitic materials, ammonia forms mainly five types of species:

- NH_4^+ ions formed as a result of ammonia protonation by acidic hydroxyls;

- $N_2H_7^+$ or $N_2H_7^+ \cdot nNH_3$ species produced by solvation of NH_4^+ in excess of ammonia;
- NH_3 coordinated to Lewis acid sites (coordinatively unsaturated metal cations);
- NH_3 H-bonded through its N-atom to weakly acidic hydroxyl groups;
- NH_3 H-bonded through its H-atoms to surface oxygen; and
- dissociated ammonia (NH_2^- groups).

Fortunately, all these forms exhibit characteristic spectral features and can usually be distinguished by IR spectroscopy (56). These traits make ammonia one of the most frequently used probe molecules for assessment of surface acidity.

All vibrations of ammonia or the ammonium ion are sensitive to the bonding of the species to the surface. In the NH stretching region, the spectral pattern is often rather complicated because of the superposition of other bands (e.g., shifted OH modes) or Fermi-resonance phenomena. Therefore, the region of the deformation modes is usually analyzed. The body of data in the literature concerning ammonia adsorption on various materials was summarized and analyzed by Davydov in 2003 (56), and general ranges for the spectral signature of the different adsorption forms were specified (see Table 2.18). The free ammonia molecule has two deformation bands: $\delta_s(NH_3)$ at 950 cm^{-1} and $\delta_{as}(NH_3)$ at 1628 cm^{-1}. The symmetric deformations are highly sensitive to the coordination and are shifted up to a position of 1260 cm^{-1} when a strong bond is formed. For NH_3 H-bonded to hydroxyl groups through the N-atom, the $\delta_s(NH_3)$ modes are observed around 1150–1100 cm^{-1}. This possibility to estimate the strength of the Lewis acidity on the basis of spectral data is often decisive in the choice of NH_3 as a probe molecule. However, this spectral property does not come to bear for silica and zeolitic materials, because they are not transparent in the relevant spectral region.

Table 2.18 Deformation modes of adsorbed NH_3 and NH_4^+ ions (56)[a]

Adsorbed species	$\delta_s(NH_x)$, cm^{-1}	$\delta_{as}(NH_x)$, cm^{-1}
NH_3 coordinated to Lewis acid site	1280–1150	1630–1600
NH_3 H-bonded to OH groups through the N-atom	1150–1100	1625
NH_3 H-bonded through H-atoms	1070–1020	1610–1600
NH_4^+ ions	ca. 1680	1485–1405

[a]Some authors have reported values outside the specified regions.

When ammonia is protonated, the symmetric deformations (IR–silent) are shifted to about 1680 cm^{-1} and the antisymmetric ones to about $1450–1400 \text{ cm}^{-1}$. In fact, the symmetric deformations are detected in the spectra of ammonium ions on surfaces, which indicates that the symmetry has been lowered. When ammonia dissociates on acid–base pairs, the deformation modes of the formed NH_2^- groups are detected around 1550 cm^{-1}.

IR spectra of ammonia adsorbed on oxovanadium species supported on porous silica are shown in Figure 2.33. Two kinds of interactions with hydroxyl groups are evident *(282)*. The broad band at about 1470 cm^{-1} corresponds to the bending vibrations of ammonium ions formed by reaction with acidic vanadia hydroxyls. The band at 1635 cm^{-1} is ascribed to the bending mode of ammonia molecules H-bonded to silanols. Simultaneously, a shift of the OH stretching modes of the silanol groups (external silanols at 3745 cm^{-1} and terminal and H-bonded silanols in hydroxyl chains at 3710 and 3547 cm^{-1}, respectively) by about -800 cm^{-1} is observed. The bands at 3390 and 3330 cm^{-1} are assigned to N–H modes.

A drawback of using ammonia as a probe molecule is the large variety of possible adsorption forms. For example, an ammonium ion formed in the adsorption process could be bound to lattice oxygen ions through two, three, or four hydrogen atoms. A similar variety of interactions can also stabilize coordinated ammonia or ammonia H-bonded to weakly acidic hydroxyls. In the latter case, measurement of the hydroxyl acidity through the H-bond method should be made with another probe for which such

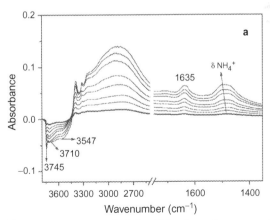

Figure 2.33 IR difference spectra recorded after NH_3 adsorption at room temperature on a sample of oxovanadium species supported on porous silica outgassed at 423 K. NH_3 equilibrium pressure in the range: 0.01–20.0 mbar. *Reproduced from Ref. (282).*

secondary interactions are limited. Another complication in regard to pro-tonated ammonia is that at high coverages, positively charged dimers or polymers can be formed *(427,428)*; that is, one needs to control the adsorbed amount to not exceed a 1:1 stoichiometry. Finally, coordinated ammonia can affect the acidity of the surface hydroxyls, as demonstrated by adsorption of weak bases on TiO_2 and on TiO_2 partly covered by ammonia *(269,357)*.

Unfortunately, the position of the $\delta_{as}(NH_4)$ modes of the ammonium ion strongly depends on the adsorption geometry and does not permit any definite conclusions on the acid strength. To determine the acid strength, one needs to combine IR spectroscopy with temperature-programmed desorption experiments. In conclusion, ammonia as a probe gives information on the number of acidic hydroxyls (that are able to pro-tonate it) and, by combination with other techniques, on the strength of these sites.

In the following, we provide a brief survey of measurements of the acid-ity of surface hydroxyl groups. Such data can be found in hundreds of papers, and we shall limit the considerations to investigations focusing on funda-mental changes in the OH region and/or giving new insight into the observed phenomena.

Silanol groups on silica *(429)* and porous silica *(430)* are weakly acidic, and many measurements demonstrate that they are not able to protonate ammonia. Bordiga *et al.* *(429)* detected a band around 1520 cm^{-1} when a high NH_3 equilibrium pressure was maintained over a silica sample. A similar band (at 1480 cm^{-1}) was detected in the spectra of a silicalite sam-ple. However, these bands were not assigned to ammonium ions. Many authors reported that silanols in zeolites form only H-bonds with ammonia.

In contrast, the bridging hydroxyls of zeolites react with ammonia to give ammonium ions ($\delta_{as}(NH_4)$ at 1490–1405 cm^{-1}) *(28,99,133,215,225,226, 416,427,431–439)*. Ammonium ion formation occurs also with hydroxyls of related materials, such as porous phosphates *(133,210,440,441)*. A careful analysis of the symmetry of the species formed in various zeolites was pub-lished by Zecchina *et al.* *(427)*. The authors found that mainly bidentate and tridentate species with various local symmetries were formed in ZSM-5, MOR, BEA, SAPO-34, and FAU zeolite and zeotype materials. In addi-tion, some tetradentate species were found, with the most significant con-centrations present in a MOR zeolite. At higher ammonia coverage, the authors noticed the solvation of the ammonium ion and formation of $N_2H_7^+$ or $N_2H_7^+ \cdot nNH_3$ species. Bands at about 3380 cm^{-1} *(427)* and 1712 and 2750 cm^{-1} *(442)* were proposed to be indicative of these species.

More recently *(428)*, the bands responsible for dimeric structures were identified for a sample of TON zeolite, as labeled in Figure 2.34.

In general, protonated ammonia species in zeolites are stable. According to Ref. *(437)*, ammonia in ZSM-5 zeolite desorbs at a temperature of 473 K, whereas in Ref. *(432)* it was reported that after evacuation at 478 K only a small fraction of the intensity at 3605 cm^{-1} (bridging hydroxyls) is recovered and the near-complete desorption needs a temperature of 688 K. A temperature of 473 K was reported for the onset of decomposition of the more strongly bound complexes in HY and HZSM-22 zeolites; the more weakly bound complexes started to decompose at 373 K *(438)*. The hydroxyls in FER zeolite characterized by a band at 3601 cm^{-1} were restored after evacuation of adsorbed ammonia at 470 K, whereas the hydroxyls absorbing at 3591 cm^{-1} retain ammonia even at 670 K *(434)*. Three temperature-programmed desorption peaks of ammonia were found

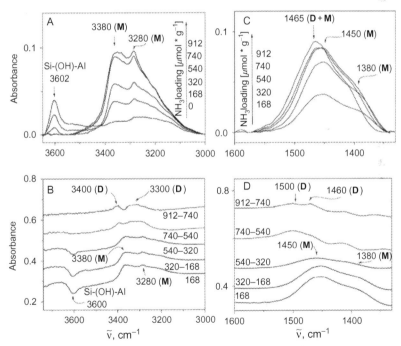

Figure 2.34 (A and C) IR spectra recorded after the adsorption of increasing amounts of NH$_3$ on zeolite TON (Si/Al = 27) at a temperature of 400 K, with the ammonia loadings given in the figure; (B and D) the difference spectra (differences between the spectra of two consecutive additions of ammonia). The symbols M and D indicate the bands of monomeric ammonium ions and dimers, respectively. *Reproduced from Ref. (428).*

for SAPO-34 at 507, 623, and 763 K and were associated with species protonated by hydroxyls of different acid strength (210). Ammonium ions in $AlPO_4$-5 were found to decompose at 373 K, whereas the respective complexes of ZrAPO-5 with ammonia decomposed during evacuation at 473 K. It was concluded that ZrAPO-5 is characterized by a higher number of more acidic hydroxyls, proposed to be bridging OH groups with participation of zirconium (441). Differences in the decomposition temperature have been used to distinguish different types of bridging hydroxyls in MOR (23) and USY zeolites (24).

Incorporation of various cations into mesoporous silica often leads to the creation of Brønsted acid sites. For example, adsorption of ammonia on Ti/MCM-41 (430,443), [Al]MCM-41 (444), and [Al]SBA-15 (182), as well as on titanium, zirconium, and hafnium species supported on mesoporous silica (445), and on TiO_2–SiO_2 (110) leads to the appearance of bands of protonated ammonia. However, the adsorption is reversible (430) or partly reversible (182,443,445) at ambient temperature, indicating a weaker acidity of these materials in comparison to zeolites.

It has been reported that hydroxyls able to protonate ammonia exist on various alumina surfaces. Two types of Brønsted acid sites were found on γ-Al_2O_3 by Wang et al. (446). Ammonium ions stable toward evacuation at 473 K are also formed on β-Al_2O_3 (447). However, the possibility of ammonia disproportionation on alumina to NH_4^+ and NH_2^- has been pointed out (59). NH_4^+ species that are stable at ambient temperature were also reported to be formed with Al–OH groups on silica–alumina (248). According to several authors (215,433,436), the aluminol groups on extraframework alumina material in zeolites (characterized by hydroxyl bands at about 3665 cm^{-1}) are also able to protonate ammonia and are restored by evacuation at 723 K. However, similar species in DAY characterized by a band at 3615 cm^{-1} do not react with NH_3 (439). It was also found that the so-called VHF (very high-frequency) band in zeolites (located at 3785–3780 cm^{-1}, associated with nonframework aluminum) disappeared after interaction of the samples with strong bases such as NH_3 (215,431,436). Initially, this behavior was taken as evidence for strong acidity of the respective hydroxyls (436), but later on it was noticed that the disappearance of the OH band coincides with the rise of a band at 1330 cm^{-1}. This band is typical of coordinated ammonia and reveals interaction of the hydroxyls with NH_3 coordinated to Lewis acid sites (431). The band at 3782 cm^{-1} reappears after thermal desorption of ammonia at 723 K (431,436).

Most authors agree that ammonia adsorption does not reveal protonic acidity on pure TiO_2 *(356)* or ZrO_2 *(276)*. However, strong Brønsted acid sites were detected on sulfated samples *(276,356,448,449)*. By means of successive adsorption of small doses of ammonia on sulfated titania, Desmartin-Chomel *et al. (448)* found that initially NH_3 dissociates, thus creating SOH groups which then protonate ammonia.

Bulk and supported oxides with pronounced covalent character of the M–O bond, for example, vanadia *(119,450,451)* and tungsten oxide *(450,452,453)*, also manifest Brønsted acidity. The acid strength depends on the sample, and ammonium ions are usually stable up to temperatures of 423–523 K *(450–452)*. After reduction, the covalency of the M–O bond is reduced, and these materials lose their Brønsted acidity *(56)*.

Fiorilli *et al. (454)* investigated an SBA-15 material containing carboxylic groups and reported that the —COOH groups were able to protonate ammonia at ambient temperature in a reversible process.

Evidently, there is no straight correlation between the acidity of hydroxyls estimated by using ammonia as a probe and the acidity measured by the hydrogen-bond method. As a rule of thumb, hydroxyls experiencing a CO-induced red shift of the OH modes of more than approximately 150 cm^{-1} are able to protonate ammonia.

Substituted amines are characterized by even higher basicity than ammonia and can thus be protonated by hydroxyls that are not able to form $NH_4{}^+$. For example, Brønsted acidity has been found on titania when using trimethylamine (TMA) as a probe *(455)*. TMA has also been successfully used for revealing the heterogeneity of the bridging hydroxyls in FAU zeolites *(456,457)*.

4.1.2 Pyridine

Pyridine (Py, C_5H_5N) is the most commonly used IR probe for measuring protonic acidity according to the ion-pair method. It is a very strong base (PA $= 930.0 \text{ kJ mol}^{-1}$ *(148)*, $pK_a = 5.2$ *(134)*) and is easily protonated by acidic surface hydroxyl groups. It has been pointed out *(56,134)* that the two scales, PA and pK_a, do not coincide when comparing pyridine and ammonia: pyridine is characterized by a lower pK_a and a higher PA than ammonia. These values indicate that ammonia is more basic than pyridine in water, but less basic in the gas phase. This seemingly conflicting behavior is explained by the possibility of solvation of the ammonium ion in water: it can form four H-bonds with water molecules, which additionally stabilizes the ion. In fact, theoretical investigations *(366)* have indicated that a similar stabilization can occur with the pyridinium ion, by interaction between hydrogen atoms from its C–H moieties with surface oxygen. However, this interaction is much weaker than the interaction of the NH groups with framework oxygen.

Although the pyridine molecule is larger than the ammonia molecule, it is still a useful probe for porous materials because its kinetic diameter (0.533 nm) allows entry into a range of pores.

When pyridine is adsorbed on oxide surfaces or in porous materials, the following species are commonly observed: (i) pyridine coordinated to Lewis acid sites, (ii) pyridine H-bonded to weakly acidic hydroxyls, and (iii) protonated pyridine. At high coverage, physisorbed pyridine and protonated dimers can also be observed. All these adsorbate structures can be easily distinguished on the basis of their spectral signature. There are two main reasons for which pyridine is the standard probe for assessing surface acidity: Reactive pyridine adsorption is very rare, and it is possible to selectively detect Lewis and Brønsted acid sites (and also to estimate the Lewis acid strength). It is generally accepted that hydroxyl groups able to protonate pyridine may be designated as Brønsted acid sites. However, this designation is arbitrary because bases stronger than pyridine can be protonated by less acidic hydroxyls.

Pyridine is a relatively complex molecule and exhibits a number of different bands in IR spectra. Among others, the bands characterizing the ν_{8a} and ν_{19b} modes have been found to be sensitive to the coordination or protonation of the molecule (see Table 2.19). Note that the band that is diagnostic for the PyH^+ ion at about 1545 cm^{-1} (ν_{19b} mode) does not overlap with any of the other bands.

Table 2.19 Dependence of the ν_{8a} and ν_{19b} modes of Py and PyH^+ on the bonding to the surface (138,458)

Adsorbed species	$\tilde{\nu}_{8a}$, cm^{-1}	$\tilde{\nu}_{19b}$, cm^{-1}
Gas-phase Py	1584	1439
Py H-bonded to OH groups through the N-atom	1595	1445
Py coordinated to Lewis acid site	1630–1600	1450
PyH^+	1640	1545 (1552–1540)

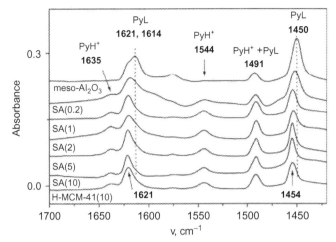

Figure 2.35 IR spectra of pyridine adsorbed on Al$_2$O$_3$, on aluminosilicates of various Si/Al ratios, and on HMCM-41 (Si/Al = 10) at a temperature of 420 K. Bands of pyridine adsorbed on Lewis acid sites are labeled PyL, and bands of pyridinium ions are labeled PyH$^+$. *Reproduced form Ref. (230).*

Typical spectra of adsorbed pyridine are presented in Figures 2.35 and 2.36; the bands corresponding to coordinated and protonated species are marked.

Although the C–H modes of adsorbed pyridine are also sensitive to the bonding, they are rarely utilized in practice. It has been proposed that these modes could help to unambiguously distinguish between H-bonded and weakly coordinated species *(459)*.

To obtain reliable results with pyridine as a probe, one should pay attention to the experimental details, in particular one needs to ensure that weakly adsorbed species are removed. A standard procedure in the investigation of zeolites and other porous materials starts with the introduction of pyridine at a temperature of 423 K and a partial pressure of about 100 Pa. The goal of this step is to allow diffusion of pyridine vapor into the small pores. As exemplified in Figure 2.37, the pressure is essential for pyridine diffusion into the pores of FER zeolite, even at 573 K.

Next, in order to remove physisorbed species, the apparatus with the sample is evacuated at the same temperature. There is no standard temperature for the evacuation or the measurement, and some variations can be found in the literature. It is important to know that for some samples, the described procedure can lead to the loss of important information. For example, it was found that on gallia–silica mixed oxides, the formation of

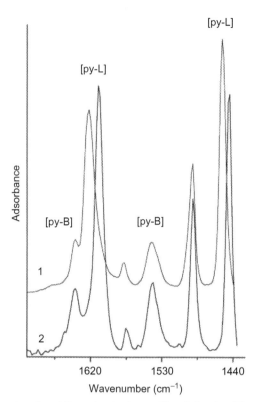

Figure 2.36 IR spectra of pyridine adsorbed on sulfated alumina (1) and sulfated zirco-nia (2) at a temperature of 423 K. Bands of pyridinium ions formed on Brønsted acid sites are labeled [py-B]. Bands of pyridine adsorbed on Lewis acid sites are labeled [py-L]; their position is sensitive to the nature of the cation (Al^{3+} vs. Zr^{4+}). *Reproduced from Ref. (261).*

PyH$^+$ ions is reversible at ambient temperature and under a certain equilib-rium pressure *(461)*. In such cases, experiments with gradually decreasing equilibrium pressure followed by evacuation at stepwise increasing temper-atures can give more detailed information.

Note that the ν_{8a} and ν_{19b} bands are highly sensitive to the Lewis acid strength (see the examples in Figure 2.36). Regarding Brønsted acid sites, the spectra merely provide information on their presence, and to assess the acid strength one needs to conduct temperature–programmed desorp-tion experiments. For example, the acidic bridging hydroxyls (3608 cm^{-1}) in ITQ–33 zeolite are completely consumed after interaction with pyridine at a temperature of 423 K (see Figure 2.38). Subsequent evac-uation at 523 K leads to a substantial decrease in intensity of the bands

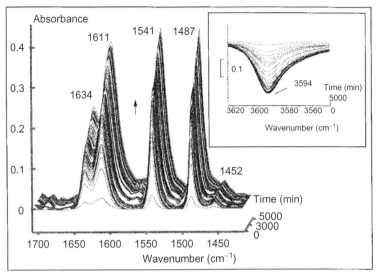

Figure 2.37 Evolution of IR spectra with time after adsorption of pyridine on FER at a temperature of 573 K and pyridine partial pressures of 10^{-2} (bottom spectrum) to 10^{-1} mbar (top spectrum). Inset shows the consumption of the bridging hydroxyl groups (difference spectra). *Reproduced from Ref. (460).*

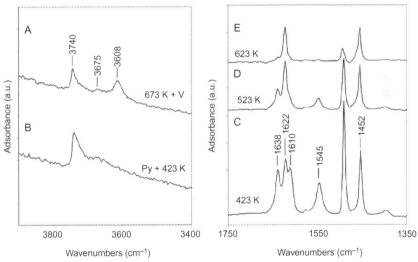

Figure 2.38 Acid properties of Al-ITQ-33 zeolite measured by pyridine adsorption and subsequent (attempted) desorption at increasing temperatures. On the left is plotted the hydroxyl stretching region, where (a) is the IR spectrum after thermal treatment at 673 K under vacuum and (b) is the spectrum after adsorbing pyridine followed by desorption at 423 K. On the right is shown the C–C stretching region of the adsorbed pyridine after evacuation at (c) 423, (d) 523, and (e) 623 K. *Reproduced from Ref. (462).*

indicative of PyH^+ (1638 and 1545 cm^{-1}), and they ultimately disappear after evacuation at 623 K.

Under certain conditions, including those of high coverage, positively charged dimeric species (viz., bis(pyridinium) ions) may be formed after adsorption of pyridine. These species are characterized by the absence of the typical band for PyH^+ at 1545 cm^{-1}; instead, a doublet appears in the region of 1446–1420 cm^{-1} (463). According to Buzzoni et al. (464), dimeric species are easily produced when the hydroxyl groups donating protons are highly acidic. The conjugate bases of strong acids, which in this case are the oxygen ions of the zeolite framework, are weak, and the bond between them and the PyH^+ ions is not strong. In the presence of a pyridine excess, these weakly bound PyH^+ ions can form $PyH^+\cdots Py$ adducts. The adducts are characterized by $NH^+\cdots N$ hydrogen bonds that produce a broad band, which is split into two components (with a minimum around 2300 cm^{-1}) as a result of Fermi resonance.

It was reported (465) that the bands of PyH^+ (1545 cm^{-1}) on HY are less intense if the evacuation after adsorption is conducted at ambient temperature rather than at 373 K. An additional band at 1446 cm^{-1} was registered at 373 K. At the time, the interpretation of the results was that a high temperature is needed to achieve proton transfer to H-bonded pyridine (1446 cm^{-1}). However, theoretical investigations (466) indicated that the proton-transfer barrier is quite small, and the hydrogen-bonded complexes are expected to have a very short lifetime. A contemporary interpretation is that the increase in intensity of the PyH^+ bands during evacuation at higher temperatures is associated with decomposition of bis(pyridinium) ions.

A band at 1462 cm^{-1} is sometimes observed after desorption of pyridine. Morterra and Magnassa (458) have assigned this band to pyridine bound simultaneously to a Lewis acid site and a proton.

The number of published articles that include characterization of surface acidity by pyridine adsorption is immense. Therefore, we are not able to present a comprehensive analysis of the available data on the pyridine–solid interaction. In the next paragraphs, we shall briefly review the main conclusions that were obtained through the use of pyridine as a probe and will contrast the results of pyridine adsorption with those of ammonia adsorption.

Hydroxyl groups that are not able to protonate ammonia also do not protonate pyridine. However, hydroxyls exist that are able to protonate ammonia but not pyridine. For example, no PyH^+ is detected after adsorption of pyridine on alumina, whereas protonated ammonia has been detected on this oxide by several authors. The same situation is encountered for mixed titania–silica. In this context, one should pay attention to the different conditions at which the surface acidity is probed by ammonia or pyridine. As noted earlier, reversible formation of PyH^+ can occur at ambient temperature *(461)*.

In agreement with the results of NH_3 adsorption, adsorption of pyridine indicates the presence of Brønsted acid sites on sulfated oxides. However, there are reports suggesting that the protonic acidity of sulfated titania detected by ammonia may be induced through ammonia dissociation *(448)*. Adsorption of pyridine revealed protonic acidity of sulfated zirconia, and this acidity was related to an OH band at 3650 cm^{-1} which was associated with disulfates *(467)*.

Results consistent with those of ammonia adsorption were obtained in the analysis of supported vanadia and molybdena. The development of protonic acidity with the increase of the vanadia loading on dealuminated BEA zeolite can be easily traced by adsorption of pyridine (Figure 2.39).

On the basis of pyridine adsorption results, it has been established that the substitution of aluminum by iron in ZSM-23 zeolite leads to a considerable decrease in the strength of the Brønsted acid sites *(469)*.

4.1.3 Substituted pyridines

Although they are not as widely used as pyridine, substituted pyridines have found their own place in the characterization of hydroxyl groups. It was proposed *(470)* that, for steric reasons, 2,6-dimethylpyridine (DMP, lutidine) does not interact with Lewis acid sites and is thus a proton-specific probe. Later, it was demonstrated that DMP *(247,256,471–473)* (as well as 2,4,6-trimethyl pyridine or collidine *(474)*) still forms coordination bonds with coordinatively unsaturated surface cations. This bond is only weakened by the steric interference of the methyl groups with the surface but not prevented. In any case, the steric hindrance leads to preferential interaction of DMP with hydroxyl groups *(471–473)*. DFT calculations suggest that the proton transfer is promoted by the stabilization of the lutidinium ion on the deprotonated site, rather than by the intrinsic acidity of the acid site itself *(475)*.

Table 2.20 summarizes the spectral parameters of differently bound DMP. Compared to pyridine, lutidine seems to have some advantages as a probe. The most important aspect is that the spectra of protonated lutidine seem to contain information on the strength of the Brønsted acid sites. It was demonstrated *(256)*, with a series of FAU zeolites, that the positions of the ν_{8a} and $\nu(NH)$ bands (2975–2494 cm^{-1}) of $DMPH^+$ depend on the acid

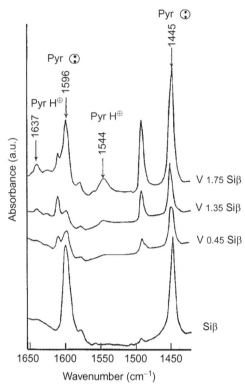

Figure 2.39 FTIR spectra of samples [Si]BEA, $V_{0.45}$[Si]BEA, $V_{1.35}$[Si]BEA, and $V_{1.75}$[Si]BEA after adsorption of pyridine (room temperature, 1 h) and evacuation (423 K, 1 h, 10^{-3} Pa). Bands of pyridinium ions are located at 1637 and 1544 cm^{-1}, and bands of pyridine coordinated to Lewis acid sites are located at 1596 and 1445 cm^{-1}. *Reproduced from Ref.* (468).

Table 2.20 Dependence of the ν_8 and ν_{19} modes of DMP and DMPH$^+$ on the bonding to the surface *(138,256)*

Species	$\tilde{\nu}_{8a}$, cm^{-1}	$\tilde{\nu}_{8b}$, cm^{-1}	$\tilde{\nu}_{19a}$, cm^{-1}	$\tilde{\nu}_{19b}$, cm^{-1}
Gas-phase DMP	1584	1580	1480	1410
Physisorbed DMP	1595	1600	–	–
DMP H-bonded to OH groups	1600	1580	1480	1410
DMP coordinated to Lewis acid site	1620–1595	1573	1477	1410
DMPH$^+$	1655–1627	1630	1473	1415

strength of the hydroxyl groups: the higher the $\nu(NH)$ wavenumber and the lower the ν_{8a} wavenumber, the stronger is the acidity. Regarding such measurements, caution needs to be applied with the exact determination of the $\nu(NH)$ maximum, because the band is split into two components as a result of Fermi resonance.

DMP is more basic than pyridine ($pK_a = 6.7$ vs. a pK_a of 5.2 for pyridine) and can thus be protonated by weaker Brønsted acid sites. Consistent with the difference in pK_a, protonic acidity was detected on γ-Al_2O_3 by DMP *(256)* but not by pyridine. Similar results were obtained for TiO_2–ZrO_2 mixed oxides *(476)*.

Onfroy *et al.* *(477)* investigated DMP adsorption on various oxides and zeolites with the goal of determining the absorption coefficients. Figure 2.40 presents typical spectra of protonated DMP formed on acidic solids (phosphated silica and HY zeolite), as well as the relationship used to calculate the absorption coefficient.

A drawback of using DMP is the relatively large size of this molecule (characterized by a kinetic diameter of 0.67 nm), which prevents it from accessing sites in small pores.

Another probe molecule that is of particular interest for investigating surface hydroxyls is 2,6-di-*tert*-butylpyridine (DTBP) *(159,462,474,478,479)*. It is assumed that this molecule is selective for Brønsted sites because of the steric impossibility to react with coordinatively unsaturated surface cations. However, the assignment of the bands in the 1700–1300 cm^{-1} region is still under debate, and the N–H$^+$ stretching band around 3370 cm^{-1} has been proposed to be diagnostic for protonation *(478)*. As seen in Figure 2.41, DTBP essentially interacts with all bridging hydroxyls in BEA zeolite (absorbing at 3610 cm^{-1}), and the produced N–H$^+$ band at 3370 cm^{-1} is narrow and well defined. Because of its relatively large kinetic diameter (0.79 nm *(159,479)*), the DTBP molecule is not able to enter the pores of zeolites like ZSM-5 or MCM-22.

4.2. Proton transfer after adsorption of acidic molecules

As already mentioned, some hydroxyls (e.g., Mg–OH) are too basic to be affected by CO. However, basic hydroxyls are expected to interact easily with acidic molecules. Lavalley *(480)* reported that the strongly basic groups are mainly of type I and proposed the following order of basicity:

$$Mg-OH > Zn-OH > La-OH > Al-OH.$$

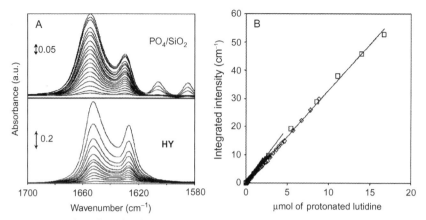

Figure 2.40 Phosphated silica and HY zeolite: (A) Infrared spectra recorded after increasing amounts of lutidine were adsorbed; upper part: phosphated silica (20 mg pellet) from 0.068 to 2.845 µmol lutidine; lower part: HY zeolite (20 mg pellet) from 0.087 to 8.874 µmol lutidine; (B) evolution of the integrated band intensity of protonated species as a function of the amount of protonated lutidine on phosphated silica (filled triangles), HY zeolite (diamonds), and HY after thermodesorption (squares). *Reproduced from Ref. (477).*

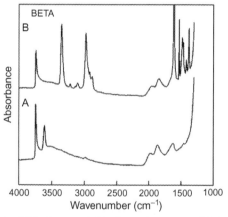

Figure 2.41 IR spectra (a) before and (b) after adsorption of 2,6-di-*tert*-butylpyridine (DTBP) on beta zeolite. The band of bridging hydroxyl groups at 3610 cm^{-1} disappears completely, and characteristic signals of the DTBPH$^+$ ion are formed at 3370, 1616, and 1530 cm^{-1}. *Reproduced from Ref. (478).*

In Sections 4.2.1, 4.2.2, and 4.2.3, we shall consider the interaction of surface hydroxyls with three molecules of increasing acid strength: CO_2, NO_2, and SO_2. As will be shown, some of these molecules can also interact with highly acidic surface hydroxyls.

4.2.1 CO_2

Reactive adsorption of CO_2 with the participation of surface hydroxyls that are basic in nature results in the formation of bicarbonate species. The OH modes of surface bicarbonates are detected in the interval between 3625 and 3605 cm^{-1}, while the deformation COH modes are positioned around 1236–1220 cm^{-1}. Both modes are sensitive to H/D exchange. Additional bands reported are the symmetric and antisymmetric C–O stretches at 1670–1595 and 1490–1370 cm^{-1}, respectively (481). Although the positions of these bands have been proposed to contain information on the basicity of the hydroxyls, it was remarked that they depend on too many factors to allow unambiguous conclusions (56,480).

It is established that indeed the higher-frequency hydroxyls (believed to be more basic) interact preferentially with CO_2. This behavior has been observed for alumina (482) and zirconia (483). However, on some other oxides, as, for example, on thoria, hydroxyls of type II also interact with CO_2 (480). By using oxide samples with ^{18}O-labeled OH groups, integration of the hydroxyl oxygen into the formed bicarbonate structures was proven (484).

The intensity of the $\delta(COH)$ band around 1225 cm^{-1} has been used as a measure of the concentration of basic OH species on various oxides (485).

By means of CO_2 adsorption, the basicity of the so-called VHF hydroxyls in zeolites (located at about 3780 cm^{-1}) was revealed (486).

4.2.2 NO, NO_2, and N_2O_3

The nonreactive interaction of OH groups with NO was considered in Section 3.5. Here, we shall summarize the data on the interaction of NO with surface hydroxyls in the presence of O_2 (including also interaction with NO_2). In addition, redox reactions with NO will be discussed in Section 4.4.

When NO is coadsorbed with O_2 on HZSM-5, it interacts with the acidic hydroxyls to form NO^+ ions (N–O stretching modes at 2133 cm^{-1}) which are localized in the zeolite's exchange cation positions (140). The reaction mechanism is complicated and consists of several steps. It is known that co-condensation of NO and NO_2 results in the formation of asymmetric N_2O_3, which can be represented as a $[NO][NO_2]$ complex with a very weak N–N bond (487). In the first step, $[NO][NO_2]$ approaches the OH group, the proton is exchanged by a NO^+ ion, and nitrous acid (HONO) is formed. In the second step, HONO reacts with a proton to water and NO^+. The overall reaction can be written as follows:

$$2\,NO + \tfrac{1}{2}O_2 + 2\,O_{lattice} - H^+ \rightarrow H_2O + 2\,O_{lattice} - NO^+ \qquad (2.18)$$

The same mechanism was later on adopted for NO_x interaction with other zeolitic OH groups, including those in BEA *(488)*, FER *(489,490)*, MOR *(491,492)*, and SSZ-13 *(493)*. It was also reported that the N–O stretching frequency depends on the basicity of the oxygen to which the NO^+ cation is bound, and it was proposed that $\nu(NO)$ could be used to estimate surface basicity *(494)*.

Concerning the interaction of NO_2 with basic surface hydroxyls, no clear picture emerges from the literature. The disappearance of the OH groups in the presence of NO_2 has been reported many times *(495–498)*. By using a partly deuterium-exchanged TiO_2 surface, it was proven that water evolves during NO_2 adsorption *(495)*. Therefore, it was proposed that surface nitrate species replace the OH groups according to the following disproportionation reaction:

$$3\,NO_2 + 2\,M - OH \rightarrow 2\,M - NO_3 + NO + H_2O, \qquad (2.19)$$

which is known to proceed in solutions. A similar mechanism has been suggested for pure and sulfated zirconia *(497)* and for alumina *(496,499)*. In the latter case, a slightly modified reaction was formulated *(499)*:

$$2\,NO_2 + 2\,M - OH \rightarrow M - NO_3 + M - NO_2 + H_2O. \qquad (2.20)$$

4.2.3 SO₂

The SO_2 molecule is characterized by pronounced acidic properties and is thus expected to interact readily with basic hydroxyls. However, many of the IR spectroscopic investigations dealing with SO_2 adsorption or $SO_2 + O_2$ coadsorption on various surfaces do not discuss the changes in the OH stretching region. More than 15 years ago, Lavalley *(480)* remarked that SO_2 interaction with basic hydroxyls is not well understood, and this statement seems to be still valid. Based on the disappearance of the bands characterizing the surface hydroxyls on alumina in a stream of SO_2 and O_2, Liu et al. *(500)* suggested that hydroxyl groups play an important role in the formation of surface HSO_3^- species.

Similar conclusions, but concerning the formation of sulfate or bisulfate species, were drawn for Fe_2O_3 *(501)*. Various authors postulated that OH groups donate protons to sulfates and SOH groups are formed *(502,503)*. Grassian et al. *(504)* investigated the mechanism of SO_2 adsorption on titania with the help of isotopically labeled samples which contained either $Ti^{18}OH$

surface hydroxyls produced by treatment with $H_2{}^{18}O$, or TiOD groups. Two important conclusions were drawn from the interpretation of the IR spectra: (i) water was formed during the process and its formation did not involve oxygen from the surface OH groups, and (ii) labeled oxygen went into the sulfur-containing surface products. On the basis of these results, the authors proposed a reaction scheme for the interaction of SO_2 with TiO_2, with surface sulfites and water as the products.

4.3. Adsorption of alcohols

In their now classical work on the classification of different types of surface OH groups, Tsyganenko and Filimonov (28) noted that for a series of oxides, the number of bands of methoxy groups that appear after CH_3OH adsorption corresponds to the number of originally present bands of hydroxyl groups. The authors observed this correlation for MgO, CaO, TiO_2, ZrO_2, and CeO_2; whereby for CeO_2, the $\nu(CO)$ modes were sensitive to the geometry of the adsorption complex.

Following these findings, Lavalley and coworkers expanded the data set in a series of papers (505–510) and pursued the idea to use alcohols as probe molecules for identifying the number of cations that can be simultaneously bound to one adsorbed molecule. It was found that methoxy species are formed according to two principal routes. The first route is the dissociative adsorption of methanol on Lewis acid–base pairs:

$$
\begin{array}{ccc}
 & H_3CO & H \\
 & | & | \\
M^{n+} - O^{2-} + CH_3OH & \rightarrow M^{n+} & -O^{2-}
\end{array}
\qquad (2.21)
$$

and the second route is the interaction of the surface hydroxyl groups with the alcohol:

$$
M^{n+} - OH^- + CH_3OH \rightarrow M^{n+} - OCH_3 + H_2O. \qquad (2.22)
$$

By using $CH_3{}^{18}OH$, it was established that the O–H bond of methanol is broken in both reactions (505–507).

It is evident that the reaction in Equation (2.22) will be favored by basicity of the surface hydroxyls. Indeed, this reaction has been observed to proceed even at ambient temperature on oxides that are characterized by nonacidic hydroxyls, specifically MgO (28,507,511), CaO (28), Al_2O_3 (512,513), Fe_2O_3 (113,514), CeO_2 (28,509,515), TiO_2 (28,270,516), ZrO_2 (28,517,518), and CeO_2–ZrO_2 (508,510). Moreover, ZrO_2 is a typical example demonstrating that the more basic hydroxyls on a surface react more readily with methanol *(517,518)*. The hydroxyl band at 3763 cm^{-1} is preferentially consumed upon methanol dosage, followed by the band at 3665 cm^{-1}.

Equation (2.22) implies that water is formed during the reaction, which, when retained on the surface, complicates the spectra. Therefore, it is necessary to remove H_2O by evacuation. The evacuation temperature depends on the hydrophilicity of the surface and is different for different materials, for example, a temperature of 473 K is suitable for ZrO_2 *(517)*.

Equations (2.21) and (2.22) indicate that the number of surface methoxy species formed should correspond to the number of hydroxyl groups on a fully hydroxylated surface, that is, a surface where the acid–base pairs have been converted to two hydroxyl groups through water addition. It is practically impossible to distinguish the methoxy products formed according to the two reactions because they are virtually identical.

Typical IR spectra of methoxy groups are presented in Figure 2.42. The informative vibrations are the $\nu(CO)$ modes that appear in the 1150–1000 cm^{-1} region. Generally, the methoxy groups of type I are

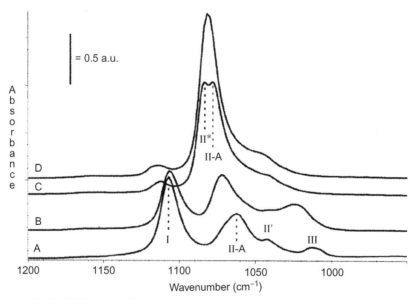

Figure 2.42 $\nu(OC)$ bands for methoxy groups on ceria reduced by H_2 at 573 K (a), 623 K (b), 673 K (c), and 773 K (d). Roman numerals refer to methoxy species on different types of hydroxyl groups. *Reproduced from Ref. (509).*

observed at wavenumbers above 1100 cm^{-1}, whereas those of types II and III are observed at lower frequencies.

Although silanols are slightly acidic, SiO_2 was also found to form alkoxy groups. Whether or not the reaction occurs depends on the interaction temperature and the nature of the alcohol. It was reported that evacuation after methanol adsorption did not result in the appearance of methoxy groups, whereas some ethoxy species were formed after ethanol adsorption and subsequent evacuation [519]. However, after interaction of silica with CH_3OH at elevated temperatures (>423 K), surface silanols are converted into methoxy species. It was suggested [519] that the reaction mechanism involves protonation of methanol by the silanol groups. This mechanism is in contradiction with results of experiments using silica and isotopically labeled compounds, which indicate that the methanol O–H bond is broken during methoxylation [520]. It was shown that silanols on the external surface of zeolites can also be methoxylated [521].

It is well established that at ambient temperature and low coverage, methanol is H-bonded to the acidic hydroxyls in zeolites (PA of $CH_3OH =$ 754.3 kJ mol^{-1} [148,229]). Thus, the practical use of this probe is restricted mainly to nonacidic oxides. Unfortunately, even in the absence of acidity, additional complications arise from the diversity of the adsorption forms. Coordinated methanol is often obtained as an additional species, and various methanol conversion products (formates, aldehydes) may be observed [56], especially on reactive surfaces such as those of Fe_2O_3 [514] or CeO_2 [515].

Notwithstanding these complications, some important insights have been gained by probing surface OH groups with methanol. For example, the conversion of type II methoxy species (i.e., methoxy formed on type II hydroxyls) on hydrogen-reduced ceria to type I species during oxidation of the sample with O_2 has been observed [509]. We suggest that CH_3OH could be a useful probe, especially for the basicity of the surface hydroxyls, provided the adsorption is performed by gradually increasing coverage through dosing small amounts.

4.4. Redox processes involving OH groups

4.4.1 Redox reactions with release of hydrogen (H₂)

It is known that strong acids act as oxidants. Consequently, acidic surface hydroxyls can also be oxidizers. This reaction potential of surface OH groups is well expressed when they interact with supported metals. For example, it was reported that adsorption of carbon monoxide on dispersed supported metallic rhodium leads to formation of geminal dicarbonyls of Rh^+ [163].

It was established that the oxidation of metallic rhodium proceeds in parallel with the decrease in intensity of the bands characterizing the hydroxyl groups of the support. On the basis of these observations, the following reaction has been proposed (Equation 2.23) *(522–526)*:

$$Rh^0 + OH^- + 2\,CO \rightarrow O^{2-} - Rh^+\,(CO)_2 + \tfrac{1}{2}\,H_2 \qquad (2.23)$$

The reaction is promoted by the high stability of the $Rh^+(CO)_2$ species and has been confirmed by DFT calculations *(527–529)*.

Similar observations were made for supported ruthenium catalysts, and again surface hydroxyls were proposed as oxidizing agents *(530,531)*. In principle, one would expect that surface hydroxyls can oxidize only metals with an electron affinity lower than that of hydrogen. However, oxidation of metallic copper by zeolite hydroxyls has also been reported *(532,533)*. This oxidation process can be easily explained by considering the effect of the present CO: as in the case of rhodium, the reaction is favored by the strong heat of adsorption of CO on isolated Cu^+ ions. Even in the absence of CO, redox chemistry has been observed. The oxidation of metallic copper in ZSM-5 was investigated in an argon stream, and it was established that the band characterizing bridging hydroxyls ($3611\ cm^{-1}$) started to decrease in intensity when the temperature reached 573 K *(532)*. While Cu^+ ions were again the main product, some Cu^{2+} species were also detected.

Recently, even oxidation of metallic gold by some hydroxyl groups of MgO was reported *(534)*. It was concluded that the hydroxyl groups absorbing at $3710\ cm^{-1}$ (corresponding OD groups at $2735\ cm^{-1}$) were involved in the reaction because of their consumption during gold deposition. The gold oxidation state was estimated by CO adsorption experiments.

4.4.2 Redox reactions involving the cation binding sites of surface OH groups

Another possibility of redox reactions involving surface hydroxyl groups is the change of the oxidation state of the cations to which they are bound.

In CoAPO-18 catalysts, a fraction of the framework aluminum is isomorphously substituted by tetrahedral Co^{3+} ions. These ions are not associated with charge defects. However, reduction of Co^{3+} to structural Co^{2+} ions creates the need for charge compensation, which is accomplished by adding a proton and formation of a bridging Co–(OH)–P hydroxyl *(535)*. This redox reaction can be monitored by IR spectroscopy; Figure 2.43

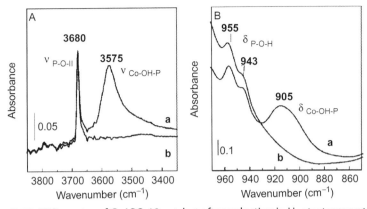

Figure 2.43 FTIR spectra of CoAPO-18 catalyst after reduction in H_2 at a temperature of 673 K (curve a) and after calcination in O_2 at a temperature of 823 K (curve b). *Reproduced from Ref. (535).*

shows that only reduced CoAPO-18 is characterized by a vibration at 3575 cm^{-1} assignable to Co–(OH)–P groups. The respective deformation modes are evidenced by a band at 905 cm^{-1}.

Fe^{3+}–OH groups in iron-exchanged zeolites are another example of OH groups associated with a cation that can serve as an oxidant *(217,536,537)*. It was reported that these groups are strong oxidizing agents. Even at ambient temperature, these hydroxyls oxidize NO, and at 673 K, they are reduced by CO *(217,536,537)*. It has been proposed that Fe^{3+}–OH species are intermediates in the selective catalytic reduction of N_2O by CH_4 *(536)*.

4.5. Ion exchange

The proton in an OH group can be exchanged with another cation. The other cation may be a metal cation, but can also be cation such as NH_4^+ or NO^+. It is obvious that the more acidic the OH group, the easier the exchange. Because the bridging hydroxyls in zeolite are among the most acidic surface OH groups, ion exchange is widely used for modifying the properties of zeolites by modifying their constitution.

4.5.1 Solid-state ion exchange

When the exchange reactions are performed in the solid state, they can be monitored by IR spectroscopy. Lamberti *et al. (538)* proposed a procedure of preparation of Cu ZSM-5 by fully exchanging the acidic protons by Cu^+ through reaction of HZSM-5 with CuCl vapor. Similar procedures have been applied to introduce not only copper cations *(538,539)* but also other

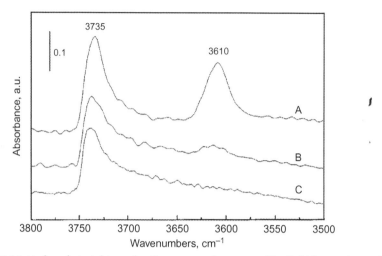

Figure 2.44 Hydroxyl stretching vibration range of spectra of In_2O_3/H-beta mixture I after activation in high vacuum at a temperature of 670 K for 1 h (A) and of mixtures I (B) and II (C) after subsequent reduction with H_2 at 670 K for 0.5 h. Mixtures I and II contain 0.346 and 0.691 mmol g^{-1} indium, respectively. The bridging hydroxyl groups at 3610 cm^{-1} react in a reducing atmosphere, whereas the silanol groups at 3735 cm^{-1} remain almost unaffected. *Reproduced from Ref. (542).*

cations, including cobalt *(540,541)*, nickel *(334,541)*, indium *(542)*, and lanthanum cations *(543,544)*. In some cases, reduction is necessary to assist the exchange procedure. For example, heating of In_2O_3/HBEA mixtures barely affects the zeolite hydroxyls (see Figure 2.44). In contrast, reduction with H_2 (aiming at the creation of In^+ ions) results in consumption of the acidic zeolite hydroxyls while the silanol groups remain largely unaffected.

4.5.2 Ion exchange in liquid phase

When the ion exchange is performed in solution (typically aqueous), the spectra of the starting and the final materials are compared, and conclusions are made on the basis of the consumption of different OH groups.

First, we consider the liquid-phase exchange of bridging hydroxyls in zeolites with monovalent cations. This reaction is simple, because one proton is exchanged with one metal cation. It has been found by many authors that the bands of the bridging hydroxyls of various zeolites are consumed or strongly reduced in intensity during ion exchange *(193,215,221,545,546)*. The reverse reaction is also used to prepare the H-forms of zeolites *(547,548)*.

When the introduced cation is divalent, the situation is more complicated. General stoichiometric considerations show that one cation should replace two zeolitic protons. This reaction requires two vicinal hydroxyls and is therefore not always possible. The availability of such hydroxyls depends on the aluminum content and distribution in the zeolite. Alternatively, a monovalent moiety may be exchanged, for example, $[M-OH]^-$. A similar scenario is possible for trivalent ions, whereby the charge-balancing moiety could also be a $[M=O]^+$ ion *(542)*. Consistent with the proposed structures, hydroxyls on exchanged cations are sometimes observed, but they are not common. Their absence is probably due to their low thermal stability. Interaction between two exchanged $[M-OH]^-$ cations may result in the creation of associated cations $[M-O-M]^{2+}$, which would in fact replace two protons.

Another peculiarity of exchanged divalent cations is their possibility to hydrolyze according to the Hirschler–Plank mechanism *(139)*:

$$O^{2-} - M^{2+} - O^{2-} + H_2O \rightarrow OH + M^{2+}(OH) - O^{2-} \qquad (2.24)$$

Indeed, after ion exchange, new OH groups were detected in M^{2+}BEA samples *(549)*, which were characterized by bands at 3692 cm^{-1} (magnesium), 3682 cm^{-1} (calcium), 3679 cm^{-1} (strontium), and 3672 cm^{-1} (barium). Two assignments of these new hydroxyls were proposed: $M^{2+}-OH$ species or original zeolite hydroxyls interacting with M^{2+}. The CaBEA sample was probed by pyridine, and it was found that the new OH species, absorbing at 3682 cm^{-1}, disappeared and then were restored after evacuation at 623 K. The results were taken as evidence of the creation, by exchange, of new protonic sites with weaker acidity than the bridging zeolite hydroxyls. We would like to remark that $Ca^{2+}-OH$ groups should not exhibit any acidity.

As expected for both mechanisms, it was found that the intensity of the bands of the bridging hydroxyls of zeolites decreases when divalent cations are introduced through ion exchange *(193,220,395,549–553)*. However, in a few investigations, concerning mainly BEA zeolite, the opposite trend was observed. For NiBEA *(219)* and ZnBEA *(404)*, the bands of the bridging hydroxyls are more intense in the cation-exchanged samples than in the H-forms. It was reported *(404)* that the concentration of isolated Brønsted acid sites (3612 cm^{-1}) in ZnBEA initially decreased slightly with increasing Zn^{2+} content (0–0.15 Zn/Al), whereas at higher loading, their concentration increased *(404)*. Similar results were obtained with a Ni/BEA sample, although the nickel was introduced by deposition–precipitation *(554)*. It has already been mentioned that BEA zeolite is easily dealuminated. The above observations can be rationalized assuming that some aluminum occupies cation positions (of the exchangeable type) and is preferentially exchanged with other metal cations and also with protons *(219,341,555)*. This scenario is consistent with the spectral behavior. The so-called VHF Al–OH band around 3780 cm^{-1} is often observed in spectra of H-forms of BEA zeolite. However, this band is reportedly absent or of strongly reduced intensity after ion exchange with cations of various elements: sodium *(545,546)*, cesium *(215,221)*, magnesium, calcium, strontium, barium *(540)*, zinc *(334)*, cobalt *(433,540)*, nickel *(219,334)*, iron *(556)*, and lanthanum *(543,557)*. These observations indicate that the VHF band could be associated with aluminum cations in exchange positions *(219)*.

Because of their weak acidity, terminal silanols are usually not involved in ion exchange. However, under particular conditions these hydroxyls can, to a limited extent, participate in the cation exchange *(540)*.

It was found that highly siliceous zeolites have a cation-exchange capacity that is higher than expected from the content of tetrahedral aluminum in the framework. This aluminum-independent exchange capacity was attributed to the presence of internal silanol groups *(558–560)*. By now, it is well established that silanol nests that are created as a result of the extraction of aluminum atoms from tetrahedral positions easily exchange with metal cations from solution. The metal cations then occupy the vacant tetrahedral positions in the zeolite. Using this procedure, a series of materials modified by titanium *(52)*, vanadium *(216,283,468,548,561)*, cobalt *(220)*, and nickel *(219)* have been synthesized. With the same method, aluminum can be reintroduced into the tetrahedral positions *(562)*. It has also been reported that the band of H-bonded hydroxyls in HBEA decreased after cation exchange with lanthanum *(543,544)* or cesium *(215,221)*.

Although mesoporous silica–alumina materials usually have no bridging hydroxyls, cations can be easily incorporated into these materials by conventional ion exchange procedures *(236,248)*. Moreover, it has been reported that these cations behave like the cations in zeolites *(236)*. A possible explanation for this behavior is the participation of pseudo-bridging hydroxyls in the exchange process.

It should be emphasized that hydroxyls that are not acidic do not participate in ion exchange, contrary to what is sometimes suggested. Indeed, it has been reported that the hydroxyl groups on titania are unaffected after deposition of copper, cobalt, or nickel ions by a procedure similar to that used for ion exchange *(563)*. To explain the observation, it was proposed that the deposited cations are located on coordinatively unsaturated oxygen ions.

5. LOCATION AND ACCESSIBILITY OF OH GROUPS IN POROUS MATERIALS

To be active in a catalytic reaction, a surface hydroxyl group must be accessible to the reactants, and the reaction products should be able to diffuse away into the reaction medium. The accessibility of the hydroxyls is related to their location: OH groups on the external surface of a solid are generally accessible, whereas the accessibility of hydroxyls in cages (or pores) depends on the dimension of both, the cage and the guest molecule. IR spectroscopy provides direct information on the accessibility, while information on the location of the hydroxyls in pores is mainly indirect.

It should be emphasized that the accessibility depends on the experimental conditions. It needs to be considered that the size of pores in crystals and hence the accessibility of hydroxyls located in these pores depend on the temperature. When the size of the guest molecule is similar to that of the pore, kinetic factors should be taken into account. Busca and coworkers *(106,564)* reported that pyridine, when applied for a short time of contact at room temperature and at 0.5 Torr (0.67 mbar) partial pressure, does not perturb the bridging hydroxyls of HFER. In contrast, it was established that about 60% of these groups are accessible to pyridine at a temperature of 400 K and about 80% at a temperature of 570 K *(372)*. It was also demonstrated that the value of 80% could not be overcome at a pyridine pressure of 10^{-2} mbar *(460)* (see Figure 2.37).

To assess the accessibility of families of hydroxyls to a given probe molecule, the perturbation of their OH stretching modes can be monitored and the integrated intensities of the bands can be used for quantitative analysis. Note that the probe may form H-bonds with the hydroxyls or may be protonated. By using several probes with different sizes and testing their access to a site, it is possible to determine the critical size of the opening that permits or prevents access. This information allows to (indirectly) draw conclusions on the location of the hydroxyls.

Different hydroxyls of the same family may have different accessibility, for various reasons. Some pores may be partly blocked by amorphous material. The real size of the pore is also affected by the presence of exchange cations. Dealumination or desilication of zeolites (which create mesopores) facilitate the access to OH groups *(159,185,345,565–567)*.

Corma *et al.* *(478)* investigated the adsorption of DTBP on a series of zeolites and found that different fractions of the Brønsted sites were affected (see Table 2.21). Later on, Thibault-Starzyk *et al.* *(565)* introduced the so-called accessibility index, ACI, for substituted alkylpyridines (pyridine, 2,6-lutidine, and 2,4,6-collidine). The ACI was defined as the fraction of acid sites that can interact with a given probe molecule, in reference to the total number of acid sites. It is possible to use the total number of OH groups or the total number of bridging hydroxyls as a reference. Variations of the definition were made for specific case studies. Sadowska *et al.* *(566)* determined the accessibility factor for ZSM-5 zeolite as a ratio of concentrations of acid sites, those accessible to pivalonitrile relative to those accessible to pyridine. To calculate the total ACI (i.e., accessible sites referenced to the total number of acid sites), it is necessary to know the molar absorption coefficients of the probe for interaction with both, Lewis and Brønsted acid sites. However, for OH groups, the ACI can be easily determined by analyzing the ν(OH) band intensity.

There are several serious problems regarding the determination of the accessibility of hydroxyl groups. First, a probe adsorbed at the pore mouth (of, e.g., a zeolite) can block further access to hydroxyls that are principally accessible. For example, Zecchina *et al.* *(568)* reported that tetrahydrofuran and diethyl ether do not enter the MOR channels because the strongly adsorbed species formed at the pore entrances block further molecules from entering. When strong bases are used as probes, it is possible that hydrogen atoms from inaccessible position are "extracted" into the accessible space. For example, CO does not interact with the OH groups located in the sodalite cages of FAU zeolites because it cannot physically come into contact

Table 2.21 Accessibility of protonic acid sites of different standard zeolites to 2,6-di-*tert*-butylpyridine (DTBP) *(479)*

Sample	Si/Al	Percentage interaction
MOR	10.0	36
BEA	12.5	100
Y (FAU)	2.5	57
ZSM-5 (MFI)	25.0	3
ZSM-11 (MEL)	34.0	43
M50 (MWW)	50.0	6
SSZ-24 (AFI)	50.0	30
SSZ-26 (CON)	20.0	33

with them when adsorbed at low temperatures. However, stronger bases such as pyridine or acetonitrile interact with these hydroxyls although they are larger probes than CO; presumably the bases strongly attract the protons, forcing them out of the sodalite cages *(172,173)*.

Another problem that is still under discussion concerns the size of the guest molecule. Researchers mostly use the kinetic diameters for guidance (see Table 2.22). However, it has been remarked *(574,580)* that the kinetic diameter is not the best criterion for the molecular dimension. The flexibility and the shape of the molecule in relation to the shape of the pore openings should be taken into account. Consideration of these properties explains some results that indicated penetration of relatively large molecules into smaller pores. Armaroli *et al. (331)* observed differing behavior of *m*-xylene and 2,6-lutidine (similar in size) in accessing the ZSM-5 pores and suggested that chemical effects also play a role for accessibility.

Various other quantities can be used instead of the kinetic diameter, for example the molecular dimensions (see Table 2.23). The use of the dimensions is particularly recommended when examining sorption of nonspherical molecules into materials with cylindrical pores. Sigma–Aldrich *(569)* has compiled information on the so-called critical diameter for several commonly used sorptives (Table 2.22). Molecules are not able to enter a pore with dimensions smaller than their critical diameter.

Busca *(581)* has recently published an excellent review on the accessibility and the location of hydroxyls in porous materials. He divided the molecular probes into several classes for this purpose. The molecules from each class are characterized by the same functional groups but by different alkyl substituents that can cause sterical hindrance. Thus, molecules from one class have similar chemical properties while having different dimensions.

A widely used class of compounds are pyridine and alkylpyridines, including 2,6-lutidine, 2,4,6-collidine, DTBP, quinoline, 4-methylquinoline, and 2,4-dimethyl-quinoline *(565,582)*. These molecules interact with acidic hydroxyls through their N-atoms with formation of protonated species. Because they are strong bases, the possibility to extract protons from their original positions cannot be ruled out. An advantage (but in some cases disadvantage) of many pyridines with bulky substituents is that, for steric reasons, they do not interact with Lewis acid sites through the N-atoms, or this interaction at least is hindered.

Amines with substituents of various sizes are not as widely investigated as pyridines and include methylamine, dimethylamine, TMA, and trimethylsilyl-diethylamine. Like pyridine, they are protonated by the acidic OH groups.

Table 2.22 Critical and kinetic diameters of some molecules used as IR probes to determine accessibility

Molecule	Kinetic diameter, nm	Ref.	Critical diameter, nm	Ref.
NH_3	0.26	*(334)*	0.36/0.165	*(569)/(130)*
H_2O	0.265	*(570,571)*	0.32/0.189	*(572)*
H_2	0.29	*(573)*	0.24	*(569)*
Ar	0.34	*(302)*	0.38	*(569)*
O_2	0.346	*(571)*	0.28	*(569)*
N_2	0.364	*(571)*	0.30	*(569)*
CO	0.376	*(144)*	0.28/0.328/ 0.073	*(569)/(572)/ (130)*
CH_4	0.38	*(570,571)*	0.40	*(569)*
CH_3CN	0.38–0.42	*(574)*		
C_2H_4	0.39/0.44	*(575)/(570)*	0.402	*(569)*
CO_2	0.40	*(573)*	0.28/0.318	*(569)/(572)*
Xe	0.40	*(302)*		
n-Hexane	0.43	*(576,577)*		
1-Butanol	0.43	*(577)*		
C_3H_6	0.45	*(575)*		
1-Butene	0.45	*(575)*	0.51	*(569)*
2-Propanol	0.47	*(577)*		
Acetone	0.47	*(576)*		
Propionitrile	0.49	*(578)*		
i-Hexane	0.50	*(571)*		
i-Butane	0.50	*(571)*		
i-Butene	0.50	*(575)*	0.56	*(569)*
Pyridine	0.533	*(77)*		
Isobutyronitrile	0.56	*(578)*		
Toluene	0.585	*(572)*	0.67	*(569)*
p-Xylene	0.585	*(572)*	0.67	*(569)*

Table 2.22 Critical and kinetic diameters of some molecules used as IR probes to determine accessibility—cont'd

Molecule	Kinetic diameter, nm	Ref.	Critical diameter, nm	Ref.
Benzene	0.585	*(572)*	0.68	*(569)*
Cyclohexane	0.60	*(577)*	0.61	*(569)*
2,2-Dimethylbutane	0.60–0.62	*(574)*		
Ethylbenzene (styrene)	0.60/0.62	*(572)/(570)*	6.7	*(572)*
Pivalonitrile	0.6, >0.62	*(98)/(578)*		
1,2,3-Trimethyl-benzene	0.66	*(572)*	0.764	*(572)*
Dimethylpyridine	0.67	*(479)*		
m-Xylene	0.68/0.74	*(572)/(577)*	0.71/0.68	*(569)/(572)*
o-Xylene	0.68	*(572)*	0.74	*(569)*
Quinoline[a]	0.71	*(579)*		
2,4,6-Trimethyl-pyridine (colidine)	0.74			
1,2,4-Trimethyl-benzene	0.76	*(572)*	0.725	*(572)*
2,6-Di-*tert*-butylpyridine	0.79	*(159,479)*		
Mesitylene	0.84	*(329)*		
1,3,5-Triiso-propylbenzene	0.85	*(580)*		
1,3,5-Trimethyl-benzene	0.86	*(572)*	0.818	*(572)*

[a]2-Azabicyclo[4.4.0]deca-1(6),2,4,7,9-pentaene.

Aliphatic and aromatic nitriles are another widely used class of compounds for testing accessibility. They are less strongly adsorbed on acid sites than pyridines or amines. This class of probes includes the very small acetonitrile and propionitrile, isobutyronitrile, pivalonitrile, benzonitrile, *ortho*-tolunitrile, and 2,2-diphenylpropionitrile. Interaction with both Lewis and Brønsted acid sites typically occurs and can be monitored by IR spectroscopy. The nitriles are relatively weakly bound to acidic hydroxyls by

Table 2.23 Molecular dimension (*a, b, c*) of some molecules

Molecule	*a*, nm	*b*, nm	*c*, nm	Ref.
CH$_3$CN	0.17	0.18	0.39	*(574)*
n–Hexane	0.19	0.25	0.82	*(574)*
n–Hexane	0.39	0.43	0.91	*(580)*
2,2-Dimethylbutane	0.335	0.45	0.575	*(574)*
Benzene	0.34	0.62	0.69	*(580)*
p-Xylene	0.37	0.62	0.69	*(580)*
Mesitylene	0.37	0.78	0.85	*(580)*
(CH$_3$)$_3$C-CN (pivalonitrile, PN)	0.375	0.432	0.51	*(574)*
o-Xylene	0.41	0.69	0.75	*(574)*
Pyridine	0.435	0.425	–	*(574)*
3-Methyl-pentane	0.46	0.58	0.86	*(580)*
C$_6$H$_6$	0.46	0.515	–	*(574)*
Cyclohexane	0.47	0.62	0.69	*(580)*
2,2-Dimethylbutane	0.59	0.62	0.67	*(580)*

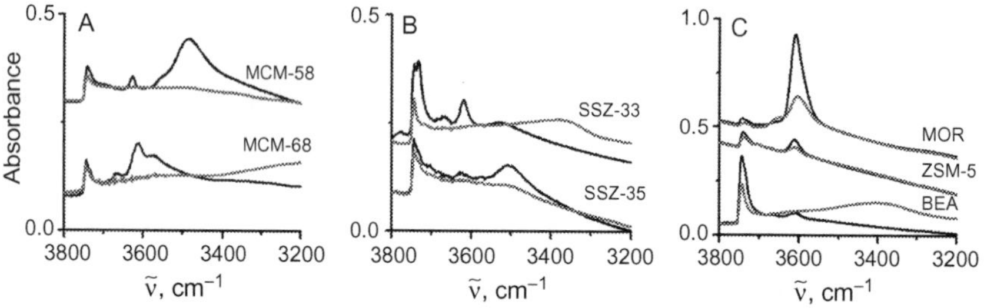

Figure 2.45 IR spectra in the OH region of various zeolites before (black lines) and after (red lines) pivalonitrile adsorption at room temperature and subsequent desorption by evacuation for 20 min at the same temperature. *Reproduced from Ref.* (583). (See the color plate.)

H-bonds, thus shifting the OH stretching modes. Among the nitriles, pivalonitrile is an often used probe for characterization of the external hydroxyls of zeolites with small and medium-sized pores. Figure 2.45 shows the accessibility of the bridging hydroxyls in various zeolites to pivalonitrile (note that the weak complexes with silanol groups have been decomposed by evacuation). It is evident that only a fraction of the bridging hydroxyls in HMOR and HZSM-5 are affected by this probe.

A variety of hydrocarbons have also been used, and a substantial amount of data exists. When aliphatic hydrocarbons are utilized, the interaction is relatively weak. Stronger interaction is achieved with alkenes and aromatics but it is still weaker as compared to nitriles.

An often used approach is the consecutive exposure of the surface to two probes. Initially, the sample is saturated with a bulky molecule and then the remaining, unperturbed OH groups are probed with a small molecule, typically by means of low-temperature CO adsorption *(185,194,567)*. With this type of experiment, precise information on the properties of the difficult-to-access hydroxyls can be obtained.

Small probe molecules are also used, especially when investigating materials with small pores. For example, according to Ref. *(334)*, the bridging OH groups in the side channels of mordenite (characterized by a band at 3585 cm^{-1}) do not interact with CO, whereas they protonate the smaller ammonia molecule.

As already discussed, information on the location of the OH groups can be obtained by the OH stretching frequency alone (see Figure 2.15). In general, because of H-bonding, the OH groups located in smaller cages vibrate with lower frequencies. Using this approach, Gil *et al.* *(98)* concluded that OH groups in MCM-58 zeolite characterized by a band at 3485 cm^{-1} are located in five-membered rings.

Analysis of the published results reveals the following general picture of the location of hydroxyls in zeolites. The bridging hydroxyls are located in the zeolite pores. On the contrary, the silanol groups are mainly located at the external surface (typical band at 3748 cm^{-1}). Hydroxyls associated with EFAL species (bands around 3780 *(159)* and 3650 cm^{-1} *(196)*) are also found to be located at the internal surfaces (in contrast to the EFAL species responsible for Lewis acidity).

In Figure 2.46, the spectra of acetonitrile and pivalonitrile adsorbed on HFER are compared. It can be seen that the smaller-sized acetonitrile interacts with both, bridging hydroxyls and surface silanols (spectrum d). However, pivalonitrile can only reach the external silanols and essentially does not affect the bridging hydroxyls located at the internal surface.

There are many experimental observations revealing that the accessibility of hydroxyls depends on a variety of factors:

A. The fraction of bridging OH groups in zeolites accessible to large molecules increases with the relative increase of the external surface area.

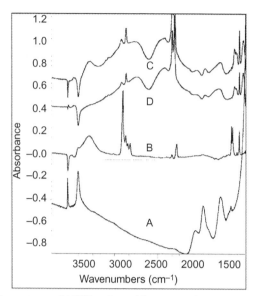

Figure 2.46 FTIR spectrum of HFER activated by outgassing at a temperature of 723 K (a), and difference FTIR spectra (with the spectrum of the activated sample subtracted) recorded after adsorption of pivalonitrile (b) or acetonitrile (c) on HFER at room temperature (3 min contact, 0.5 Torr) and after outgassing at room temperature (d). *Reproduced from Ref. (564).*

Thus, collidine does not affect the bridging hydroxyls in conventional HZSM-5 materials. However, a large fraction of these hydroxyls are accessible to collidine in a nanosheet ZSM-5 material *(194)*. It was concluded that in this material most of the acidic hydroxyls, while located inside the zeolite channels, are in close proximity to the channel mouths. Similar high accessibility was found for a two-dimensional zeolite *(188)*. It was reported that the ACI linearly correlated with the external surface area of a hierarchical zeolite *(584)*.

B. Creation of mesopores (which to some extent is equivalent to an increase of the external surface area) improves the accessibility of the bridging hydroxyls *(138,159,185,329,343,479,565,566)*. Partial destruction of micropores through dealumination also enhances accessibility *(567)*.

C. Some zeolites (of the MWW type: MCM-22, MCM-49) are characterized by the presence of acidic hydroxyls on the external surface *(109)*. According to Ref. *(581)*, these hydroxyls are located in the hemisupercages that are exposed at the (001) surface of these zeolites.

D. Some silanol groups (of zeolites and related materials) are located inside the pores. Their presence is assumed to be associated with defects. Because of the pore effect, the OH stretching frequency of these groups is slightly lower than that of external silanols; different authors report different values between 3740 *(486)* and 3680 *(581)* cm^{-1}. However, it should be taken into account that strong evidence exists for vibrations of external silanols on silicalite in the 3740–3730 cm^{-1} region *(585)*.

In the following paragraphs, we shall briefly summarize the IR spectroscopic results on the location and accessibility of different hydroxyls in zeolites that are important from a commercial point of view. The results apply for intact (defect-free) materials and are ordered by framework type. The dimensions of the pores can be found in Table 2.4.

FAU (HX, HY) and EMT. There are two main families of OH groups in FAU zeolites. The first kind of hydroxyls, also called HF (high-frequency) hydroxyls (with an IR band around 3650–3625 cm^{-1}), are located in the supercages (1.3 nm in diameter) and the access is through 12-membered rings (0.74 nm in diameter). Thus, these hydroxyls can interact with relatively large molecules. Corma *et al.* *(579)* reported that quinoline is too large to enter the supercages and does not penetrate deep into Y zeolite. It follows that this molecule may be used to probe the external surface of FAU zeolites.

The other kind of hydroxyls ("LF" for low frequency, detected around 3550 cm^{-1}) are located in the small sodalite cages and are not accessible even to CO (at least not at low temperatures). However, as already stated, these hydroxyls can interact with stronger bases, whereby the proton is displaced from its original position. Alternatively, it was proposed *(172)*, for the case of pivalonitrile adsorption, that the C–N moiety of nitriles penetrates the hexagonal six-ring window separating the supercage from the sodalite cage, with the rest of the molecule still being located in the supercage. Upon dealumination, the LF hydroxyls become more accessible.

The wavenumbers of these two types of OH groups are consistent with their location in large and small cages, respectively.

Romero Sarria *et al.* *(456,457)*, by investigating the adsorption of NH$_3$ and TMA, provided evidence of the existence of a third type of bridging OH groups in the hexagonal prisms of FAU zeolites.

BEA. This zeolite is characterized by large pores. There are two types of channels, "small" 12-membered ring channels in the [001] direction with a cross-section of 0.56 × 0.56 nm and "large" 12-membered ring channels in the [100] direction with a cross-section of 0.66 × 0.67 nm. In a series of investigations, Busca and coworkers *(586,587)* obtained data on the location of the OH groups by adsorption of pivalonitrile. Bridging OHs located in the smaller channels (inaccessible to pivalonitrile) were found to be homogeneous and characterized by a sharp band at 3609 cm^{-1}. In contrast, the OH groups of the larger channels (accessible to pivalonitrile) were heterogeneous (with three components at 3628, 3608, and 3590 cm^{-1}).

It was found that the principal part of the 3746 cm^{-1} silanol band was shifted by −204 cm^{-1} after adsorption of hexamethylbenzene *(588)*. Part of the band was shifted only by −42 cm^{-1}. The results were taken to indicate that a fraction of the silanol groups were barely accessible to hexamethylbenzene, leading to only a weak interaction with the methyl groups.

HBEA is a zeolite that is easily dealuminated and various aluminol groups are often observed in samples. The band at 3782 cm^{-1} disappears fully after adsorption of pyridine or pivalonitrile, which indicates that the respective species are not located in the small channels *(587)*. However, only half of the band is affected by adsorption of DTBP *(159)*, indicating some variation in the location of these species.

MFI (HZSM-5). This zeolite is one of the most extensively investigated, and there is a large amount of data on the accessibility of the hydroxyl groups. There is no definite proof of chemical heterogeneity of the bridging hydroxyls in HZSM-5. The MFI structure is characterized by two types of channels: (i) straight channels with a nearly circular opening (0.53 × 0.56 nm) and (ii) sinusoidal channels with an elliptical opening (0.51 × 0.55 nm). Therefore, one could expect differences in the accessibility of the hydroxyl groups.

It was reported that acidic hydroxyls in HZSM-5 do not interact with large molecules, including quinoline *(589)*, adamantane–carbonitrile *(590)*, *o*-xylene *(311)*, mesitylene *(51)*, 1,3,5-trimethylbenzene *(591)*, and pivalonitrile *(311,585,586)*.

A small fraction of the acidic hydroxyls interact with cyclohexane *(592)*, W(CO)$_6$ *(204)*, and only a minor fraction of them are affected by toluene *(331)*. Larger fractions are affected by benzene and *p*-xylene *(311)*. 4-Methylquinoline affects about 35% of the acidic hydroxyls *(591)*. 2,2-Dimethylbutane slowly accesses the pores and shifts the OH band from 3612 to 3485 cm^{-1} *(581)*. The residual OH groups are characterized by a band at 3619 cm^{-1}. Although a relatively large molecule (with a kinetic diameter of 0.6 nm), 2,2-dimethylbutane interacts with acidic hydroxyls, thereby shifting the 3605 cm^{-1} band to 3465 cm^{-1} *(333)*. Two types of interactions were observed with *i*-butane and explained with steric arguments *(311)*. Small molecules such as CO or CD$_3$CN have access to all of the OH groups *(585)*.

Interestingly, in the presence of benzene, pivalonitrile interacted with some acidic hydroxyls *(333)*. The results were tentatively explained by the formation of a complex between benzene and pivalonitrile with an effective diameter smaller than the diameter of pivalonitrile.

Data on the location of aluminol groups (3665 cm^{-1}) were reported in Ref. *(196)*. These hydroxyls were not affected by 2,4,6-collidine adsorption, and it was concluded that their location must be at the internal surface.

FER. Ferrierite has two kinds of channels, one with dimensions of 0.42 × 0.54 nm and the other with dimensions of 0.35 × 0.48 nm. The spectrum of HFER in the OH stretching region shows one main band with at least three components (the reported positions vary). Montanari *et al. (586)* investigated the adsorption of isobutyronitrile, which presumably enters only the large channels, and concluded that in these channels, hydroxyl groups vibrating at about 3590 cm^{-1} are located. A band around 3600 cm^{-1} that was unaffected by isobutyronitrile adsorption was associated with OH groups in the smaller channels. It was also found that 2-methyl-propionitrile only slightly affects the bands of the bridging hydroxyls. The maximum of the unperturbed band of the bridging hydroxyls is at 3598 cm^{-1} (hydroxyls located in the 8-membered ring channels), while the maximum of the band of the perturbed OH is at 3591 cm^{-1} (hydroxyls located in the 10-membered ring channel) *(586)*. Because of its large size, pivalonitrile does not affect the acidic hydroxyls in this zeolite *(564,586,593)*. It needs to be pointed out that this interpretation is not consistent with the rule that lower-frequency bands correspond to hydroxyls in smaller cavities.

The above results are confirmed by adsorption of different alkanes on HFER *(594,595)*. It was reported that *n*-hexane interacts with 20% of the OH groups responsible for a band at 3609 cm^{-1}, with 50% of the OH groups responsible for a band at 3601 cm^{-1}, and with 100% of the OH groups responsible for a band 3587 cm^{-1}. It was also found that only 23% of the bridging OH groups are accessible to *i*-butane *(594)*, whereas all of them are inaccessible to xylenes *(331)*.

Yoda *et al. (596)* adsorbed 2-methylpropane on HFER at a temperature of 203 K and observed that only 2% of the OH groups absorbing at 3610 cm^{-1} were affected, whereby their vibrations were shifted to 3531 cm^{-1} (a shift typical for hydroxyl interaction with saturated hydrocarbons). It was concluded that 2-methylpropane is unable to diffuse into the pores and interacts with the bridging hydroxyls close to the pore mouth through the methyl groups. The fraction of affected bridging OH groups was higher when 2-methylbutane or 2-methylpentane were adsorbed at 273 K, because the part of the molecule able to enter the pores, namely the linear alkyl chain, was longer. Furthermore, coadsorption of 2-methylpropane and propene showed that not all pores are blocked by 2-methylpropane. A model was proposed according to which only the pores having OH groups near the mouths are blocked.

MOR. Mordenite has two types of channels (Figure 2.47): 12-membered ring straight channels (0.70 × 0.65 nm) and 8-membered ring compressed channels (0.26 × 0.57 nm). The channels are interconnected by eight-membered side pockets (openings of 0.37 × 0.58 nm).

The vibrations of the bridging hydroxyls of this material are similar and form one band centered at about 3605 cm^{-1} that can be broken down into several components The hydroxyls in the large channels are characterized by a contribution at 3612 cm^{-1} to the band of the bridging hydroxyls. They are

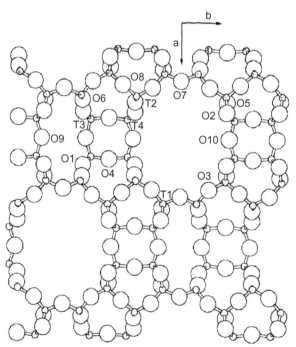

Figure 2.47 Projection of the mordenite framework along the [001] direction. The different positions of the T- and O-atoms are labeled. *Reproduced from Ref.* (597).

accessible to pyridine *(184,597,598)*. Only few of these hydroxyls interact with pivalonitrile.

Another component at about 3585 cm^{-1} is considered to correspond to bridging OH groups in the side pockets and smaller channels. The data on the accessibility are controversial. It was reported that these hydroxyls interact with ammonia *(334,597)* whereas they are not affected by the slightly larger molecules CO and N$_2$ *(334)*. According to several authors *(184,597)*, the hydroxyls absorbing at 3595 cm^{-1} are not accessible to pyridine, but others report that they are partly accessible *(574,598)*, although not all of them protonate pyridine *(574)*.

Busca and coworkers *(581,586)* have proposed three families of hydroxyls in mordenite:

- hydroxyls located inside the side pockets (3588 cm^{-1}, position O6 in Figure 2.47); these hydroxyls are accessible to linear nitriles but not to branched and aromatic nitriles;
- hydroxyls located at the intersection between side pockets and main channels (3609 cm^{-1}, position O5 in Figure 2.47). They can interact with linear and monobranched nitriles (e.g., isobutyronitrile) but not with the doubly branched pivalonitrile and with aromatic nitriles;
- hydroxyls pointing to the center of the main channels (3606 cm^{-1}, positions O2, O3, O7, and O10 in Figure 2.47). These OH groups are the most accessible and can interact with isobutyronitrile but not with 2,2-diphenyl-propionitrile.

No hydroxyls are located in the eight-membered ring channels.

Other zeolites. MCM-22 (MWW) is characterized by hemisupercages at the external surface, and the bridging hydroxyls (at approximately 3620 cm^{-1}) located in these cages are highly accessible *(581)*. Other hydroxyls are less accessible because they are located at the internal surface: inside nontruncated supercages (3620 cm^{-1}) or in the sinusoidal channels (≈ 3610 cm^{-1}). It was found that *p*-cumene can reach about half of the bridging hydroxyls *(386)*. Pivalonitrile only interacts with the bridging hydroxyls on the external surface and possibly those very close to the pore entrances *(109)*.

Theoretically, pivalonitrile should access all bridging hydroxyls in MCM-58 and MCM-68. This expectation was confirmed by experiments *(98,583)* (see Figure 2.45). Only a very small fraction of OH groups in MCM-58 remained unperturbed after pivalonitrile adsorption.

The bridging hydroxyls in SUZ-4 are more easily accessible than those in FER. They are partly accessible to *i*-butane and fully accessible to *n*-hexane *(595)*.

All OH groups (3611 and 3575 cm^{-1}) in ZSM-12 are accessible to 1,2,4-trimethylbenzene *(165)*.

Hydroxyls located in small pores present in bi-pillared montmorillonite are inaccessible to pivalonitrile *(599)*.

6. OH ISOTOPOLOGUES

6.1. OD groups

6.1.1 Chemical difference between ^{1}H hydrogen (protium) and ^{2}H hydrogen (deuterium)

Hydrogen has two stable isotopes, protium and deuterium. According to the IUPAC classification, deuterium is denoted as ^{2}H or D. The natural abundance of deuterium is very low, about 156 ppm, and restricts the large-scale application of deuterium-exchanged compounds. However, deuterium-exchanged compounds are very useful and often easily obtainable model molecules for various investigations.

It is generally assumed that the chemical properties of different isotopes of one element are identical. However, the considerable difference between the masses of protium and deuterium leads to measurable differences in the chemical properties of the elements themselves and of their compounds that are much more pronounced than for the isotopes of other elements *(600–602)*. The slight but significant toxicity of D_2O is also a result of this phenomenon *(603)*.

In a recent investigation *(604)*, the OH bond distance in H_2O was found to be by 3% larger than the OD bond distance in D_2O, whereas the distance of the H-bond formed to another H_2O molecule is by 4% shorter than for the deuterated molecules. This difference has important consequences and is higher than expected previously.

The common explanation of these effects is the difference in zero-point energy between the O–H and O–D bonds *(601,602,605)*. The zero-point energy corresponds to the vibrational energy of the bond in the ground state and is equal to the system energy at absolute zero temperature. In fact, most of the bonds have this energy also at room temperature. Because of the substantial difference in the masses of protium and deuterium, the OH and OD stretching frequencies differ significantly. An important consequence of the difference in zero-point energies is that the dissociation energies of the OH and OD bonds are different: they are equal to the difference between the energy E_1 and the zero-point energy (Figure 2.48), that is, O–D bonds are stronger than the respective O–H bonds. Also, the O–D bond is more

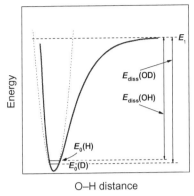

Figure 2.48 Schematic representation the Morse potential curve of OH and OD bonds.

harmonic than the O–H bond because the zero-point level for OD lies closer to the bottom of the Morse potential curve, where the deviations from the harmonic oscillator model are smaller.

The energy states are the reason for the different reactivity of H- and D-containing compounds. Many investigations have been devoted to kinetic isotopic effects, which we will not discuss in detail because several reviews exist *(606–608)*. We note that one should take into account the effect of the difference in the masses of H and D on the velocity of passage over the potential energy barrier, and the possibility for nonclassical penetration of the energy barrier (tunneling).

6.1.2 Acidity of OH and OD groups

As one would expect, the substitution of hydrogen by deuterium affects the acidity of chemical compounds. At first, there were different opinions on the direction of the effect. According to Duncan and Cook *(600)*, D_2O is a stronger acid than H_2O. However, most authors are now of the opposite opinion. It was reported *(609)* that the difference in the acid dissociation constants of H_2O and D_2O, ΔpK_a, is 0.958, where

$$\Delta pK_a = -\log[(D_3O^+)(OD^-)] - (-\log[(H_3O^+)(OH^-)]) \quad (2.25)$$

Slightly lower acidities of deuterated liquid acids as compared to the respective H-forms have also been documented *(609,610)*. However, the situation in solutions is complicated because the solvents (i.e., H_2O for H-acids and D_2O for D-acids) exhibit different basic properties, and it is difficult to distinguish between the different contributions. Moreover, theoretical considerations are usually limited to the formation of H_3O^+, whereas in reality the proton is bound to several molecules of water *(601,611)*. It was also reported that, in solutions, the isotopic effect on the hydroxyl acidity decreases when the acidity increases *(601)*.

Very recently, experimental proof was presented indicating that surface OD groups are less acidic than the respective OH groups *(376)*. The conclusion was made by analyzing the C–N stretching modes of CD_3CN H-bonded to bridging OH and OD groups in ZSM-5 zeolite. A difference of $\tilde{\nu}(CN)$ larger than 1 cm^{-1} was reported which corresponds to about 0.2 units of the Hammett acidity function. It was also found that the isotopic effect (in this case the difference in C–N stretching frequency of CD_3CN interacting with OH or OD groups, respectively) increased with the hydroxyl acidity. This behavior is the opposite of the behavior observed in solutions *(601)*. In the latter case, the presence of the solvent complicates the picture and the measured difference is a sum of various effects. The observation is important because the acidity of surface OH groups determines, to a high extent, the surface and catalytic properties of the solids *(59)*.

6.1.3 OH/OD isotopic shift factor

In the ideal case, the bands of OH groups should be shifted, after exchange with deuterium, by the isotopic shift factor, *i*, which is easily calculated on the basis of the following formula *(612)*:

$$i = \frac{\nu_{OH}}{\nu_{OD}} = \sqrt{\frac{\mu_{OD}}{\mu_{OH}}}, \tag{2.26}$$

where μ is the reduced mass.

The force constant is assumed to be the same for OH and OD bonds. According to Equation (2.26), the OH/OD isotopic shift factor should be 1.3736. Sometimes, the reciprocal value of *i* is used, which for the harmonic OH oscillator is equal to 0.728.

However, analysis of published data shows that the measured isotopic shift factors for surface hydroxyls substantially deviate from this value. Reported OH and OD frequencies for various materials are summarized in Table 2.24. In most cases, the isotopic shift factor is around 1.356–1.358 and thus smaller than anticipated. For isolated hydroxyls, the factor is almost independent of $\nu(OH)$; occasionally, a slight decrease with decreasing $\nu(OH)$ is seen. A few significant deviations from the anticipated value stand out. For convenience, the values of *i* lower than 1.353 are highlighted by italics in Table 2.24. Most of these low values are associated

Table 2.24 Reported stretching frequencies of isolated surface OH and respective OD groups and calculated isotopic shift factors i

OH group	Sample	$\tilde{\nu}(OH)$, cm^{-1}	$\tilde{\nu}(OD)$, cm^{-1}	i	i^{-1}	Ref.
SiOH	SiO$_2$	3751	2762	1.358	0.736	(613)
SiOH	SiO$_2$	3746	2762	1.356	0.737	(21)
		3748	2763	1.356	0.737	
SiOH	SiO$_2$	3738	2756	1.356	0.737	(614)
		3670	2710	1.354	0.738	
		3530	2600	1.358	0.736	
		3520	2610	*1.349*	0.742	
SiOH	SiO$_2$–Al$_2$O$_3$	3744	2764	1.355	0.738	(615)
SiOH	SiO$_2$–Al$_2$O$_3$	3745	2757	1.358	0.736	(616)
SiOH	MCM-41	3745	2760	1.357	0.737	(430)
SiOH	HY	3748	2759	1.358	0.736	(302)
SiOH	HY	3740	2760	1.355	0.738	(617)
SiOH	HNaY	3740	2755	1.358	0.737	(618)
SiOH	HY, NaHX	3745	2760	1.357	0.737	(619)
		3738	2756	1.356	0.737	
SiOH	HZSM-5	3745	2746	1.360	0.735	(197)
SiOH	HZSM-5	3747	2761	1.357	0.737	(152,153)
SiOH	HZSM-5	3740	2760	1.355	0.738	(620)
SiOH	HZSM-5	3740	2757	1.357	0.737	(421)
SiOH	Zn/HZSM-5	3745	2758	1.358	0.736	(621)
SiOH	[Si]BEA	3738	2755	1.357	0.737	(21)
SiOH	HBEA	3737	2760	1.355	0.739	(486)
SiOH	[Si]BEA	3525	2609	*1.351*	0.740	(21)
SiOH	HMOR	3752	2766	1.356	0.737	(293)
SiOH	HFER	3746	2761	1.357	0.737	(594)
SiOH	HFER	3744	2758	1.358	0.737	(622)

Continued

Table 2.24 Reported stretching frequencies of isolated surface OH and respective OD groups and calculated isotopic shift factors *i*—cont'd

OH group	Sample	$\tilde{\nu}(OH)$, cm^{-1}	$\tilde{\nu}(OD)$, cm^{-1}	*i*	*i*$^{-1}$	Ref.
SiOH	SUZ-4	3746	2760	1.357	0.737	*(594)*
SiOH	HERI	3745	2760	1.357	0.737	*(623)*
SiOH	MCM-58, MCM-68	3743	2758	1.357	0.737	*(98)*
Si(OH)Al	HY	3644	2691	1.354	0.738	*(302)*
		3555	2624	1.355		
Si(OH)Al	HY	3637	2682	1.356	0.737	*(624)*
		3634	2681	1.355	0.738	
		3629	2677	1.356	0.738	
		3565	2632	1.355	0.738	
		3550	2622	1.354	0.739	
		3547	2619	1.354	0.738	
Si(OH)Al	HY	3641	2687	1.355	0.738	*(169)*
		3546	2620	1.353	0.739	
Si(OH)Al	HY, NaHX	3634	2681	1.355	0.738	*(619)*
Si(OH)Al	HY, NaHX	3631	2677	1.356	0.737	*(619)*
		3658	2698	1.358	0.737	
		3648	2687	1.357	0.737	
		3639	2682	1.357	0.737	
Si(OH)Al	HNaY	3645	2690	1.355	0.738	*(618)*
		3550	2615	1.358	0.737	
Si(OH)Al	HNaY	3645	2688	1.356	0.737	*(624)*
		3558	2627	1.354	0.738	
		3551	2622	1.354	0.738	
		3546	2619	1.354	0.739	
Si(OH)Al	HNaCsY	3659	2698	1.356	0.737	*(624)*
		3536	2610	1.355	0.738	

Table 2.24 Reported stretching frequencies of isolated surface OH and respective OD groups and calculated isotopic shift factors i—cont'd

OH group	Sample	$\tilde{\nu}(OH)$, cm^{-1}	$\tilde{\nu}(OD)$, cm^{-1}	i	i^{-1}	Ref.
Si(OH)Al	Pd/HY	3645	2688	1.356	0.737	(625)
		3558	2623	1.356		
Si(OH)Al	USY	3625	2670	1.358	0.737	(626)
		3600	2655	1.356	0.738	
		3560	2630	1.354	0.739	
Si(OH)Al	USY	3628	2676	1.356	0.736	(627)
		3548	2621	1.354	0.739	
Si(OH)Al	DAY	3625	2670	1.358	0.737	(628)
		3540	2610	1.356	0.737	
Si(OH)Al	AgA	3619	2668	1.356	0.737	(629)
		3613	2664	1.356	0.737	
		3602	2665	*1.352*	0.740	
		3591	2649	1.356	0.738	
		3576	2632	1.359	0.736	
Si(OH)Al	HZSM-5	3627	2676	1.355	0.738	(630)
		3619	2670	1.355	0.738	
		3606	2660	1.356	0.738	
Si(OH)Al	HZSM-5	3620	2675	1.353	0.739	(631)
Si(OH)Al	HZSM-5	3616	2667	1.356	0.738	(152,153)
Si(OH)Al	HZSM-5	3617	2669	1.355	0.738	(98)
Si(OH)Al	HZSM-5	3611	2665	1.355	0.738	(624)
Si(OH)Al	HZSM-5	3610	2662	1.356	0.737	(197)
Si(OH)Al	HZSM-5	3610	2660	1.357	0.737	(421)
Si(OH)Al	HZSM-5	3605	2670	*1.350*	0.741	(620)
Si(OH)Al	ZSM-5, MOR	3610	2665	1.355	0.738	(592)
		3250	2450	*1.327*	0.754	

Continued

Table 2.24 Reported stretching frequencies of isolated surface OH and respective OD groups and calculated isotopic shift factors i—cont'd

OH group	Sample	$\tilde{\nu}(OH),$ cm^{-1}	$\tilde{\nu}(OD),$ cm^{-1}	i	i^{-1}	Ref.
Si(OH)Al	BEA	3610	2665	1.355	0.738	(624)
Si(OH)Al	BEA	3606	2660	1.356	0.738	(486)
Si(OH)Al	HMOR	3616	2668	1.355	0.738	(293)
		3586	2646	1.355	0.738	
Si(OH)Al	HMOR	3612	2664	1.356	0.738	(344)
		3584	2646	1.355	0.738	
Si(OH)Al	HMOR	3612	2664	1.356	0.737	(624)
Si(OH)Al	HFER	3609	2662	1.356	0.738	(154)
		3596	2653	1.355	0.738	
		3555	2623	1.355	0.738	
Si(OH)Al	HFER	3609	2663	1.355	0.738	(632)
Si(OH)Al	HFER	3602	2658	1.355	0.738	(594)
Si(OH)Al	HFER	3586	2644	1.356	0.737	(622)
Si(OH)Al	MCM-58	3628	2679	1.354	0.738	(98)
		3556	2628	1.353	0.739	
		3485	2576	1.353	0.739	
Si(OH)Al	MCM-68	3617	2669	1.355	0.738	(98)
Si(OH)Al	ERI	3612	2670	1.353	0.739	(623)
		3565	2640	*1.350*	0.741	
Si(OH)Al	SAPO-11	3630	2680	1.354	0.738	(624)
Si(OH)Al	SAPO-17	3626	2673	1.357	0.737	(624)
		3616	2667	1.356		
		3593	2652	1.355		
Si(OH)Al	SAPO-34	3626	2674	1.356	0.737	(624)
		3600	2657	1.355	0.738	
Si(OH)Al	ZK-5 (KFI)	3610	2664	1.355	0.738	(624)
		3563	2630	1.355	0.738	

Table 2.24 Reported stretching frequencies of isolated surface OH and respective OD groups and calculated isotopic shift factors i—cont'd

OH group	Sample	$\tilde{\nu}(OH),$ cm^{-1}	$\tilde{\nu}(OD),$ cm^{-1}	i	i^{-1}	Ref.
Si(OH)Al	SUZ-4 (SZR)	3603	2659	1.355	0.738	*(594)*
AlOH	Al$_2$O$_3$	3785	2790	1.357	0.737	*(10)*
		3770	2775	1.359	0.736	
		3730	2750	1.356	0.737	
		3675	2710	1.356	0.737	
		3590	2650	1.355	0.738	
AlOH	HZSM-5	3680	2710	1.358	0.736	*(631)*
AlOH	HZSM-5	3667	2703	1.357	0.737	*(152,153)*
AlOH	HZSM-5	3653	2693	1.356	0.737	*(421)*
AlOH	MCM-68	3670	2704	1.357	0.737	*(98)*
MgOH	MgO	3740	2765	1.353	0.739	*(10)*
		3615	2675	*1.351*	0.740	
		3550	2650	*1.340*	0.747	
		3400	2600	*1.308*	0.765	
FeOH	Fe-ZSM-5	3674	2712	1.355	0.738	*(20)*
		3635	2681	1.356	0.738	
ZnOH	ZnO	3672	2707	1.356	0.737	*(111,633)*
		3656	2698	1.355	0.738	
		3639	2683	1.356	0.737	
		3620	2669	1.356	0.737	
		3564	2626	1.357	0.737	
		3448	2551	*1.352*	0.740	
ZnOH	ZnO	3502	2591	*1.352*	0.740	*(634)*
ZnOH	Zn/H-MFI	3675	2705	1.359	0.736	*(621)*
TiOH	TiO$_2$	3720	2748	1.354	0.739	*(495)*
		3650	2703	*1.350*	0.740	

Continued

Table 2.24 Reported stretching frequencies of isolated surface OH and respective OD groups and calculated isotopic shift factors *i*—cont'd

OH group	Sample	$\tilde{\nu}(OH),$ cm^{-1}	$\tilde{\nu}(OD),$ cm^{-1}	i	i^{-1}	Ref.
TiOH	TiO$_2$	3733	2751	1.357	0.737	(504)
		3717	2741	1.356	0.737	
		3691	2719	1.358	0.737	
		3672	2708	1.356	0.737	
		3642	2676	1.361	0.735	
ZrOH	ZrO$_2^a$	3775	2782	1.357	0.737	(635)
		3738	2752	1.358	0.737	
		3671	2706	1.357	0.737	
		3400	2520	*1.350*	0.741	
ZrOH	ZrO$_2$	3774	2780	1.358	0.737	(636)
		3733	2751	1.357	0.737	
		3673	2709	1.356	0.737	
ZrOH	MOF: UiO-66	3666	2708	1.354	0.739	(637)
LaOH	La$_2$O$_3$	3662	2700	1.356	0.737	(638)
CeOH	CeO$_2$	3657	2694	1.357	0.737	(639)
		3610	2598	*1.351*	0.720	
CeOH	CeO$_2$ (reduced)	3675	2716	1.353	0.739	(640)
		3650	2680	1.362	0.734	
CeOH	Cu/CeO$_2$	3663	2702	1.356	0.738	(641)
CeOH	Pt–CeO$_2$	3650	2685	1.359	0.736	(640)
POH	SAPO-5, SAPO-11	3676	2710	1.356	0.737	(348)
COH	Citosan^b	3435	2529	1.358	0.736	(642)
		3290	2432	1.353	0.739	
		3180	2372	*1.341*	0.746	
COH	HCO$_3$/Fe$_2$O$_3$	3619	2671	1.355	0.738	(484)

^{a}Average values.
^{b}Poly-β-(1,4)-glucosamine.
Values of *i* lower than 1.353 are highlighted by italics.

with H-bonded hydroxyls. Thus, analysis of the literature data strongly suggests that the H/D isotopic shift factor decreases upon formation of an H-bond, reaching values as low as 1.308 *(10)*.

These conclusions are further supported by data on the isotopic shift factor of hydroxyl groups forming H-bonds with adsorbate molecules (Table 2.25). These data are compiled in Figure 2.49; the decrease of the isotopic shift factor upon H-bond formation is plotted vs. $\Delta\tilde{\nu}/\tilde{\nu}_0$ for the OH groups. Although the data were collected in different laboratories with different equipment and a relatively large uncertainty is expected with such small shifts, a clear trend emerges. One can definitely conclude that the isotopic shift decreases with increasing strength of the OH–adsorbate interaction.

Similar variations of the isotopic shift factor have been reported for bulk hydroxyls *(161,643–644)*. Various models were proposed and the anomalous isotopic shift factor was associated with the anharmonicity of the OH bond *(161,644)*. In a recent review *(643)*, three main causes of the anomalous OH/OD isotopic effects were identified: (i) higher anharmonicity of the OH vibrations as compared to OD; (ii) contribution of the bending modes via coupling, and (iii) tunneling effects (taking place with strongly hydrogen-bonded systems). It was also pointed out that weak H-bonds complicate the situation because under these circumstances, the anharmonicity decreases.

The OH and OD stretching modes of free OH radicals are at 3568.0 and 2632.1 cm^{-1}, respectively *(645)*, which corresponds to an isotopic shift factor of 1.3556. This value is again lower than the theoretical one and is similar to the values reported in Table 2.24 for isolated surface hydroxyls. These measured frequencies of the stretching modes are known to be affected by anharmonicity. The harmonic OH and OD frequencies (calculated on the basis of overtone modes) are at 3735.2 and 2720.9 cm^{-1}, respectively. Thus, the isotopic shift factor for the harmonic frequencies is 1.3728, which is very close to the theoretical value of 1.3736. The same value for the hydroxyl anion, OH^-, is 1.3726. Therefore, one can conclude that the deviations of the isotopic shift of free OH radicals and ions are mainly the result of anharmonicity. Similar calculations for isolated surface OH groups *(21)* show that also in this case, the deviations are mainly caused by anharmonicity (isotopic shift factor of the harmonic frequencies of 1.372–1.373).

The significant deviations of *i* from the theoretical value for H-bonded hydroxyl groups deserve further exploration. In most cases concerning

Table 2.25 OH/OD isotopic shift factors i for surface hydroxyl groups before and after formation of H-bonds with weak bases

Sample	OH group/complex	$\tilde{\nu}(OH)$ cm^{-1}	$\tilde{\nu}(OD)$ cm^{-1}	i	Ref.
SiO$_2$	SiOH	3748	2763	1.3565	*(21)*
	SiOH$\cdots$CH$_4$	3714	2739	1.3560	
	SiOH$\cdots$CO	3660	2700	1.3556	
	SiOH$\cdots$CD$_3$CN	3429	2552	1.3437	
SiO$_2$	SiOH	3751	2768	1.3551	*(234)*
	SiOH$\cdots$TMP[a]	3506	2608	1.3443	
SiO$_2$	SiOH	3750	2764	1.3567	*(632)*
	SiOH$\cdots$*cis*-2-butene	3525	2615	1.3480	
HZSM-5	Si(OH)Al	3616	2667	1.3558	*(152)*
	[Si(OH)Al]$\cdots$N$_2$	3496	2584	1.3529	
	[Si(OH)Al]$\cdots$CO	3306	2460	1.3439	
2D zeolite	Si(OH)Al	3594	2654	1.3542	*(188)*
	[Si(OH)Al]$\cdots$CO	3215	2409	1.3345	
	[Si(OH)Al]$\cdots$C$_2$H$_4$	3107	2330	1.3335	
HY	[Si(OH)Al]	3641	2687	1.3550	*(169)*
	[Si(OH)Al]$\cdots$CO	3354	2495	1.3443	
HY	Si(OH)Al	3660	2664	1.3739	*(317)*
	[Si(OH)Al]$\cdots$C$_2$H$_4$	3285	2506	1.3664	
HFER	[Si(OH)Al]	3608	2662	1.3554	*(234)*
	[Si(OH)Al]$\cdots$CO	3317	2468	1.3440	
HFER	[Si(OH)Al]	3612	2663	1.3564	*(632)*
	[Si(OH)Al]$\cdots$C$_2$H$_6$	3513	2591	1.3558	
HFER	[Si(OH)Al]	3609	2663	1.3552	*(632)*
	[Si(OH)Al]$\cdots$CO	3317	2468	1.3440	
HFER	[Si(OH)Al]	3606	2661	1.3551	*(632)*
	[Si(OH)Al]$\cdots$C$_2$H$_4$	3213	2403	1.3370	

[a]2,4,4–Trimethyl-2-pentene.

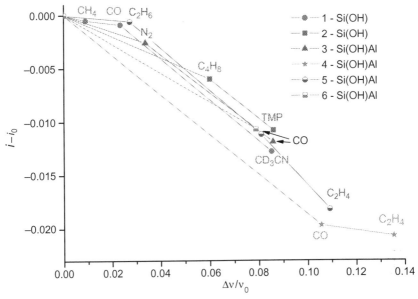

Figure 2.49 Correlation between the shift of the O–H modes induced by adsorption of weakly basic molecules ($\Delta\nu/\nu_0$) and the decrease of the OH/OD isotopic shift factor (i_0 stands for the isotopic shift factor, ν_{OH}/ν_{OD}, in absence of adsorbate). Data for individual samples are connected by straight lines. (1) Interaction of SiOH groups on silica with CH_4, CO, and CD_3CN *(21)*; (2) interaction of SiOH groups on silica with 2,4,4-trimethyl-2-pentene *(234)* and *cis*-2-butene *(632)*; (3) interaction of bridging hydroxyls in HZSM-5 with N_2 and CO *(152)*; (4) interaction of bridging hydroxyls in a two-dimensional zeolite with CO and C_2H_4 *(188)*; (5) interaction of bridging hydroxyls in HFER with C_2H_6, CO, and C_2H_4 *(632)*; and (6) interaction of bridging hydroxyls in HY with CO *(169)*. (See the color plate.)

surface hydroxyls, the reasons for the deviations of i from the theoretical value have not been discussed. Recently, an empirical equation was proposed for isolated hydroxyls *(98)*:

$$\tilde{\nu}_{OD} = 0.709(\tilde{\nu}_{OH}) + 103.7 \qquad (2.27)$$

Equation (2.27) is not valid for H-bonded hydroxyls *(98)*. Boscoboinik *et al.* *(188)* assumed that the difference between the isotopic shift factors of free and H-bonded hydroxyls arises from the coupling of the OH modes with intramolecular vibrations. However, coupling is generally assumed to affect the width and not the position of the OH band and therefore should not affect the isotopic shift factor.

Zhdanov *et al. (646)* proposed that the additional decrease of the isotopic shift factor for H-bonded hydroxyls is related to an increase of the anharmonicity of the O–H bond upon H-bond formation. Unfortunately, it is usually difficult to experimentally determine the harmonic frequencies of these species because the intensity of the overtones decreases dramatically with the formation of the H-bond *(21,41,42)*. Only when the H-bond is weak, the harmonic frequencies may be obtained. As already noted, the harmonicity of the O–H bond increases with the formation of weak H-bonds and decreases with the formation of strong H-bonds *(43,44,647)*. Therefore, formation of weak H-bonds should lead to an increase of the isotopic shift factor, contrary to the experimental observations. This discrepancy implies that other factors must have a stronger influence.

It is instructive to consider what would happen with the ν(OH) and ν(OD) bands upon formation of H/D-bonds if the oscillators were harmonic and the acidity of the OH and OD groups was identical. Both the OH and OD bands would be shifted toward lower frequencies and the shift values, corrected with the isotopic shift factor, should be identical. However, because the OD groups possess lower intrinsic acidity, the OD band should be shifted to a lesser extent than the OH band. Consequently, the isotopic shift factor of H-bonded hydroxyls should be lower than the isotopic shift factor of isolated (unperturbed) groups. This effect can be seen in Figure 2.5, where the changes of the spectra of SiO_2 (spectra c) and deuteroxylated SiO_2 (spectra f) caused by adsorption of CD_3CN are compared in the OH and OD regions. The x-axis scale of the panel presenting the OD modes (panel B) is expanded by the experimental isotopic shift factor for the isolated OH/OD groups. Thus, if the OH and OD groups exhibited the same acidity, the shifted OD band should have been observed at 2528 cm^{-1}, instead of 2552 cm^{-1}. On the basis of these results, it was proposed that the experimental OH/OD isotopic shift factor can be used to estimate the extent of H-bonding of a given OH group *(21)*.

Finally, there are isotopic effects on the band intensities. The linear intensities (absorbance values) of the OH and OD groups are almost the same. However, the integral intensity of ν(OD) bands is smaller than that of ν(OH) bands *(648)*. This difference arises from the fact that the OD bands are more "compressed" with respect to the x-axis scale. Because the isotopic shift is not a constant value, the ratio between the absorption coefficients of different OH/OD pairs of bands should also slightly differ. The absorption coefficient of ν(OD) for surface OD groups on silica was determined to be 0.63–0.73 times the value of the ν(OH) absorption coefficient *(617)*.

6.1.4 Deuterium exchange

The most convenient way to exchange OH with OD groups is to use D_2O. The reaction proceeds even at ambient temperature, but subsequent evacuation is necessary in order to remove molecularly adsorbed water. In principle, more acidic OH groups are more readily exchanged with D_2O less acidic ones. D_2 has also been applied for OH/OD exchange but in these cases the D_2 molecule needs to be activated and the reaction temperature is generally higher than when D_2O is used *(344,348,594,622,624–626,629,631–633)*. H/D exchange also occurs after interaction of OH groups with various deuterated compounds *(616,627)*. The rate of H/D exchange is one of the possible measures of surface acidity because more acidic hydroxyls are exchanged more rapidly *(57)*. H/D exchange is also used to draw conclusions on the origin of hydrogen in surface species or reaction products, and on the reaction pathway in general *(57)*.

There are several reasons to perform H/D exchange for the spectral characterization of surface hydroxyl groups. H/D exchange is of unique specificity to check whether a vibration involves a hydrogen atom or not. Indeed, although the 3800–3200 cm^{-1} region is mainly occupied by O–H stretching vibrations, other modes can also appear there. These modes are mainly overtones and combinations of various vibrations, or N–H stretching vibrations. For more details on the bands in the OH region that do not involve hydrogen, see Section 7.

It was proven, by using exchange with deuterium, that bands around 1000 cm^{-1} characterize the δ(OH) modes of hydrogen-bonded hydroxyls on oxide surfaces *(10)*. The use of a deuterated zeolite sample allowed the detection of the in-plane O–D bending modes around 880 cm^{-1} *(624)*, whereas the respective O–H modes are generally masked by the zeolite framework vibrations.

H/D exchange is a common approach to obtain information on the accessibility of hydroxyl groups, that is, to elucidate whether they are on the surface or not. In this context, one should pay attention to the experimental conditions because with a rise in temperature, occluded hydroxyls may become accessible.

Another reason to perform an H/D exchange is to obtain high-quality spectra *(615,622)*. Because of scattering of the IR radiation (depending on the nature of the sample), the ν(OH) region can be too noisy and the hydroxyl spectra are of low quality; sometimes the ν(OH) bands are not even observable. In addition, the shape of the spectrum may be affected by atmospheric water outside the IR cell or (during low-temperature experiments) water condensed on the cell windows. The $\nu(\text{O–D})$ modes are located around 1000 cm^{-1} below the respective $\nu(\text{O–H})$ modes in a region where noise is limited.

By performing deuterium exchange preferentially with the most acidic hydroxyl groups (using C_6D_6), it was possible to detect small amounts of bridging hydroxyls on amorphous silica–alumina *(616,627)*. The authors registered weak bands at 2632 and 2683 cm^{-1}, with the latter having a low-frequency shoulder. After low-temperature CO adsorption, the band at 2683 cm^{-1} shifted to 2530 cm^{-1} ($\Delta\tilde{\nu}(\text{OD}) = -152 \text{cm}^{-1}$), and the shoulder around 2658 cm^{-1} shifted to 2483 cm^{-1} ($\Delta\tilde{\nu}(\text{OD}) = -175 \text{cm}^{-1}$). It was stated that these shifts correspond to $\Delta\tilde{\nu}(\text{OH})$ shifts of -238 and -272 cm^{-1}, respectively, which confirmed the high acidity of the hydroxyls.

Molecular water is characterized by a deformation mode around 1630 cm^{-1}. This band cannot always be discerned because many other surface species (carbonates, nitrates) manifest bands in this region. The band should be shifted to about 1200 cm^{-1} when using deuterated compounds. Moreover, when the H/D exchange is only partial, the deformation of the HOD molecule should be observed around 1415 cm^{-1} *(495)*.

H/D exchange is a good method to check for the existence of spectral effects *(152–155,169,613)*. In the absence of such effects, the spectra of OH and the respective OD groups should be analogous. However, spectral phenomena such as Fermi resonance may be different for OH and OD groups. Recently, using this approach, it was shown that the bands of acidic hydroxyls in some zeolites (HZSM-5 *(152,153)*, HFER *(154)*) when shifted to lower frequencies upon formation of OH$\cdots$CO complexes are split as a result of Fermi resonance with the second overtone of the $\delta(\text{OH})$ mode. This kind of Fermi resonance occurs when the maximum of the shifted OH band is at around 3300 cm^{-1}, implying that no resonance takes place when the acidity of such hydroxyls is measured with a weaker base, for example, N_2 (OH band shifted to ≈ 3500 cm^{-1}), or when hydroxyls of weaker acidity are tested by CO. The bridging OH groups of HDY are shifted from 3641 to 3370 cm^{-1} upon CO adsorption, and the corresponding OD groups are shifted from 2687 and 2506 cm^{-1} *(169)*. In this case, the shift is small enough to avoid the Fermi resonance. On the contrary, resonance occurs when benzene interacts with the acidic hydroxyls of HZSM-5, because the induced shift of the $\nu(\text{OH})$ modes is similar to those induced by CO *(155)*. It is important to exclude spectral effects and properly determine the acidity and the possible heterogeneity of hydroxyls in zeolites, because these sites are relevant for catalytic reactivity.

The appearance of two combination bands for SiOH groups, at about 4563 and 4530 cm^{-1}, has been debated, and it was recently shown that only one corresponding band is observed with a deuteroxylated sample *(613)*. It was proposed that two surface modes with bending character arise from the coupling of the bending Si–OH modes with Si–O–Si stretching vibrations. No such coupling can occur with deuterated samples because the Si–OD bending modes appear at frequencies well below the Si–O–Si stretching vibrations.

We would like to remark that surface complexes represent good model systems for investigations of OH/OD exchange phenomena. They are typically well-defined species, and usually there is no interference from other molecules, as there can be in solutions. In addition, one can perform all experiments at low temperature, thus reducing the temperature effect.

6.2. ^{18}OH groups

Oxygen has three stable isotopes, ^{16}O, ^{17}O, and ^{18}O. The natural abundance of ^{17}O and ^{18}O is low, about 380 and 2050 ppm, respectively. ^{18}O is more suitable for vibrational spectroscopy because the difference in mass relative to ^{16}O is larger than it is for ^{17}O, and thus isotopic effects are more pronounced and better observable. Because of the higher abundance of ^{18}O as compared to ^{17}O, ^{18}O-containing compounds are also less expensive.

Exchange of OH groups with ^{18}O is a less commonly used approach than OH/OD exchange. The shift of the OH stretching vibration is much smaller, by two orders of magnitude, than the shift upon H/D exchange and amounts to approximately -10 cm^{-1}. The theoretical isotopic shift factor is 1.0033. Because of the small value of the shift, its experimental determination is accompanied by a relatively large uncertainty. Moreover, for reasons of chemical reactivity, this exchange is much more difficult to perform than H/D exchange. Silica OH groups are exchanged to a high degree with H$_2^{18}$O at a temperature of 673 K (whereas H/D exchange occurs at room temperature) *(649)*. Table 2.26 shows the observed frequencies after ^{18}O exchange and the respective isotopic shift factors.

The exchange of ^{16}O with ^{18}O can be used to distinguish between OH and NH groups if their bands appear in the spectral region where both kinds of species can absorb. In principle, the H/D exchange should provide the same information, because the OH/OD and NH/ND isotopic shift factors are different (1.3744 and 1.3693, respectively, for harmonic oscillators). However, the deviations from theory (see above) make the results

Table 2.26 Reported stretching frequencies and calculated isotopic shift factors i of isolated surface OH and respective ^{18}OH groups

OH group	Sample	$\tilde{\nu}$(OH), cm^{-1}	$\tilde{\nu}(^{18}$OD), cm^{-1}	i [a]	Ref.
SiOH	SiO$_2$	3749	3738	1.003	(649)
SiOH	MCM-41	3744	3735	1.002	(650)
SiOH	DAY	3739	3727	1.003	(650)
TiOH	TiO$_2$	3663	3653	1.003	(651)
		3569	3559	1.003	
TiOH	TiO$_2$	3733	3720	1.003	(504)
		3717	3708	1.002	
		3691	3677	1.004	
		3672	3665	1.002	
		3642	3630	1.003	
FeOH	Fe–ZSM-5	3674	3663	1.003	(20)
FeOH	Fe–ZSM-5	3635	3621	1.004	(20)
COH	HCO$_3$/Fe$_2$O$_3$	3619	3609	1.003	(484)
ZrOH[b]	ZrO$_2$	953	948	1.005	(10)
TiO$_2$[b]	ZrO$_2$	953	948	1.005	(10)

[a]Isotopic shift factor, ν(OH)/$\nu(^{18}$OH).
[b]Deformation modes.

ambiguous. On the contrary, a shift of a questionable band after exchange of ^{16}O with ^{18}O proves that the band characterizes vibrations involving oxygen. For example, adsorption of ammonia on TiO$_2$ resulted in the appearance of a band at 3569 cm^{-1}. When ammonia was adsorbed on an ^{18}O-enriched titania surface, the respective band was observed at 3559 cm^{-1}. An N–H mode could thus be excluded and the band was ascribed to O–H modes (651).

Lavalley and coworkers (505,506) investigated the adsorption of ^{18}O-labeled CH$_3$OH on different oxides to prove that methanol dissociates on the surfaces through breaking of the O–H bond. They observed ^{18}O-containing methoxy species and not ^{18}OH groups.

By combining deuterium and ^{18}O exchange, it was demonstrated that the deformation δ(MOH) mode of H-bonded hydroxyls on oxides is located at 1100 cm^{-1} for Al$_2$O$_3$, 953 cm^{-1} for ZrO$_2$, and around

900 cm^{-1} for MgO *(10)*. The Zr–^{18}OH band was detected at 948 cm^{-1}, and the other bands were not detectable after exchange with D or ^{18}O because they were shifted to frequencies too low to be observed.

^{18}OH groups are often used to trace the origin of oxygen in surface species or catalytic reaction products. For example, it was concluded that the surface hydroxyls on titania were involved in the reaction with SO$_2$ to yield sulfite and bisulfite *(504)*. Using the same method, it was decided that oxygen from the OH groups on Fe$_2$O$_3$ and Al$_2$O$_3$ participated in the formation of bicarbonates when the samples were contacted with CO *(484)*. A similar approach was used to determine the products of cyanogen chloride (ClCN) decomposition on the γ-Al$_2$O$_3$ surface *(652)*. The appearance of ^{18}O-labeled NCO species indicated that surface hydroxyl groups of γ-Al$_2$O$_3$ were involved in the ClCN decomposition through Ti–NCO intermediates. As a note of caution, it should be pointed out that the procedure to exchange surface hydroxyls to ^{18}OH groups normally also leads to exchange of coordinatively unsaturated O^{2-} surface ions (invisible to IR spectroscopy), and these ions could also have participated in the investigated reactions.

7. OTHER TOOLS FOR IDENTIFICATION OF HYDROXYL GROUPS AND PRACTICAL HINTS

7.1. Distinguishing between adsorbed water and free hydroxyls

Water is a product of many reactions and can adsorb on the catalyst surface, thus affecting the hydroxyl stretching region in IR spectra. H$_2$O can be produced even after adsorption of a simple adsorbate as a result of surface reactions. Most of these reactions were already considered in the sections above, among them: formation of water during adsorption of NO$_2$ (through production of NO$^+$ or nitrates, see Eqns. (2.19) and (2.20)), SO$_2$, or alcohols. Water on activated samples can also appear as a result of readsorption from the apparatus walls.

High surface densities of adsorbed water give rise to a broad band in the OH region as a result of hydrogen-bonded OH groups. At low coverage, the adsorbed water molecules are usually isolated and manifest two OH stretching and one deformation mode (around 1640–1620 cm^{-1}). When the adsorbed molecules have C_{2v} symmetry, the OH modes are split into symmetric and antisymmetric vibrations. Often, one of the water hydrogen atoms is H-bonded to surface oxygen *(11,422,423)* and the OH modes are no longer degenerate. Depending on the strength of the H-bond, which depends on the sample, the modes appear at different frequencies. The variability of the spectral pattern impedes the distinction between surface hydroxyl groups and water hydroxyls. In principle, water can be identified by its deformation mode. Unfortunately, the intensity of the $\delta(H_2O)$ band is low, and the band is sometimes difficult to detect. Moreover, the deformation modes may be masked by other strong bands in the region (e.g., from nitrates or carbonates *(495)*).

A powerful technique for establishing whether a band or a component of a band around $1650–1600$ cm^{-1} is associated with adsorbed water is D/H isotopic exchange. The exchange results in a red shift of the water deformation modes (for more details, see Section 6.1). Another possibility is to analyze the overtones and combination modes (see Section 7.3).

7.2. IR bands of other compounds absorbing in the OH region

Although the region between 3800 and 3000 cm^{-1} is mainly occupied by $\nu(OH)$ modes, there are other vibrations that can appear in this region, which may lead to misinterpretations of the spectra. These bands are usually overtones and combination modes, but sometimes $\nu(NH)$ modes appear in the lower part of the $\nu(OH)$ region. If the assignment of a particular band to an OH stretching mode is in doubt, a D/H exchange experiment will help clarify the interpretation. However, to distinguish between O–H and N–H modes, it is necessary to perform experiments with labeled oxygen or nitrogen.

Below, we shall briefly describe the most frequently observed non-hydroxyl bands that appear in the same spectral region as $\nu(OH)$.

Adsorbed CO_2 is characterized by bands at about 1380 (ν_1) and 2350 cm^{-1} (ν_3). The ν_2 mode around 670 cm^{-1} is usually not detected because it is masked by bulk absorption of the sample. However, the combinations $(\nu_1 + \nu_3)$ and $(\nu_3 + 2\nu_2)$ are observed around $3760–3710$ and $3605–3595$ cm^{-1}, respectively, and the intensity may be similar to that of the bands of the OH groups *(364,653)*. Galhotra *et al.* *(653)* reported a shift of these bands when $C^{18}O_2$ was adsorbed on zeolites instead of CO_2: the band originally positioned at 3713 cm^{-1} was detected at 3632 cm^{-1}, and the band originally positioned at 3605 cm^{-1}, at 3527 cm^{-1}.

Overtones in the $\nu(OH)$ region were also reported for adsorbed nitrogen oxides. N_2O_4 is characterized by a strong band at approximately 1750 cm^{-1} *(487)*. The overtone of this band is reported to appear near 3500 cm^{-1} *(496)*. A $(\nu_1 + \nu_3)$ combination mode at $3480–3450$ cm^{-1} can be observed after adsorption of N_2O *(487)*. Low-temperature NO adsorption usually results in formation of a *trans*-$(NO)_2$ dimer. This dimer is characterized by well-defined N–O overtones and combination modes, which for species adsorbed on HZSM-5 zeolite absorb at 3764 and 3470 cm^{-1} *(300)*. The bands shifted to 3698 and 3405 cm^{-1}, respectively, when ^{15}NO was used in the experiments.

Bands at 3700–3620 cm^{-1} have been recorded for deeply reduced ceria *(117,654)*. These bands were not affected by deuterium exchange and were assigned to electronic transitions.

The $(\nu + \delta)_{OH}$ combination modes of deuterium-exchanged samples appear in the OH region. The modes have been detected at 3366 cm^{-1} for SiOD *(21,409)* groups and around 3560 cm^{-1} for bridging Si(OD)Al groups of DZSM-5 *(152)*. At high exchange degrees, these bands may become more intense than the residual OH bands.

Attention should also be paid to the nature of the hydroxyl groups. Formation of bicarbonates leads to the appearance of a C–OH band around 3600 cm^{-1} *(482,483)*, which should not be considered as a surface hydroxyl of M–OH type. Similarly, S–OH and N–OH groups can be formed after adsorption of S- and N-containing molecules, respectively *(502,503)*.

7.3. Overtones and combination modes of surface OH groups

In the majority of reports on the spectroscopic characterization of surface hydroxyls, the analysis of the spectra is restricted to the ν(O–H) region. Occasionally, the assignments are supported by analysis of the deformation modes, but these are usually masked by bulk absorption of the sample. Additional helpful information can be obtained from the overtones and combination modes. Unfortunately, the according bands are of low intensity and occur at high frequencies where noise becomes relevant in transmission spectra. Advantageous in this respect is the diffuse reflectance technique but it renders quantification extremely difficult.

Overtones appear in the spectra because of the anharmonicity of the O–H bond. It is important to consider that the ν_{02}(OH) bands appear always at a wavenumber lower than twice that of the fundamental hydroxyl stretching mode. The difference is used to calculate the anharmonicity of the O–H oscillator according to Equation (2.1) and also the harmonic frequencies according to Equation (2.2). In principle, accurate calculations should be based also on the next overtones but it was reported that, for OH oscillators, calculations using the first overtone always give correct results *(613)*. In the absence of any other factors, the anharmonicity decreases with decreasing wavenumber because the energy of the motion decreases and the level is shifted down to the bottom of the Morse curve. Consequently, the anharmonicity of O–D oscillators is much lower than that of the respective O–H oscillators.

The intensity (in transmission IR experiments) of the first overtone of the O–H stretching modes may be roughly estimated to be approximately 1–1.5% of the intensity of the fundamental band. The actual intensity depends on many factors, including the anharmonicity. The intensity of the first overtone of the OD stretching modes for SiOD groups on silica is about 1% of the intensity of the fundamental band, whereas for the OH bands, the intensity of the first overtone is 1.5% (21). An important phenomenon that was already discussed is the strong intensity decrease of the overtone bands upon formation of H-bonds. As a result, the overtones of H-bonded hydroxyls are not observable or their intensities are negligible (21,41,42,411). The intensity of the second overtone is very low and was estimated for SiOH groups to amount to only 2.5% of the intensity of the first overtone band (613). Like the fundamental bands, the first overtones (when observable) are shifted through adsorption of molecular probes (21). The shift is slightly smaller than the doubled value of the shift of the fundamental modes.

Table 2.27 summarizes literature data on the overtone frequencies of selected surface hydroxyl groups.

Combination modes are widely used to derive the wavenumbers of vibrations that are not directly observable because they are masked by the strong absorbance of the bulk of the solid. As may be expected, these modes are shifted as a result of H-bond formation with adsorbed molecules, but the shift is smaller than the shift of the fundamental OH stretching vibrations (183). The smaller shift results because H-bonding shifts the stretching and deformation modes in different directions and the two effects are partly canceled in the combination mode.

A detailed analysis of the combination bands of hydroxyl groups on silica (3748 cm^{-1}) has been provided by Burneau and Carteret (613). They ranked the different combination modes by their intensity in relation to the intensity of the first overtone. The only combination band with an intensity comparable to the intensity of the first overtone (40–50%) is that of the ($\nu(OH) + \delta(OH)$) combination mode. The SiOH bending mode is coupled with Si–O–Si stretching vibrations, thus inducing two surface modes with bending character, at 760 and 835 cm^{-1}. As a result, the combination mode is split into two components, at 4516 and 4582 cm^{-1}. For the

Table 2.27 Fundamental and overtone OH and OD stretching frequencies of hydroxyl groups on silica and related materials

Sample	Hydroxyl	$\tilde{\nu}_{01}(OH)$ or $\tilde{\nu}_{01}(OD)$, cm^{-1}	$\tilde{\nu}_{02}(OH)$ or $\tilde{\nu}_{02}(OD)$, cm^{-1}	$\tilde{\nu}_{harm}$, cm^{-1}	Anharmonicity (X_{12}), cm^{-1}	Ref.
SiO_2	SiOH	3745	7325	3910	82.5	(312)
SiO_2	SiOH	3748	7331	3913	82.5	(613)
		3751[a]	7337[a]	3916	82.5	
SiO_2	SiOH SiOH···CH$_4$ SiOH···CO	3746	7325	3913	83.5	(21)
		3748[a]	7329[a]	3915	83.5	
		3714[a]	7265[a]	3877	81.5	
		3660[a]	7170[a]	3810	75	
[Si]BEA	SiOH	3738[a]	7310[a]	3904	83	(21)
	SiOH···CO	3644[a]	7126[a]	3806	81	
Minerals[b]	SiOH	3740	7316	3904	82	(32)
	Si(OH)$_2$	3650	7126	3824	87	
Minerals[b]	Al(OH)Fe	3600	7064	3736	68	(32)
	Fe(OH)Fe	3560	6956	3724	82	
	Mg$_3$OH	3680	7188	3852	86	
HZSM-5	SiOH	3745	7323	3912	83.5	(411)
	Si(OH)Al	3613	7077	3762	74.5	
HZSM-5	Si(OH)Al	3610	7065	3765	77.5	(312)
HFER	Si(OH)Al	3600	7048	3752	76	(323)
HMOR	SiOH	3745	7318	3917	86	(411)
	Si(OH)Al	3609	7066	3781	76	
H$_2$O on HZSM-5	H–O–H[c]	3696	7157	3931	117.5	(411)
		3560–3500	6860	3730	~100	
H$_2$O on HMOR	H–O–H[c]	3660	7170	3810	75	(411)
		3560–3500	6850	3740	~105	
SiO_2	SiOD	2762	5438	2848	43	(613)

Continued

Table 2.27 Fundamental and overtone OH and OD stretching frequencies of hydroxyl groups on silica and related materials—cont'd

Sample	Hydroxyl	$\tilde{\nu}_{01}(OH)$ or $\tilde{\nu}_{01}(OD)$, cm^{-1}	$\tilde{\nu}_{02}(OH)$ or $\tilde{\nu}_{02}(OD)$, cm^{-1}	$\tilde{\nu}_{harm.}$, cm^{-1}	Anharmonicity (X_{12}), cm^{-1}	Ref.
SiO$_2$	SiOD	2762	5434	2852	45	(21)
	SiOD···CH$_4$ SiOD···CO	2763[a]	5436[a]	2853	45	
		2739[a]	5391[a]	2826	43.5	
		2700[a]	5321[a]	2779	39.5	
[Si]BEA	SiOD	2755[a]	5422[a]	2483	44	(21)
	SiOD···CO	2689[a]	5287[a]	2780	45.5	

[a]At 100 K.
[b]Bulk hydroxyls.
[c]Because of the reduced symmetry of H$_2$O, two independent ν(OH) bands are detected. Data concerning weakly H-bonded hydroxyls are also included.

same reason, the very weak $(2\nu(OH) + \delta(OH))$ mode is also split into two components, at 8100 and 8165 cm^{-1}. No coupling was observed for a deuterated sample, because the δ(OD) mode at 605 cm^{-1} is well separated from the Si–O–Si stretching vibrations. The intensity of all other combinations was less than 1% of the intensity of the first overtone band.

Similar $(\nu_{OH} + \delta_{OH})$ combination bands have been detected for silanol groups in zeolites in the 4573–4540 cm^{-1} region (183,312,411).

The bridging hydroxyls in zeolites (3610–3600 cm^{-1}) also manifest a pronounced $(\nu_{OH} + \delta_{OH})$ mode, which is observed at higher frequencies than the combination modes of the silanol groups, namely in the region of 4654–4661 cm^{-1} (183,312,323,411). Thus, like the overtones, the combination modes of these different bands are well separated (see Figure 2.50).

The analysis of the combination modes is very useful for the detection of adsorbed water. The δ(H$_2$O) mode is around 1640–1600 cm^{-1}, and consequently, the $(\nu(OH) + \delta(H_2O))$ modes of adsorbed water are detected in the 5330–5000 cm^{-1} region, that is, at higher frequencies than the analogous mode of surface OH groups (409,411,614,655). Moreover, it was suggested that the exact frequency provides information on the adsorption state of water (409). Water molecules adsorbed only through the oxygen atoms (S$_0$ state) will absorb at the highest frequencies (Figure 2.51B and C). For this state, the band is assigned to the $(\nu_{as}(OH) + \delta(H_2O))$ combination. When one of the hydroxyls of the water molecule is engaged in H-bonding with surface oxygen (S$_1$ state), the combination mode is shifted

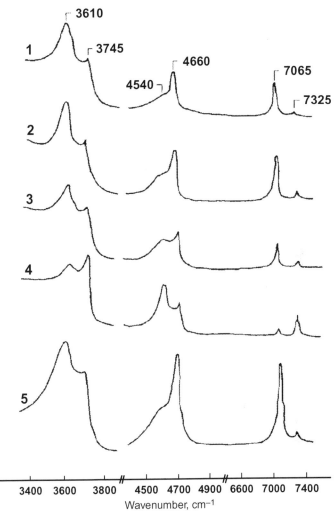

Figure 2.50 Diffuse reflectance IR spectra of zeolites with a high silica content in the range of 3300-7500 cm^{-1}: (1) HZSM-5, obtained by NH$_4$-form decomposition; (2) HZSM-5, obtained from NaZSM-5 by HCl treatment; (3) Na-ZSM-5; (4) silicalite; and (5) hydrogen form of mordenite. *Reproduced from Ref.* (312).

to lower frequencies. Indeed, a band at 5279 cm^{-1} was reported for water adsorbed on bridging hydroxyls in zeolites *(411)*. When both hydroxyls of water form H–bonds (S$_2$ state), the band shifts further to lower frequencies *(409,411,655)*. Finally, polymeric chains of H$_2$O molecules (S$_n$ state) will manifest a band at the lowest frequency (among these states). For

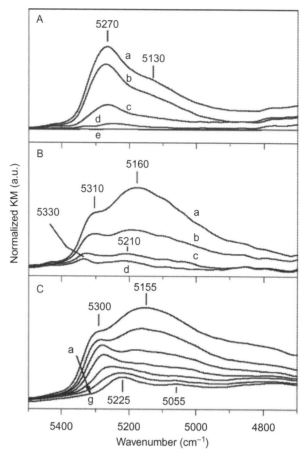

Figure 2.51 Near-IR spectra in the range of 5500-4700 cm^{-1} of: (A) SiO$_2$, (B) Al$_2$O$_3$, and (C) TiO$_2$ initially contacted with water vapor at room temperature and then progressively outgassed until high vacuum conditions were reached. Lettering in each panel is in the order of decreasing water coverage. *Reproduced from Ref. (409).*

comparison, the $(\nu(OH) + \delta(H_2O))$ combination mode of liquid water is observed as a band at 5200 cm^{-1} that extends toward lower frequencies. According to Takeuchi *et al.* *(655)*, H$_2$O adsorbed on TiO$_2$ is characterized by the following combination bands, depending on the mode of adsorption: S$_0$, 5308 cm^{-1}; S$_1$, 5188 cm^{-1}; S$_2$, 5027 cm^{-1}; and S$_n$, 4841 cm^{-1}.

7.4. Coefficients of molar absorptivity (absorption coefficients)

For quantitative investigations, one needs to know the molar absorptivity, that is the absorption (or in the older literature extinction) coefficients of the bands of OH groups and of the adsorbate bands that are typical of the interaction between the respective probe and the surface hydroxyls. It should be noted that the absorptivity of ν(OH) bands strongly depends on the presence of H-bonds.

Table 2.28 summarizes literature data on the absorption coefficients of hydroxyls on silica and zeolites. The most commonly used dimension is cm μmol^{-1}. The decadic absorption coefficient is more prevalent than the Napierian and is usually used in IR spectroscopy. In older references, one can encounter that the linear intensity was analyzed and the absorption coefficient was expressed in cm^2 μmol^{-1}. To avoid any confusion, we shall not present such data here.

While there are substantial differences between the reported values, the main deviations concern older investigations. The data in Table 2.28 strongly indicate that a decrease in the frequency of hydroxyls induced by H-bonding is associated with an increase of the absorption coefficient. Using this concept, Makarova et al. (146) determined the absorption coefficients of the high- and low-frequency bands that characterize two types of bridging hydroxyls in HY zeolite.

According to Ref. (434), the absorption coefficient of the OH groups in FER zeolite samples gradually increases from 0 to 3 cm μmol^{-1} with increasing exchange degree, that is, replacement of alkali metal cations with protons (which is also reported to lead to an increase in acid strength).

Vimont et al. (614) determined the absorption coefficient for the ($\nu + \delta$) OH combination band of silanol groups on silica (4590–4515 cm^{-1}) to be 0.16 cm μmol^{-1}. A slightly larger value, 0.22 cm μmol^{-1}, was found for the ($\nu + \delta$) combination band of water adsorbed on silica.

For quantitative measurements, it is necessary to know the absorption coefficients of selected bands of the adsorbate. Although these coefficients are allowed to vary for different solids, they should be of the same order of magnitude. The differences in the reported values for the same probe and different solids are larger than seems reasonable. Common problems with such measurements concern the determination of the adsorbed amount; for example, sorptive can also be retained on the apparatus walls (pyridine) or on resin-containing parts of the apparatus (ammonia).

Ammonia is often used to quantify the number of Brønsted acid sites, and the band representative of NH$_4^+$ ions around 1450 cm^{-1} is analyzed. However, the data for the absorption coefficient provided by different groups scatter widely. Davydov (56) reported a value of $\varepsilon_{1430} = 0.215$ cm μmol^{-1}. Significantly higher values of $\varepsilon_{1430} = 13.0$ cm μmol^{-1} and $\varepsilon_{1430} = 14.8$ cm μmol^{-1}

Table 2.28 Decadic integral molar absorption coefficients ε (O–H stretching modes) of some surface hydroxyls

Sample	$\tilde{\nu}$(OH), cm^{-1}	ε, cm µmol^{-1}	Note	Ref.
SiO$_2$	3745	~3		(656)
HY	3635	3.2		(146)
	3645	8.5		
HY	3625	3.1	Determined assuming the same ε for the HF and LF bands	(657)
	3550	3.1		
HY	3650	5.28		(658)
	3550	3.52		
HY	3635	7.5		(659)
	3547	5.6		
	3596 + 3522	4.7		
HY	3640	12.2		(660)
	3550	19.9		
HY	HF	5.28		(661)
	LF	9.5		
HMOR	3605	3.5–4.65		(657)
HMOR	3605	3.00		(343)
HMOR	ca. 3650	2.5		(598)
	ca. 3600	3.8		
HFER	3604	4.05		(372)
HZSM-5	3605	3.7		(657)
MAZ	3600	4.68		(343)
HSAPO-34	3630	3.9		(210)
	3614	3.9		
	3600	6.0		

were reported by Bortnovsky *et al. (662)* and Datka *et al. (434)*, respectively, and used by other authors *(210,232)*. However, recently, Datka and coworkers *(213,216,230)* determined a value of 0.11 cm µmol^{-1} (in one instance, the dimension used was cm^2 µmol^{-1} *(213)*). The same value (in cm^2 µmol^{-1})

was used by other authors *(182)*. An intermediate value of 1.47 cm μmol^{-1} has also been proposed *(663)*.

Pyridine is the conventionally used probe for quantification of different types of acid sites, and thus the knowledge of the absorption coefficients of bands characterizing protonated and coordinated pyridine is important. Again, there is no full agreement between the reported data. Datka *et al.* *(664)* reported a value of 0.73 cm μmol^{-1} for the absorption coefficient of the band at 1545 cm^{-1} (PyH$^+$), and this value was used by others *(103)*. In a series of later publications by the same author *(216,230)*, a corrected value of 0.070 cm μmol^{-1} was proposed, which was also used by others *(98)*.

Emeis *(657)*, Khabtou *et al.* *(659)*, Take *et al.* *(665)*, and Guisnet *et al.* *(343)* measured higher extinction coefficients for the 1545 cm^{-1} PyH$^+$ band than Datka: 1.13 *(343)*, 1.36 *(138)*, 1.5 *(665)*, 1.6 *(657)*, and 1.8 cm μmol^{-1} *(659)*, and these values were used by other authors in a number of investigations *(182,184,339)*.

The following values were reported for absorption coefficients of other probe molecules: ν(CN) band at 2297 cm^{-1} for adsorbed CD$_3$CN, 2.05 cm μmol^{-1} *(372)*; ν_{8a} and ν_{8b} bands of protonated 2,6-dimethylpyridine (at 1644 and 1628 cm^{-1}, respectively), 6.81 cm μmol^{-1} *(477)*; and the 2278-2258 cm^{-1} band of pivalonitrile, 0.11 cm μmol^{-1} *(566)*.

The discrepancies between different sources suggest that absorption coefficients reported in the literature may be associated with (sometimes significant) errors. A solution to this problem is to analyze and interpret changes of absorption coefficients. Through this approach, deviations of the absolute absorption coefficients from the correct values cancel each other. Most investigations applying this approach concern changes upon H–bond formation *(21,136,146)* or compare the absorption of different hydroxyl bands *(68,146)*. Some sources provide the relative absorption coefficients of adsorbates *(144,202)*.

8. MATERIALS OF PRACTICAL INTEREST

In this section, we will consider the hydroxyl population on some materials of practical interest including oxides, zeolites, and porous phosphates, which have found application as catalysts and catalyst supports as well as metal organic frameworks.

8.1. Magnesia

The hydroxyl spectra of MgO are relatively simple. After evacuation at high temperature (e.g., 973 K), a single sharp band in the region of

3750–3740 cm^{-1} dominates in the spectra *(54,351,361,382,666,667)* (see Figure 2.11). Another band at lower frequencies, that is broader and consists of several components, is detected when the evacuation temperature is lower *(27,28,54,114,301,666,668,669)*. The maximum of this band is coverage-dependent as shown in Figure 2.11 and the reported position varies depending on the source, but is generally between 3640 and 3515 cm^{-1}. As may be expected, the hydroxyls on magnesia are highly basic, and their O–H stretching modes are not shifted by adsorption of CO *(114)* or CH$_4$ *(301)*. A very small shift was reported after benzene adsorption (-30 cm^{-1}) *(351)*, which, as already discussed, may, at least partly, be caused by interaction of benzene with the oxygen atoms of the hydroxyl groups.

Although the simple structure of MgO facilitates modeling, there is still serious disagreement in the interpretation of the hydroxyl spectra. In 1965, Anderson *et al.* *(670)* investigated water adsorption on MgO and proposed that H$_2$O dissociates on the (001) plane, thus producing hydroxyls of type I (3740 cm^{-1}) and type V (≈ 3500 cm^{-1}). The authors pointed out that the bridging hydroxyls were more perturbed by interaction with their neighbors than the terminal hydroxyls. An interpretation proposed later by Tsyganenko and Filimonov *(27,28)* is very similar and was accepted by many researchers *(114)*.

Coluccia *et al.* *(666)* noticed that the high- and low-frequency bands did not change in concert which contradicted the model proposed by Anderson. The authors reported that the high-frequency band could also be produced after dissociative adsorption of H$_2$. To explain this phenomenon, they assigned the high-frequency band to OH groups on corners and edges (where H$_2$ dissociates), and the low-frequency band to OH groups on flat crystal planes.

A careful investigation of the MgO hydroxyls was performed by Knözinger *et al.* *(668)*. The authors considered four types of possible OH groups on the MgO surface; the structures are depicted in Figure 2.52.

The following key experiments helped devising an advanced model of the magnesia OH groups:
- The high-frequency band consists of at least three components. During evacuation at temperatures between 473 and 673 K, a component in the interval between 3740 and 3690 cm^{-1} disappears. A component at 3750 cm^{-1} is much more stable and only disappears during evacuation at 773-1030 K, leaving a third component at 3712 cm^{-1}.
- The low-frequency band disappears in parallel with the 3740-3690 cm^{-1} component of the high-frequency band.
- Adsorption of protonic acids (expected to produce only bridging hydroxyls) results in the formation of a band at 3712 cm^{-1}.

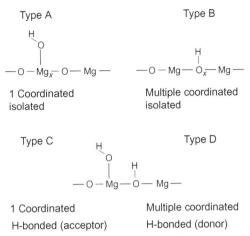

Figure 2.52 Classification of isolated and H-bonded OH groups resulting from hydroxylation of a MgO surface; x denotes the coordination number of Mg^{2+} and is 3 for corner, 4 for edges, and 5 for regular planes. *Reproduced from Ref.* (668).

On the basis of these observations, it was proposed that the type A hydroxyls (see Figure 2.52) are characterized by a band at 3750-3737 cm^{-1} and the type B hydroxyls, by a band at 3737–3700 cm^{-1}. The hydroxyls of the types C and D are H-bonded and absorb in the regions of 3740-3690 and 3695–3400 cm^{-1}, respectively. The latter two features change in concert, and the respective species recombine to give water at temperatures below 673 K. Later, the model was expanded, and the frequencies of B-type hydroxyls in contact with different species produced after adsorption of H-acids were specified *(667)*.

Chizallet *et al. (671)* critically reviewed the different models of magnesia hydroxyls. On the basis of DFT calculations, these authors formulated a model similar to that of Knözinger *et al. (668)*. However, the C species were proposed to be the species with the highest frequencies. We note that this detail disagrees with the observation that the highest frequency species are still present on the surface after the complete removal of the D-species, whereby the model requires the C and D species to disappear simultaneously with production of H_2O (see Figure 2.52).

Other DFT calculations *(19)* indicated that the frequency of the bridging hydroxyls on magnesia is slightly lower than the frequency of the type I OH groups, consistent with Knözinger's model. It was further inferred that, for type I species, an increase of the magnesium ion coordination should lead to a small decrease in the O–H frequency.

8.2. Zinc oxide

The surface of zinc oxide has been characterized by vibrational spectroscopy by a number of research groups *(111,263,310,633,634)*. The crystal planes terminating ZnO particles are usually stabilized by OH groups. This termination is illustrated in Figure 2.53, which shows hydrogen atoms on the $(000\bar{1})$ plane. Because ZnO single crystals are available, high-resolution electron energy loss spectroscopy (HREELS) has been performed, and the spectra can be used for interpretation of the hydroxyl IR spectra of ZnO. Considering information from HREELS, Noei *et al. (111)* proposed the following band assignments for surface hydroxyls on powdered ZnO:

- Band at 3620 cm^{-1}, OH groups on the $(000\bar{1})$ ZnO plane;
- Bands at 3656 and 3639 cm^{-1}, OH groups on the $(10\bar{1}0)$ ZnO plane;
- Band at 3672 cm^{-1}, OH groups on the $(10\bar{1}0)$ ZnO plane interacting with coadsorbed water.

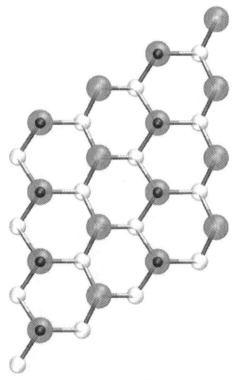

Figure 2.53 (2×1) H-terminated ideal ZnO $(000\bar{1})$ surface. Zinc atoms, white; oxygen atoms, red; and hydrogen atoms, blue (on top of oxygen). *Reproduced from Ref. (633).* (See the color plate.)

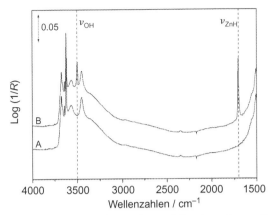

Figure 2.54 DRIFT spectra showing the products of H_2 adsorption on ZnO NanoTek at a temperature of 323 K: (a) in He and (b) in 2% H_2/He. H_2 dissociates to form an OH group and ZnH species. *Reproduced from Ref.* (633).

In addition, two broader bands that were observed only in the IR spectra, at 3564 and 3448 cm^{-1}, were associated with hydroxyls on defect sites (see Figure 2.54, spectrum a).

A band at 3572 cm^{-1} was found to be characteristic of the polar $(000\bar{1})$ ZnO plane *(672)*. Additional bands were registered in the spectra of other preparations of ZnO, for example, at 3670, 3625, and 3445 cm^{-1} *(263)*.

It was also found, by means of low-temperature CO adsorption, that the ZnO hydroxyls are characterized by weak acidity *(263)*.

The dissociative adsorption of H_2 on ZnO has been extensively investigated *(310,633,634)*. It was established that a new hydroxyl band, around 3500 cm^{-1}, is formed, together with a band at 1709 cm^{-1} that is ascribed to ZnH species (see Figure 2.54, spectrum b). The position clearly indicates that these hydroxyls are bridging. Probably, the same species may also be formed through dissociative adsorption of water, but would be very unstable and easily recombine to produce water.

8.3. Alumina

There are many investigations of the hydroxyl population on various aluminas because of the extraordinary importance of this oxide in catalysis. Noticeably, the hydroxyl spectra hardly depend on the polymorph form, although variations in the relative intensities of the OH bands are observed even with a single polymorph. Compared to oxides like MgO or ZrO$_2$, alumina is more covalent (43% covalent character of the bond vs. 33% for MgO and ZrO_2) which implies that the OH groups should generally be more acidic.

There are several bands that have been detected (with different relative intensities) in the spectra of activated aluminas (see Figure 2.55). Below is a list of the characteristics of each band.

A. Band between 3800 and 3785 cm^{-1}. This band has a low intensity but is virtually present in the spectra of all activated alumina samples, except for those of the corundum phase of alumina *(476)*. There is consensus that this band corresponds to type I hydroxyls.

B. Band at 3780–3760 cm^{-1}. This band is usually more intense than the A-band. Busca *(673)* reported that the respective hydroxyls are the most reactive ones on alumina. This band is also not observed in spectra of the corundum phase *(476)*.

C. Band in the 3745–3740 cm^{-1} region. Usually this band is of low intensity.

D. Band at 3735–3725 cm^{-1}. The "D-band" is typical of a number of different aluminas and is usually observed with high intensity. It has been suggested that there are several components to this band.

E. Band at 3710–3700 cm^{-1}. This band is sometimes not well discernible, because it is masked by other hydroxyl bands in the proximity. No

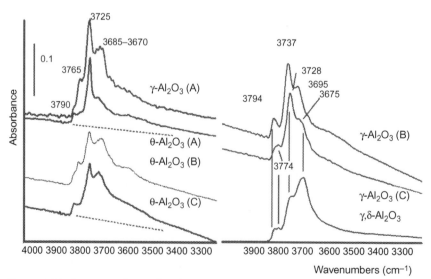

Figure 2.55 IR spectra of samples of various Al_2O_3 materials after outgassing at a temperature of 773 K for 1 h. *Reproduced from Ref. (673).*

strong acidity of the respective hydroxyls is observed. For example, the band is not affected by pivalonitrile. However, the hydroxyls could be located in small pores not accessible to this probe *(673)*.

F. Band between 3690 and 3675 cm^{-1}. The relative intensity of this band is high in the spectra of fumed alumina samples.

G. Band around 3590 cm^{-1}. This band is usually attributed to H-bonded hydroxyls. It is the first to disappear upon outgassing at high temperatures.

Unfortunately, it is difficult to selectively measure the acidity of the different hydroxyls by low-temperature CO adsorption because the shifted bands form a single feature. Some information can be extracted through analysis of the C–O stretching frequencies and it was inferred that, notwithstanding a similar OH shift, the hydroxyls characterized by the F-band are more acidic than those characterized by the D-bands *(257)*.

There have been many attempts to associate the hydroxyl groups on alumina with distinct surface structures, and the results were reviewed by Morterra and Magnassa *(458)* and very recently by Busca *(673)*. In 1965, Peri *(674)* proposed a statistical model for the distribution of hydroxyl groups considering the (001) crystal plane of alumina. Tsyganenko and Filimonov *(27,28)* attributed bands at 3800, 3740, and 3700 cm^{-1} to hydroxyls of type I, II, and III, respectively.

The most cited model was proposed in 1978 by Knözinger and Ratnasamy *(112)* who considered the structure of the (100), (110), and (111) alumina planes. The authors attributed the A-band to hydroxyls of type I formed with the participation of octahedrally coordinated aluminum on the (110) and (111) crystal faces. The B-band was also assigned to type I OH groups but attached to tetrahedrally coordinated aluminum on the (110) and (111) planes. The C and D bands were associated with bridging hydroxyls. Hydroxyls bridging two octahedrally coordinated aluminum atoms on the (110) and (111) planes manifest a band denoted by C in the above list. The D-band corresponds to hydroxyls on the (111) plane bridging one tetrahedrally and one octahedrally coordinated aluminum. Finally, the E-band was attributed to triply bridged hydroxyls on the (111) plane (involving octahedrally coordinated aluminum ions).

Busca *et al. (675,676)* argued that the coordination state of the cation should play a more important role and associated the A- and B-bands with type I hydroxyls attached to tetrahedrally coordinated aluminum. The heterogeneity of these hydroxyls was explained by the absence (A-band) or the presence (B-band) of a coordination vacancy in the vicinity of the aluminum. Correspondingly, the authors proposed that type I OH groups bound to octahedrally coordinated aluminum are responsible for the appearance of the D- and E-bands (whereby the E-band is associated with the presence of a vacancy).

A more recent model was introduced by Digne *et al. (93)* on the basis of DFT calculations. According to this model, the A- and B-bands also correspond to type I hydroxyls, but the A-band is attributed to OH on tetragonal aluminum sites on the (110) plane, and the B-band is associated with octahedral aluminum on the (100) plane. In this model, the D-band is proposed to characterize type I hydroxyls that are attached to pentacoordinated aluminum sites on the (110) plane. The doubly and triply bridged hydroxyls are associated with the E- and F-bands, respectively.

Most of the above models consider crystal planes, but the relatively low concentrations of hydroxyl groups on activated samples indicate that corner, edge, and defect sites might be responsible for the stabilization of at least a fraction of the hydroxyl groups. As noted in Ref. *(673)*, the development of a realistic model of alumina hydroxyls is still a challenge.

The extraframework alumina species in zeolites are also of significant interest. They are primarily characterized by two groups of bands, around 3780 and 3665 cm^{-1}, respectively. It was established that, in contrast to the EFAL species responsible for Lewis acidity, the hydroxylated species are located inside the zeolite pores *(196)*.

The band at 3780 cm^{-1} has mainly been observed with samples of BEA zeolite, which is known to be easily dealuminated. This band resembles in position the A- and B-bands on alumina but the analogy has to be regarded with caution because in zeolites, the respective species are isolated. In any case, it seems reasonable to assign this band to type I hydroxyls. There are many controversial reports on the properties of these hydroxyls; the picture that is emerging is that they are basic, and the interaction with bases is rather secondary: the base is adsorbed on the aluminum cation and thus affects the spectral signature of these hydroxyl groups *(431)*.

The aluminols absorbing around 3665 cm^{-1} are very often detected in the spectra of zeolites. Most authors merely report their existence, and there are not many detailed investigations of their properties. It is established that these species are moderately acidic *(58,103,177,187,190,194,205,215,219,226,227)*, and an assignment to type II hydroxyls seems plausible.

8.4. Silica

The hydroxyl IR spectra of silica activated at high temperatures are relatively simple. All authors detect a sharp band at 3750-3740 cm^{-1} that is normally extended to lower frequencies. The density of the silanols on samples evacuated at a temperature of 973 K was estimated to be around 1.2 OH nm^{-2} *(409)*. The band maximum is slightly moved to higher wavenumbers with an increase of the activation temperature *(29,410)*. At the same time, the band becomes narrower and its FWHM is reported to be 4 cm^{-1} for samples activated at 1373 K *(410)*.

The band at 3750-3740 cm^{-1} is assigned to isolated terminal SiOH groups, without any dissent. However, NMR spectra indicate the presence of geminal hydroxyls on activated silica *(108)*. This observation provoked researchers to search for spectral features corresponding to geminal silanols. In principle, $Si(OH)_2$ species should be characterized by two OH modes, symmetric and antisymmetric, and Si(OH)(OD) species should manifest one additional OH vibration. No such features were detected in IR spectra, which suggested no vibrational interaction between the two hydroxyls in the geminal species. Indirect evidence supported the idea that geminal silanols are almost indistinguishable from isolated SiOH groups in IR spectra. Namely, it has been suggested that $Si(OH)_2$ species are characterized by a band having a maximum that is by 2–3 cm^{-1} lower than the SiOH band maximum *(29,409,410)*. According to Ref. *(29)*, even $Si(OH)_3$ groups can exist on silica surfaces; their band maximum would be by 1 cm^{-1} lower than that of geminal species. Theoretical investigations have shown that no H-bonds are formed between the two OH groups of the geminal silanols *(31)*.

It is established that the low-frequency tail of the silanol band is at least partly caused by the presence of internal (occluded) silanol groups. The occluded groups resist deuterium exchange and are characterized by a broad band with a maximum centered at about 3670-3660 cm^{-1} *(21,409)*.

When silica is activated at temperatures lower than 723 K, bands of H-bonded hydroxyls are evident in IR spectra. Hoffmann and Knözinger *(410)* proposed the following assignments of the IR bands of silica in the ν(OH) region:

Si–OH groups, 3749 cm^{-1};

$Si(OH)_2$ geminal hydroxyls, 3747 cm^{-1};

Vicinal silanols that are not H-bonded, 3744-3743 cm^{-1};

H-bonded vicinal silanols, 3715 and 3510 cm^{-1}.

The band at 3715 cm^{-1} is attributed to hydroxyls acting as H-donors, while the band at 3510 cm^{-1}, to hydroxyls acting as H-acceptors. Some authors support assignment of the 3510 cm^{-1} band to Si–OH pairs *(646)*, whereas others have broadened the assignment to include H-bonded hydroxyl chains *(614)*.

Additional clarifications of the spectral behavior of H-bonded hydroxyls have been obtained by analyzing the spectra of porous siliceous materials. Dealuminated BEA zeolite shows bands of internal and external isolated silanols (3750-3730 cm^{-1}) as well as bands assigned to silanol nests (at about 3500 and 3710 cm^{-1}) *(219)*. The bands of the isolated internal silanols are detected at lower frequencies than those of the external silanols because of the effect of pores on OH stretching modes. This effect has already been discussed in detail for the bridging hydroxyls in zeolites. A perfect silanol nest (see Figure 2.2) should create one band with a frequency characteristic of H-bonded hydroxyls; and this expectation is fulfilled by the band around 3500 cm^{-1}. However, the band at 3710 cm^{-1} was found to be associated with the band at about 3500 cm^{-1} and was assigned to silanols terminating H-bonded chains. The nature of these species and the frequency are reminiscent of a case discussed above: the assignment of the 3715 cm^{-1} band of silica to vicinal hydroxyls. Note that substantial heterogeneity of the H-bonded silanols has been observed with a silicalite sample (MFI structure) *(51)*. These results indicate that the structure of the silanol nests *(677)* as shown in Figure 2.2 is too idealized and in reality, hydroxyls engaged in H-bonding with only one oxygen atom (as shown in Figure 2.6) exist.

As already discussed, silica-alumina is characterized by hydroxyl spectra similar to those of pure silica, with some silanol groups demonstrating enhanced acidity. Silanol groups on the external surface of zeolites also behave like silanols on silica; also for these groups, enhanced acidity has been reported in some instances.

8.5. Titania

The most commonly employed titania modification in catalysis is anatase, whereas there are fewer applications of rutile. In many reports, mixed anatase–rutile samples were investigated. In sum, at least 12 types of OH groups on the anatase surface have been proposed by different authors *(115)*. The most frequently reported hydroxyls are characterized by IR bands around 3720-3715 and 3675 cm^{-1} *(27,28,115,269,272–274,330,357,504,651, 678,679)*. These two bands are well discernable after evacuation at a temperature of 673 K. A band around 3640 cm^{-1} is also often reported and characterizes less stable hydroxyls *(230,273,356,504,651)*. For a Degussa P25 sample (containing rutile), an additional band at 3662 cm^{-1} was reported *(273)*. An increase in the evacuation temperature results in a complex spectrum containing a set of low-intensity bands.

Busca *et al. (356)* investigated several titania samples and reported five OH groups with bands at 3735, 3725, 3715, 3670, and 3640 cm^{-1}. For a sulfated sample, an additional hydroxyl band at 3690 cm^{-1} was detected. Interestingly, only the two hydroxyls with the highest frequencies were reported to reversibly protonate ammonia. This behavior was explained by an assignment of the bands to silica impurities.

It was established that ammonia adsorption followed by evacuation at ambient temperature barely affects the OH spectrum. However, the acidity of titania hydroxyls drastically decreases as a result of adsorbed ammonia, as shown by CO and benzene adsorption *(269,357)*.

According to Tsyganenko and Filimonov *(27,28)*, the band at 3720–3715 cm^{-1} characterizes type I hydroxyls, whereas the bands at 3675 and 3640 cm^{-1} characterize hydroxyls of type II. The type II groups are more difficult to exchange with fluorine anions than the type I OH groups *(274)*.

The first model of anatase hydroxyl groups was based on the consideration of the (001) crystal face alone *(678)*. According to another model, considering all planes to be exposed on the surface, the stable hydroxyls are of type I and are located at crystal edges *(679)*. The appearance of two types of OH groups is explained by differences in the second coordination sphere of the titanium cations. Other authors support the idea that the OH groups on anatase are located at edges, corners, and surface defects *(115,273)*. More recently, DFT calculations were performed in order to determine the genesis and location of the anatase surface hydroxyls *(680,681)*. It was proposed that dissociation of H_2O on acid–base pairs located on the (001) plane of anatase produces only type I Ti–OH groups, because a surface Ti–O bond is broken in the process *(681)*. A band at 3736 cm^{-1} was attributed to OH groups on pentacoordinated Ti^{4+} ions in the (001) plane, whereas a band at 3699–3696 cm^{-1} was assigned to hydroxyls bonded to tetracoordinated Ti^{4+} ions in the (100) plane *(680)*. Bands at about 3740 and 3700 cm^{-1} were indeed found in experiments by the same authors *(680)* and by others *(563)* but the positions do not coincide with those commonly observed for titania hydroxyl bands.

Data regarding the hydroxyl population on rutile are less abundant and controversial. It appears that, for this titania modification, the sample morphology very strongly affects the surface properties. Various OH groups were reported: 3718, 3694, and 3673 cm^{-1} *(270)*; 3685, 3655, and 3410 cm^{-1} *(678)*; 3650 and 3410 cm^{-1} *(682)*; and 3670 cm^{-1} *(683)*. A model of the rutile surface considering the (110), (101), and (100) planes was proposed more than 40 years ago *(682)*.

8.6. Zirconia

The thermodynamically stable polymorph of zirconia is the monoclinic phase, and many investigations have been performed with this material (where in some instances, traces of tetragonal zirconia were present). Most authors detect two groups of bands in the hydroxyl region of activated samples, at 3780–3760 and 3680–3670 cm^{-1}, and assign them to OH groups of type I and III (or II), respectively *(55,275,276,335,364,381,483,497,518,636,684–686)* (see Figure 2.56). An alternative interpretation was presented by Jacob *et al. (116)* who proposed that the two types of hydroxyls differ in the coordination of the oxygen atom, which is either trigonal or tetragonal. Indeed, the authors found that spectra of tetragonal zirconia (which is characterized by tetrahedrally coordinated O^{2-}) display only one OH band at 3682 cm^{-1}. The equivalent of this band is detected at 3668 cm^{-1} in spectra of the monoclinic phase, and a second band at 3774 cm^{-1} arises from OH groups involving trigonally coordinated oxygen. This model, however, does not account for the substantial differences in the chemical properties of the two types of hydroxyls.

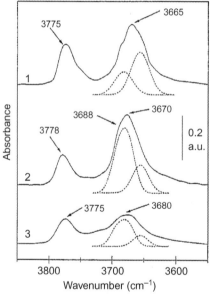

Figure 2.56 FTIR spectra of the ν_{OH} region of a ZrO_2 specimen calcined at progressively increasing temperatures and activated *in vacuo* at 673 K. 1: sample calcined at 873 K; 2: sample calcined at 1073 K; and 3: sample calcined at 1173 K (purely monoclinic zirconia). The dotted lines represent a tentative decomposition of the band of the bridged OH groups into two components by constraining the peak position (± 3 cm^{-1}) and the half-band width ($\Delta\nu_{1/2} = 31 \pm 3$ cm^{-1}). *Reproduced from Ref. (55).*

The population of type I hydroxyls (3780–3760 cm^{-1}) strongly decreases after evacuation at a temperature of 773 K *(275)*, and the bands disappear after evacuation at 873 K *(55)*. These hydroxyls are not acidic as established by low-temperature CO adsorption *(275)*. Being more basic, they interact preferentially with methanol *(518)* and CO_2 *(483)*. According to Yamaguchi et al. *(687)*, the treatment with chloroform leads to selective substitution of these hydroxyls by chloride ions.

The type III hydroxyls (3680–3670 cm^{-1}) are more stable and disappear during evacuation at 1073 K *(55)*. Some authors found that they are heterogeneous: the respective OH band consists of at least two components, as shown in Figure 2.56. It was proposed that the lower-frequency component of the band (less stable during evacuation at high temperatures) characterizes OH groups in the vicinity of other hydroxyls *(381)*. The acidity of the zirconia hydroxyls absorbing at 3660 cm^{-1}, as measured by low-temperature CO adsorption, was found to be very weak ($\Delta\tilde{\nu}_{OH} = -65\,cm^{-1}$) *(275,684)*.

Occasionally, an additional weak band around 3730 cm^{-1} has been detected and attributed to doubly bridged hydroxyls *(385,635,636,686)*. This band is often associated with traces of tetragonal zirconia that can be present in the monoclinic phase *(635,636,686)*.

Bands around 3400 cm^{-1} *(635,685)* or 3490 cm^{-1} *(688)* have been assigned to type III hydroxyls (for comparison, type II OH bands are located at 3682–3660 cm^{-1}).

In support of most of the above assignments, DFT calculations indicated that on the flat crystal planes of monoclinic zirconia, hydroxyls should exhibit bands at the following frequencies: type I in the region of 3822–3743 cm^{-1}, doubly bridged hydroxyls between 3755 and 3568 cm^{-1}, and triply bridged hydroxyls at 3647–3498 cm^{-1} *(686)*.

A few investigations address the OH groups on tetragonal zirconia. The type I hydroxyls were reported in the interval between 3765 and 3740 cm^{-1}, and their bands were of low intensity *(689,690)*. The type III OH groups were found at 3660 cm^{-1}, and doubly bridging hydroxyls were suggested to absorb at 3681 cm^{-1} *(690)*.

The ν(OH) positions of the zirconia hydroxyls are slightly red-shifted after sulfation. In general, the type I hydroxyls of sulfated samples appear with reduced intensity around 3770 cm^{-1} *(278,497,684)* and the bridging hydroxyls are detected around 3640 cm^{-1} *(261,276,278,335,497,684)*. According to Ref. *(688)*, a band at 3630 cm^{-1} indicates triply bridged OH groups in the vicinity of $S_2O_7^{2-}$ species. These hydroxyls demonstrate enhanced acidity as compared to those on nonsulfated samples. In addition, a broad band around 3400-3300 cm^{-1} is often detected (with the intensity strongly depending on the activation temperature) and is assigned to H-bonded hydroxyls from —SO_3H groups *(335,684)*.

8.7. Ceria

Ceria is an easily reducible oxide, and any redox pretreatment strongly affects the nature of the surface hydroxyl groups. Cerium oxide has been the subject of detailed DFT calculations, and the results were recently summarized *(691)*. It was noted that H_2O molecules strongly and dissociatively bind on oxygen vacancy sites and the dissociative adsorption of water is favored on defective surfaces. Even at low temperature, water dissociates on partially reduced ceria *(692)*. It was reported that surface hydroxyls on ceria are also produced as a result of H_2 dissociation *(509)*.

The following picture of the hydroxyl groups on ceria, based primarily on the findings of Lavalley and coworkers *(117,654,693)*, is broadly accepted:

- Type I, band at 3710 cm^{-1} *(117,654,685,693–696)*. These hydroxyls are of low stability and not detected in the spectra of samples activated at temperatures higher than 473 K. Bands at 3690 *(685,697)* and 3720 cm^{-1} *(698)* have also been assigned to type I hydroxyls.
- Type II, bands at 3660–3650 (II-A) and 3640–3634 cm^{-1} (II-B) *(509,639,654,695,697)*. These are the only hydroxyls detected on samples characterized by a low surface area. The position of the II-B band was reported at slightly lower wavenumbers of 3625–3620 cm^{-1} in Refs. *(654,696,697)*.
- Type III, band at 3600–3585 cm^{-1}. This band is of very weak intensity.
- A band at 3500 cm^{-1} is usually attributed to oxyhydroxy species (cerium hydroxide microphase *(654)*) or residual bicarbonates *(639)*. In some publications, the band was assigned to type III hydroxyls *(117,695,696,698)*.

The spectrum in the OH region is changed after reduction. The II-A band is gradually shifted with increasing degree of reduction to 3684 cm^{-1} and the II-B band, to 3651 cm^{-1} *(654,694)* (see Figure 2.57). According to Refs. *(693,699)*, the spectral position of OH species bridged over two Ce^{4+} ions is 3660 cm^{-1}. When an anionic vacancy affects the cerium cations, a band at 3640 cm^{-1} is observed. This band is the main feature in the spectrum after sample reduction. The type II hydroxyls rise in intensity when the reduction

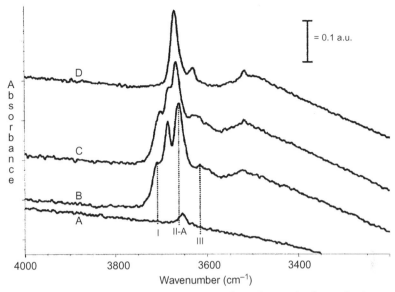

Figure 2.57 IR spectra showing the $\nu(OH)$ region of a ceria sample after activation and a series of exposures to H_2 at a temperature of 573 K (a), first exposure at 13 kPa of H_2 for 0.5 h (b), second exposure at 13 kPa of H_2 for 0.5 h (c), and third exposure at 3 kPa of H_2 for 0.5 h (d). Roman numerals refer to methoxy species on different types of hydroxyl groups. The notation II-A has been used because other type II species may be observed in addition. *Reproduced from Ref. (509).*

temperature exceeds 723 K; this phenomenon has been associated with surface reconstruction *(654).*

In addition, a new and broad band at 3450 cm^{-1} arises during reduction as a result of H_2 dissociation. The band characterizes very stable species and is attributed to H-bonded hydroxyls with a possible internal location. Other broad bands at 3620 *(117)* or 3700 cm^{-1} *(654)* are observed in the spectra of deeply reduced samples; H/D exchange experiments revealed that these bands are not associated with OH vibrations but arise from electronic transitions.

An alternative interpretation of the bands at 3675 and 3650 cm^{-1} (formed in the presence of H_2 at 523 K) to geminal hydroxyls was proposed in Ref. *(640).*

Interesting results on nanosized ceria samples with different morphology were presented by Agarwal *et al. (695).* The authors established that, although the hydroxyl spectra of all samples were similar, the reactivity of the OH groups strongly depended on the ceria nanoshape.

The type II surface hydroxyls of ceria are not acidic, according to low-temperature CO adsorption experiments *(114)*. However, a small shift of -70 to -80 cm^{-1} was detected in Ref. *(697)*, but details were not provided.

Timofeeva *et al.* *(700)* suggested that Ce–OH groups of ceria in SBA-15 are characterized by a band at 3620 cm^{-1}.

8.8. Other oxides

Spectral data of hydroxyl groups on some oxides were provided in Table 2.2. Table 2.29 summarizes data on metal oxides that were not individually discussed. It can be seen that the reported wavenumbers vary significantly for some oxides, for example, for La$_2$O$_3$. Evidently, the OH population of this oxide strongly depends on the sample origin and pretreatment.

8.9. Zeolites

Characterization of zeolite hydroxyls has been thoroughly discussed throughout the text. Instead of repeating this detail, we will provide a summary of the stretching frequencies at which the bridging hydroxyls of zeolites and related materials are observed (Table 2.30). We emphasize that the main factor determining this frequency is the size of the pore where the OH groups are located.

Table 2.29 OH stretching frequencies of hydroxyl groups on selected bulk and supported metal oxides

Oxide	$\tilde{\nu}$(OH), cm^{-1}	Note	Ref.
γ-Ga$_2$O$_3$	3693, 3660–3630 (broad)		*(379,701)*
γ-Ga$_2$O$_3$	3730, 3700, 3630, 3600	Produced by H$_2$ dissociation	*(702)*
β-Ga$_2$O$_3$	3720, 3653		*(703)*
α-Fe$_2$O$_3$	3690, 3665, 3648, 3628, 3590	After evacuation at 673 K	*(113)*
α-Cr$_2$O$_3$	3679, 3661, 3638, 3618, 3606		*(114)*
La$_2$O$_3$	3647, 3624		*(267)*
La$_2$O$_3$	3662–3658		*(638,704)*
La$_2$O$_3$	3685, 3662		*(705)*
SnO$_2$	3740–3730, 3660, 3625		*(268)*
SnO$_2$	3655, 3637, 3602, 3558		*(706)*
Nb$_2$O$_5$	3714		*(279)*

Table 2.30 OH stretching frequencies of bridging hydroxyl groups in zeolites and related porous materials

Material	Code name	Typical frequency^a, cm^{-1}	Note	Ref.
Y	FAU	3640, 3550	The bands shifted to higher frequencies with increasing Si/Al ratio	(63, 139, 166–169, 171, 394)
DAY	FAU	3625, 3600, 3560 3525		(131, 170, 173, 174, 176, 177, 191, 348, 457)
ZSM-5	MFI	3610		(102, 103, 105, 131, 152, 153, 157, 190–196, 200)
[Fe]ZSM-5	MFI	3635	Si(OH)Fe	(131, 200)
[Ga]ZSM-5	MFI	3622	Si(OH)Ga	(157, 200, 209)
FER	FER	3610, 3595, 3550		(154, 232, 434, 548)
MOR	MOR	3610, 3585		(167, 186, 369)
BEA	BEA	3610		(11, 156, 159, 322, 339, 340, 433, 436, 468, 486)
ZSM-12	MTW	3615, 3577		(98, 165, 195)
ZSM-22	TON	3606		(707)
[Fe]ZSM-22	TON	3628	Si(OH)Fe	(707)
ZSM-20	EMT/FAU	3632, 3552		(438)
EMT	EMT	3635, 3550		(171, 297, 381)
MCM-22	MWW	3627–3626		(109, 226–228)
MCM-49	–	3627		(109)

Continued

Table 2.30 OH stretching frequencies of bridging hydroxyl groups in zeolites and related porous materials—cont'd

Material	Code name	Typical frequency, cm^{-1}	Note	Ref.
MCM-58	IFR	3628, 3556		(98)
MCM-68	MSE	3617, 3574		(98)
ITQ-2	–	3628		(226)
SUZ-4	SZR			(595)
SSZ-24	AFI	3612		(104)
CHA	CHA	3620		(231)
SSZ-13	CHA	3613		(212)
		3584		
TON	TON	3598		(229)
MAZ	MAZ	3610		(343)
ERI	ERI	3612, 3565		(623)
NCL-1	–	3628, 3598		(708)
[Ga]NCL-1	–	3650, 3640		(708)
[Fe]NCL-1	–	3650, 3615		(708)
SAPO-34	CHA	3625–3614		(210,211,624)
		3630		
		3601–3600		
SAPO-5	AFI	3626		(102,319,327,348)

Material	Framework	Frequencies (cm⁻¹)		
SAPO-5	AFI	3625 / 3520		(327)
SAPO-5	AFI	3625 / 3600		(389)
SAPO-18		3626 / 3600		(397,535)
SAPO-40	AFR	3643–3628		(226,237)
SAPO-44	–	3625, 3600		(389)
SAPO-4	–	3630, 3614, 3600		(210)
SAPO-37	FAU	3645, 3575		(297)
SAPO-11	AEL	3630		(348,624)
SAPO-17	–	3626 / 3616 / 3593		(624)
CoAPO-5	AFI	3628		(348)
CoAPO-11	AEL	3628		(348)
CoAPO-18	AEI	3575	Co(OH)P	(535)

[a]Usually, the reported frequencies are within a 10 cm⁻¹ range from the typical frequency. At low temperature, the bands are slightly shifted to higher frequencies. The hydroxyls are of Si(OH)Al type unless otherwise noted.

Table 2.31 Proposed frequencies of hydroxyls formed on cations in zeolites (introduced by ion exchange)

Sample	Exchanged cation	Frequency, cm^{-1}	Note	Ref.
MgBEA	Mg	3692	Alternatively assigned to bridging OH groups affected by the respective cation	(549)
CaBEA	Ca	3682		
SrBEA	Sr	3679		
BaBEA	Ba	3672		
LTA	Ca	3555	Possibly a band at 3514 cm^{-1}	(709)
Cu-ZSM-5	Cu^{2+}	3657		(710)
Co-ZSM-5	Co	3665		(552)
Co-MOR	Co	3670, 3650		(711)
Co-MOR	Co	3654		(491)
Fe-ZSM-5	Fe	3650		(551)
Fe-ZSM-5	Fe^{3+}	3672		(537)
Fe-ZSM-5	Fe^{2+}	3665		(20)
	Fe^{3+}	3670		
FeBEA	Fe^{3+}	3686–3683		(217,537,556)
FeBEA	Fe^{3+}	3670	Alternatively, Al–OH	(550)
FeFER	Fe^{3+}	3674		(284)

Table 2.31 reports the frequencies of hydroxyl groups bound to cations that were introduced into zeolites by ion exchange. These hydroxyls can be formed according to the reaction of Equation (2.24) and, as a rule, are unstable. In some instances, the reported frequencies are very close to the frequencies of hydroxyls attached to EFAL species, and the assignments should be viewed with caution. With very few exceptions (217,709), the hydroxyl groups of this type are poorly characterized.

8.10. Metal-organic frameworks

In recent years, MOF materials have attracted increasing interest because of the prospects for various applications. Like zeolites, many of the MOFs contain structural OH groups, often bridging two or more metal atoms (712). In some cases, hydroxyls located on the external surface are found. With a few exceptions, the thorough characterization of the hydroxyl groups in MOFs has yet to be performed.

The MOF materials that have been extensively investigated by means of Fourier transform infrared (FTIR) spectroscopy of the hydroxyl groups are metal hydroxoterephthalates. Aluminum hydroxoterephthalate, denoted as MIL-53(Al), is characterized by structural bridging hydroxyls absorbing at 3704–3702 cm^{-1} *(713,714)*, and the band position is in good agreement with DFT calculations *(714)*. When aluminum is replaced by gallium, the ν(OH) modes are shifted to 3669–3668 cm^{-1} *(713,714)*. Two bridging OH groups are reported for chromium hydroxoterephthalate: at 3655 and 3610 cm^{-1} *(715)*. The OH frequencies in all materials of this family decrease to 3460–3360 cm^{-1} when the OH groups are involved in H-bonding with trapped H_2O molecules *(713)*.

An amino-functionalized Al-sample, NH_2-MIL-53(Al), was found to have the same OH groups as the amino-free sample (3700–3693 cm^{-1} *(716–718)*) as well as additional hydroxyls (3660 *(716)* or 3680 cm^{-1} *(717,718)*) attributed to OH groups interacting with the amino group via H-bonding *(716–718)*. A decrease in the OH frequency is reported in the sequence aluminum > gallium > indium *(717)*.

In analogy to zeolites, one may expect that the bridging hydroxyls in some MOFs should be highly acidic. However, the available data reveal rather weak acidity of these hydroxyls. Thus, the OH groups of MIL-53(Al) are shifted by -30 to -50 cm^{-1} after low-temperature CO adsorption *(714)*. A slightly higher acidity was observed for a MIL-53(Ga) sample: $\Delta\tilde{\nu}$(OH) $= -50$ to -100 cm^{-1} *(714)*. The OH bands of NH_2-MIL-53 (Cr) are shifted by only -180 cm^{-1} after adsorption of CD_3CN (i.e., less than silanols). Because of the basic character, it was proposed that adsorbed CO_2 is attached to the oxygen atoms from the hydroxyl groups *(715)*. To create Brønsted acidity, MIL-53(Al) MOF was sulfated and a broad band around 3000 cm^{-1} was attributed to OH stretching modes of sulfoxy groups *(719)*.

A zirconium 1,4-benzenedicarboxylate MOF (UiO-66) is characterized by triply bridged OH groups with ν(OH) at 3674–3672 cm^{-1} *(720,721)*. Full dehydroxylation occurs at 523 K, but the process is reversible and the OH groups are recovered after water adsorption *(721)*. This phenomenon has been related to the ability of zirconium cations to easily change their coordination number *(712)*. The acidity of the OH groups is weak, with a CO-induced shift of the OH modes of -83 cm^{-1} *(721)*. For an amino-functionalized UiO-66 sample, ν(OH) is observed at slightly lower frequencies, 3666 cm^{-1} (ν(OD) at 2708 cm^{-1}) *(637)*.

A nickel phosphonate MOF material (Ni-STA-12) exhibits a band at 3685 cm^{-1} related to isolated NiOH groups. These hydroxyls show no acidity according to results of low-temperature CO adsorption *(722)*.

In Ni-polypyrazol [Ni$_8$(OH)$_4$(OH$_2$)$_2$(4,4'-bis(1H-pyrazol-4-yl) bipheny)$_6$], the bridging hydroxyls are in an unusual μ_4 bridging position and are registered at 3593 cm^{-1} *(723,724)*. Also in this case, the acidity is very low: the CO-induced shift of the OH modes is only -50 to -57 cm^{-1}.

Investigations of a chromium tricarboxylate, Cr$_3^{III}$OF$_x$(OH)$_{1-x}$ (H$_2$O)$_2\cdot${C$_6$H$_3$(CO$_2$)$_3$}$_2\cdot n$H$_2$O (MIL-100), have revealed that the free Cr—OH hydroxyl groups (3585 cm^{-1}) are of very low concentration (0.05 per chromium atom) *(265,266)*. These groups are also of weak acidity; the CO-induced shift of the OH modes amounts to -90 cm^{-1}, which is similar to the shift measured for free silanols *(266)*.

It should be emphasized that MOFs often contain structural water or trapped H$_2$O molecules that may be H-bonded to structural hydroxyl groups.

9. CONCLUSIONS

The analysis of surface species by infrared spectroscopy has been developed during the past five decades, but good quality spectra have only become widely available with the advent of modern FTIR spectrometers 30 years ago. Because the equipment is relatively inexpensive and the technique is highly informative, there are already thousands of papers focused on or including infrared surface characterization. Even when an investigation is addressing another problem, usually the spectrum of the sample in the OH region is described.

This review is focused on the identification and characterization of the surface hydroxyl groups. We believe that a detailed knowledge of their properties is the basis for further investigations, including interaction of hydroxyls with different reactant molecules and elucidation of the mechanism of catalytic reactions by spectroscopy of the working catalyst.

The most commonly used mode for characterization of surface hydroxyls is the stretching frequency, ν(OH). The free OH$^-$ ion vibrates at 3555.6 cm^{-1} and its bonding to one or more atoms leads to an increase of this value up to 3800 cm^{-1}. The shift depends on the electrostatic field experienced by the OH group and on the ionic or covalent character of the bond. As a result, ν(OH) is not specific and cannot be used for unambiguous identification of the nature of the hydroxyl. Some specificity is observed with covalently bound hydroxyls such as Si—OH and P—OH.

For a detailed characterization of OH groups, it is necessary to investigate their interaction with probe molecules.

Depending on the number of surface cations to which the hydroxyl oxygen is bound, surface hydroxyls are of type I, II, and so on. Although primarily determined by the nature of the cations, the acidity of the OH groups increases from type I to types II, III, etc.

Acidity is a very important characteristic of hydroxyls. A widely used technique for assessing acidity of surface OH groups is the so-called H-bond method. Involvement of the hydroxyl in H-bonding with a probe molecule leads to a strong decrease of ν(OH) and an increase of the absorption coefficient and the FWHM of the respective bands. The shift value depends on the PAs of the OH group and of the bases involved in the formation of the H-bond and is used as a measure of the intrinsic acidity of the hydroxyl. While a large body of data on the H-bonding of different hydroxyls with various bases exists in the literature and may be consulted, the base often interacts simultaneously with other surface sites (multicenter adsorption) and the data are of limited use. The best probes are small molecules such as CO, N_2, and CD_3CN. Another way of measuring hydroxyl acidity is protonation of strong bases such as ammonia and pyridines. The protonation process is strongly favored by the stabilization of the conjugated acid on the surface. Cooperative action of two basic molecules is often required to extract a proton from the OH groups.

Basic hydroxyls are normally attached to one surface cation (type I) and interact preferentially with acidic molecules, for example, with CO_2 to form bicarbonates.

The accessibility of the hydroxyl groups in porous materials is important for acid catalysis. The accessibility and indirectly the location of the OH groups can be determined by utilizing probe molecules of varying dimensions. To characterize the probe molecules, the kinetic diameter of the molecules is typically used although it is not the best quantity to describe molecular dimension.

Deuteration of hydroxyls is a powerful methodology helping the investigations of hydroxyl groups. H/D exchange is a unique way to prove whether a given mode involves a hydrogen atom. With suitable experiments, chemical and spectral effects may be easily distinguished. Because of the low noise, OD spectra are of high quality. However, it should be taken into account that the OD groups are slightly less acidic than the respective OH groups, which hampers direct extrapolation of the results obtained with OD groups to surface OH groups.

Analysis of the deformation modes, overtones, and combination modes may also be very helpful in resolving particular questions, for example, discrimination between OH groups and adsorbed water.

Combination of IR spectroscopy with other techniques provides additional information but should be carefully made because sometimes the differences in the pretreatment conditions may affect the state of the surface hydroxyls. The most informative complementary experimental technique is NMR spectroscopy. In the recent years, much progress was attained by combining FTIR spectroscopy with DFT calculations. The role of computational chemistry for the proper assignment of IR observations increases.

We hope that this review will be helpful for the planning of experiments, choice of probe molecules, and interpretation of the results, not only for beginners but also for the experienced scientists in the field.

ACKNOWLEDGMENTS

The author thanks Prof. M. Daturi for the helpful discussions. The support from the Union Centre of Excellence (Contract No. DCVP 02-2/2009 with the Bulgarian Science Fund) is also acknowledged.

REFERENCES

1. Rosenbaum, N. H.; Owrutsky, J. C.; Tack, L. M.; Saykally, R. J. *J. Chem. Phys.* **1986**, *84*, 5308–5313.
2. Herzberg, G. *Molecular Spectra and Molecular Structure, Vol. I. Spectra of Diatomic Molecules*; Van Nostrand Reinhold: New York, 1945.
3. Cheng, B. M.; Lee, Y.-P.; Ogilvie, J.-F. *Chem. Phys. Lett.* **1988**, *151*, 109–115.
4. Kotorlenko, L. A.; Aleksandrova, V. S. *Russ. Chem. Rev.* **1984**, *53*, 1139–1153.
5. Cotton, F. A.; Wilkinson, G.; Murillo, C.; Bochmann, M. *Advanced Inorganic Chemistry*; John Wiley & Sons: New York, 1999.
6. Sathyanarayana, D. N. *Vibrational Spectroscopy: Theory and Applications*; New Age International: Delhi, India, 2007.
7. Veera Reddy, K. *Symmetry and Spectroscopy of Molecules*; New Age International: Delhi, India, 1998.
8. Weckler, B.; Lutz, H. D. *Spectrochim. Acta A* **1996**, *52*, 1507–1513.
9. Nakamoto, K. *Infrared and Raman Spectra of Inorganic and Coordination Compounds, Part B, Application in Coordination, Organometallic, and Bioinorganic Chemistry*, 6th ed.; John Wiley & Sons: Hoboken, New Jersey, 2009.
10. Lavalley, J. C.; Bensitel, M.; Gallas, J. P.; Lamotte, J.; Busca, G.; Lorenzelli, V. *J. Mol. Struct.* **1988**, *175*, 453–458.
11. Paze, C.; Bordiga, S.; Lamberti, C.; Salvalaggio, M.; Zecchina, A.; Bellussi, G. *J. Phys. Chem. B* **1997**, *101*, 4740–4751.
12. Beckenkamp, K.; Lutz, H. D. *J. Mol. Struct.* **1992**, *270*, 393–405.
13. Appelman, E. H.; Kim, H. *J. Chem. Phys.* **1972**, *57*, 3272–3276.
14. Schwager, I.; Arkell, A. *J. Am. Chem. Soc.* **1967**, *89*, 6006–6008.
15. Guillory, W. A.; Bernstein, M. L. *J. Chem. Phys.* **1975**, *62*, 1058–1060.
16. Lutz, H. D. *Struct. Bonding* **1995**, *82*, 85–103.
17. Hermansson, K. *Int. J. Quant. Chem.* **1993**, *45*, 747–758.

18. Okazaki, S.; Okada, I. *J. Chem. Phys.* **1989**, *90*, 5595–5605.
19. Almeida, A. L.; Martins, J. L.; Taft, C. A.; Longo, E.; Lester, W. A., Jr. *J. Chem. Phys.* **1998**, *109*, 3671–3685.
20. Panov, G. I.; Starokon, E. V.; Pirutko, L. V.; Paukshtis, E. A.; Parmon, V. N. *J. Catal.* **2008**, *254*, 110–120.
21. Chakarova, K.; Drenchev, N.; Mihaylov, M.; Nikolov, P.; Hadjiivanov, K. *J. Phys. Chem. C* **2013**, *117*, 5242–5248.
22. Srivastava, S. P.; Singh, I. D. *Acta Phys. Acad. Sci. Hung.* **1979**, *47*, 275–280.
23. Niwa, M.; Suzuki, K.; Katada, N.; Kanougi, T.; Atoguchi, T. *J. Phys. Chem. B* **2005**, *109*, 18749–18757.
24. Niwa, M.; Suzuki, K.; Isamoto, K.; Katada, N. *J. Phys. Chem. B* **2006**, *110*, 264–269.
25. Dyer, A. *An Introduction to Zeolite Molecular Sieves*; John Wiley & Sons: Australia, 1988.
26. Sayed, M. B.; Kydd, R. A.; Cooney, R. P. *J. Catal.* **1984**, *88*, 137–149.
27. Tsyganenko, A. A.; Filimonov, V. N. *Spectrosc. Lett.* **1972**, *5*, 477–484.
28. Tsyganenko, A. A.; Filimonov, V. N. *J. Mol. Struct.* **1973**, *19*, 579–589.
29. Takei, T.; Kato, K.; Meguro, A.; Chikazawa, M. *Colloids Surf. A* **1999**, *150*, 77–84.
30. Zhuravlev, L. T. *Colloids Surf. A* **2000**, *173*, 1–38.
31. Sauer, J.; Ugliengo, P.; Garrone, E.; Saunderss, V. R. *Chem. Rev.* **1994**, *94*, 2095–2180.
32. Madejová, J.; Pentrák, M.; Pálková, M.; Komadel, P. *Vib. Spectrosc.* **2009**, *49*, 211–218.
33. Chuang, I.-S.; Maciel, G. E. *J. Phys. Chem. B* **1997**, *101*, 3052–3064.
34. Van Der Voort, P.; White, M. G.; Mitchell, M. B.; Verberckmoes, A. A.; Vansant, E. F. *Spectrochim. Acta* **1997**, *53A*, 2181–2187.
35. Grabowski, S. J. Ed. *Hydrogen Bonding—New Insight*; Springer: The Netherlands, 2006.
36. Nibbering, E. T. J.; Dreyer, J.; Kühn, O.; Bredenbeck, J.; Hamm, P.; Elsaesser, T. *Chem. Phys.* **2007**, *87*, 619–687.
37. Pimentel, G. C.; McClellan, A. L. *The Hydrogen Bond*. W. H. Freeman: San Francisco, London, 1960.
38. Lutz, H. D. *J. Mol. Struct.* **2003**, *646*, 227–236.
39. Schuster, P.; Zundel, G.; Sandorfy, C. *The Hydrogen Bond—Recent Developments in Theory and Experiments*. North-Holland: Amsterdam/New York/Oxford, 1976.
40. Rekik, N.; Issaou, N.; Ghalla, H.; Oujia, B.; Wójcik, M. J. *J. Mol. Struct. THEOCHEM.* **2007**, *821*, 9–21.
41. Di Paolo, T.; Bourdéron, C.; Sandorfy, C. *Can. J. Chem.* **1972**, *50*, 3161–3166.
42. Perchard, J. P. *Chem. Phys.* **2001**, *273*, 217–233.
43. Berthomieu, C.; Sándorfy, C. *J. Mol. Spectrosc.* **1965**, *15*, 15–21.
44. Durocher, G.; Sándorfy, C. *J. Mol. Spectrosc.* **1965**, *15*, 22–28.
45. Sándorfy, C. *J. Mol. Struct.* **2006**, *790*, 50–54.
46. Kolomijtsova, T. D.; Shchepkin, D. N. *J. Mol. Struct.* **1994**, *322*, 211–216.
47. Rospenk, M.; Zeegers-Huyskens, T. *J. Phys. Chem.* **1997**, *101*, 8428–8434.
48. Hoelderich, W. F.; Dahlhoff, G. *Chem. Innov.* **2001**, *31*, 29–40.
49. Dzwigaj, S.; Janas, J.; Gurgul, J.; Socha, R. P.; Shishido, T.; Che, M. *Appl. Catal. B* **2009**, *85*, 131–138.
50. Dzwigaj, S.; El-Malki, E.; Peltre, M.-J.; Massiani, P.; Davidson, A.; Che, M. *Top. Catal.* **2000**, *11/12*, 379–390.
51. Bordiga, S.; Ugliengo, P.; Damin, A.; Lamberti, C.; Spoto, G.; Spano, G.; Buzzoni, R.; Dalloro, L.; Rivetti, L. *Top. Catal.* **2001**, *15*, 43–52.
52. Krijnen, S.; Sanchez, P.; Jakobs, B. T. F.; van Hooff, J. H. C. *Microporous Mesoporous Mater.* **1999**, *31*, 163–173.
53. Forni, L.; Fornasari, G.; Giordano, G.; Lucarelli, C.; Katovic, A.; Trifiro, F.; Perric, C.; Nagy, J. B. *Phys. Chem. Chem. Phys.* **2004**, *6*, 1842–1847.

54. Bailly, M. L.; Chizallet, C.; Costentin, G.; Krafft, J.-M.; Lauron-Pernot, H.; Che, M. *J. Catal.* **2005**, *235*, 413–422.
55. Cerrato, G.; Bordiga, S.; Barbera, S.; Morterra, C. *Appl. Surf. Sci.* **1887**, *115*, 53–65.
56. Davydov, A. A. *Molecular Spectroscopy of Oxide Catalyst Surfaces*; John Wiley & Sons: Chichester, England, 2003.
57. Hunger, M.; Weitkamp, J. *Angew. Chem. Int. Ed.* **2001**, *40*, 2954–2971.
58. Gabrienko, A. A.; Danilova, I. G.; Arzumanov, S. S.; Toktarev, A. V.; Freude, D.; Stepanov, A. G. *Microporous Mesoporous Mater.* **2010**, *131*, 210–216.
59. Busca, G. *Chem. Rev.* **2007**, *107*, 5366–5410.
60. Derouane, E. G.; Védrine, J. C.; Ramos Pinto, R.; Borges, P. M.; Costa, L.; Lemos, M. A. N. D. A.; Lemos, F.; Ramôa Ribeiro, F. *Catal. Rev. Sci. Eng.* **2013**, *55*, 454–515.
61. Haw, J. F.; Xu, T. *Adv. Catal.* **1998**, *42*, 115–180.
62. Hunger, M. *Handbook of Heterogeneous Catalysis*; In: Ertl, G., Knozinger, H., Schuth, F., Weitkamp, J. Eds.; Vol. 2; Wiley-VCH: Weinheim, 2008; pp 1163–1178.
63. Sauer, J. *J. Mol. Catal.* **1989**, *54*, 312–323.
64. Brunner, E.; Pfeifer, H. *Mol. Sieves* **2008**, *6*, 1–44.
65. Ernst, H.; Freude, D.; Mildner, T.; Pfeifer, H. *Proc. 12th Int. Zeolite Conf., Mater. Res. Soc.* 1999; pp 2955–2962.
66. Shenderovich, I. G.; Buntkowsky, G.; Schreiber, A.; Gedat, E.; Sharif, S.; Albrecht, J.; Golubev, N. S.; Findenegg, G. H.; Limbach, H.-H. *J. Phys. Chem. B* **2003**, *107*, 11924–11939.
67. Chuang, I.-S.; Maciel, G. E. *J. Am. Chem. Soc.* **1996**, *118*, 401–406.
68. Kazansky, V. B.; Seryk, A. I.; Semmer-Herledan, V.; Fraissard, J. *Phys. Chem. Chem. Phys.* **2003**, *5*, 966–969.
69. Sivadinarayana, C.; Choudhary, T. V.; Daemen, L. L.; Eckert, J.; Goodman, D. W. *J. Am. Chem. Soc.* **2004**, *126*, 38–39.
70. Loong, C. K.; Richardson, J. W.; Ozawa, M. *J. Catal.* **1995**, *157*, 636–644.
71. Spencer, E. C.; Levchenko, A. A.; Ross, N. L.; Kolesnikov, A. I.; Boerio-Goates, J.; Woodfield, B. F.; Navrotsky, A.; Li, G. *J. Phys. Chem. A* **2009**, *113*, 2796–2800.
72. Smrcok, L.; Tunega, D.; Ramirez-Cuesta, A. J.; Scholtzová, E. *Phys. Chem. Mirer.* **2010**, *37*, 571–579.
73. Knop-Gericke, A.; Kleimenov, E.; Hävecker, M.; Blume, R.; Teschner, D.; Zafeiratos, S.; Schlögl, R.; Bukhtiyarov, V. I.; Kaichev, V. K.; Prosvirin, I. P.; Nizovskii, A. I.; Bluhm, H.; Barinov, A.; Dudin, P.; Kiskinova, M. *Adv. Catal.* **2009**, *52*, 213–272.
74. Gonzalez-Elipe, A. R.; Espinos, J. P.; Fernandez, A.; Munuera, G. *Appl. Surf. Sci.* **1990**, *45*, 103–108.
75. Cappus, D.; Xu, C.; Ehrlich, D.; Dillmann, B.; Ventnce, C. A., Jr.; Al Shamery, K.; Kuhlenbeck, H.; Freund, H.-J. *Chem. Phys.* **1993**, *177*, 533–546.
76. McCafferty, E.; Wightman, J. P. *Surf. Interface Anal.* **1998**, *26*, 549–564.
77. Auroux, A. *Mol. Sieves* **2008**, *6*, 45–152.
78. Mekki-Barrada, A.; Aurroux, A. *Characterization of Solid Materials and Heterogeneous Catalysts: From Structure to Surface Reactivity*; In: Che, M., Védrine, J. C. Eds.; Vol. 1, Wiley-VCH: Weinheim, 2012; pp 747–852.
79. Cvetanovic, R. J.; Amenomiya, Y. *Adv. Catal.* **1967**, *17*, 103–118.
80. Rudzinski, W.; Borowiecki, T.; Panczyk, T.; Dominko, A. *Adv. Colloid Interf. Sci.* **2000**, *84*, 1–26.
81. Miyamoto, Y.; Katada, N.; Niwa, M. *Microporpus Mesoporous Mater.* **2000**, *40*, 271–281.
82. Costa, C.; Dzikh, I. P.; Lopes, J. M.; Lemos, F.; Ramôa Ribeiro, F. *J. Mol. Catal. A* **2000**, *154*, 193–201.
83. Czjzek, M.; Jobic, H.; Fitch, A. N.; Vogt, T. *J. Phys. Chem.* **1992**, *96*, 1535–1540.

84. Guisnet, M. *Acc. Chem. Res.* **1990**, *23*, 392–398.
85. Kramer, G.; McVicker, G.; Ziemiak, J. *J. Catal.* **1985**, *92*, 355–358.
86. McVicker, G. B.; Feeley, O. C.; Ziemiak, J. J.; Vaughan, D. E. W.; Strohmaier, K. C.; Kliewer, W. R.; Leta, D. P. *J. Phys. Chem. B* **2005**, *109*, 2222–2226.
87. Lercher, J. A.; Jentys, A.; Brait, A. *Mol. Sieves* **2008**, *6*, 153–212.
88. Van Santen, R. A.; Kramer, G. *J. Chem. Rev.* **1995**, *95*, 637–660.
89. Bates, S. P.; van Santen, R. *Adv. Catal.* **1998**, *42*, 1–114.
90. Van Santen, R. A. *Stud. Surf. Sci. Catal.* **1994**, *85*, 273–294.
91. Kubicki, J. D.; Sykes, D.; Rossman, J. R. *Phys. Chem. Miner.* **1993**, *20*, 425–432.
92. Vittadini, A.; Casarin, M.; Selloni, A. *Theor. Chem. Acc.* **2007**, *117*, 663–671.
93. Digne, M.; Sautet, P.; Raybaud, P.; Euzen, P.; Toulhoat, H. *J. Catal.* **2002**, *211*, 1–5.
94. Haag, W. O.; Lago, R. M.; Weisz, P. B. *Nature* **1984**, *309*, 589–591.
95. Che, M. *Turning Points in Solid-State, Materials and Surface State: A Book in Celebration of the Life and Work of Sir John Meurig Thomas*; In: Harris, K. D. M., Edwards, P. P. Eds., Royal Society of Chemistry: Cambridge, 2008; pp 588–603.
96. Loeffner, E.; Peuker, C.; Jerschkewitz, H. G. *Catal. Today* **1988**, *3*, 415–420.
97. Jacobs, P. A.; Mortier, W. J. *Zeolites* **1982**, *2*, 226–230.
98. Gil, B.; Košová, G.; Čejka, J. *Microporous Mesoporous Mater.* **2010**, *129*, 256–266.
99. Chu, C. T.-W.; Chang, C. D. *J. Phys. Chem.* **1985**, *89*, 1569–1571.
100. Strode, P.; Neyman, K. M.; Knozinger, H.; Rosch, N. *Chem. Phys. Lett.* **1995**, *240*, 547–552.
101. Pu, S.-B.; Inui, T. *Zeolites* **1997**, *19*, 452–454.
102. Kubelková, L.; Beran, S.; Lercher, J. A. *Zeolites* **1989**, *9*, 539–543.
103. Almutairi, S. M. T.; Mezari, B.; Pidko, E. A.; Magusin, P. C. M. M.; Hensen, E. J. M. *J. Catal.* **2013**, *307*, 194–203.
104. Erichsen, M. W.; Svelle, S.; Olsbyem, U. *Catal. Today* **2013**, *215*, 216–233.
105. Bevilacqua, M.; Montanari, T.; Finocchio, E.; Busca, G. *Catal. Today* **2006**, *116*, 132–142.
106. Trombetta, M.; Busca, G.; Rossini, S.; Piccoli, V.; Cornaro, U.; Guercio, A.; Catani, R.; Willey, R. J. *J. Catal.* **1998**, *179*, 581–596.
107. Chizallet, S.; Raybaud, P. *Angew. Chem. Int. Ed.* **2009**, *48*, 2891–2893.
108. Morrow, B. A.; Gay, I. D. *J. Phys. Chem.* **1988**, *92*, 5569–5571.
109. Gil, B.; Marszałek, B.; Micek-Ilnicka, A.; Olejniczak, Z. *Top. Catal.* **2010**, *53*, 1340–1348.
110. Bonelli, B.; Cozzolino, M.; Tesser, R.; Di Serioc, M.; Piumetti, M.; Garrone, E.; Santacesaria, E. *J. Catal.* **2007**, *246*, 293–300.
111. Noei, H.; Qiu, H.; Wang, Y.; Löffler, E.; Wöll, C.; Muhler, M. *Phys. Chem. Chem. Phys.* **2008**, *10*, 7092–7097.
112. Knözinger, H.; Ratnasamy, P. *Catal. Rev. Sci. Eng.* **1978**, *17*, 31–70.
113. Lorenzelli, V.; Busca, G. *Mater. Chem. Phys.* **1985**, *13*, 261–281.
114. Zaki, M. I.; Knözinger, H. *Mater. Chem. Phys.* **1987**, *17*, 201–215.
115. Hadjiivanov, K.; Klissurski, D. *Chem. Soc. Rev.* **1996**, *25*, 61–69.
116. Jacob, K. H.; Knözinger, E.; Beniez, S. *J. Mater. Chem.* **1993**, *3*, 651–657.
117. Laachir, L.; Perrichon, V.; Badri, A.; Lamotte, J.; Catherine, R.; Lavalley, J. C.; El Fallah, J.; Hilaire, L.; Le Normand, F.; Quemere, E.; Sauvion, G. N.; Touret, O. *J. Chem. Soc. Faraday Trans.* **1991**, *87*, 1601–1609.
118. Pacchioni, G.; Sousa, C.; Illas, F.; Parmigiani, F.; Bagus, P. S. *Phys. Rev. B* **1993**, *48*, 11573–11582.
119. Hadjiivanov, K.; Kantcheva, M.; Klissurski, D. *J. Catal.* **1992**, *134*, 299–310.
120. Wuttke, S.; Vimont, A.; Lavalley, J. C.; Daturi, M.; Kemnitz, E. *J. Phys. Chem. C* **2010**, *114*, 5113–5120.
121. Morrow, B. A.; Ramamurthy, P. *J. Phys. Chem.* **1973**, *77*, 3052–3058.

122. Andersson, K.; Gŕomez, A.; Glover, C.; Nordlund, D.; Öström, H.; Schiros, T.; Takahashi, O.; Ogasawara, H.; Pettersson, L. G. M.; Nilsson, A. *Surf. Sci.* **2005**, *585*, L183–L189.
123. Daturi, M. *Curr. Phys. Chem.* **2012**, *2*, 178–188.
124. Arrhenius, S. A. Z. *Phys. Chem.* **1887**, *1*, 631–648.
125. Brønsted, J. N. *Recl. Trav. Chim. Pays-Bas* **1923**, *42*, 718–728.
126. Lowry, T. M. *Chim. Ind. (London)* **1923**, *42*, 43–47.
127. Lewis, G. N. *Valency and Structure of Atoms and Molecules*; Wiley: New York, 1923.
128. Walling, C. *J. Am. Chem. Soc.* **1950**, *72*, 1164–1168.
129. Paukshtis, E. A.; Yurchenko, E. N. *Russ. Chem. Rev.* **1983**, *52*, 242–258.
130. Lercher, J. A.; Grundling, C.; Edermirth, G. *Catal. Today* **1996**, *27*, 353–376.
131. Knözinger, H.; Huber, S. *J. Chem. Soc. Faraday Trans.* **1998**, *94*, 2047–2059.
132. Zecchina, A.; Lamberti, C.; Bordiga, S. *Catal. Today* **1998**, *41*, 169–177.
133. Coluccia, S.; Marchese, L.; Martra, G. *Microporous Mesoporous Mater.* **1999**, *30*, 43–56.
134. Busca, G. *Phys. Chem. Chem. Phys.* **1999**, *1*, 723–736.
135. Zecchina, A.; Spoto, G.; Bordiga, S. *Phys. Chem. Chem. Phys.* **2005**, *7*, 1627–1642.
136. Gribov, E. N.; Cocina, D.; Spoto, G.; Bordiga, S.; Ricchiardi, G.; Zecchina, A. *Phys. Chem. Chem. Phys.* **2006**, *8*, 1186–1196.
137. Stavitski, E.; Weckhuysen, B. M. *Zeolites and Catalysis: Synthesis, Reactions and Applications*; In: Cejka, J., Corma, A., Zones, S. Eds.; Wiley, VCH: Weinheim, 2010.
138. Thibault-Starzyk, F.; Maugé, F. *Characterization of Solid Materials and Heterogeneous Catalysts: From Structure to Surface Reactivity*; In: Che., M., Vedirne, J. C. Eds.; Vol. 1; Wiley, VCH: Weinheim, 2012; pp 3–48.
139. Karge, H. G.; Geidel, E. *Mol. Sieves* **2004**, *4*, 1–200.
140. Hadjiivanov, K.; Saussey, J.; Freysz, J. L.; Lavalley, J. C. *Catal. Lett.* **1998**, *52*, 103–108.
141. Iohanssen, A. V. *Theor. Eksp. Khim. (in Russian)* **1971**, *7*, 302–327.
142. Hair, M. L.; Hertl, W. *J. Phys. Chem.* **1970**, *74*, 91–94.
143. Kustov, L. M. *Top. Catal.* **1997**, *4*, 131–144.
144. Wakabayashi, F.; Domen, K. *Catal. Surv. Jpn.* **1997**, *1*, 181–193.
145. Knözinger, H. *Handbook of Heterogeneous Catalysis*; In: Ertl, G., Knözinger, H., Weitkamp, J. Eds.; Vol. 2; Wiley-VCH: Weinheim, 1997; pp 707–732.
146. Makarova, M. A.; Ojo, A. F.; Karim, K.; Hunger, M.; Dwyer, J. *J. Phys. Chem.* **1994**, *98*, 3619–3623.
147. Datka, J.; Gil, B. *Catal. Today* **2001**, *70*, 131–138.
148. Hunter, E. P. L.; Lias, S. G. *J. Phys. Chem. Ref. Data* **1998**, *27*, 413–656.
149. Bellamy, L. J.; Hallam, H. E.; Williams, R. L. *J. Chem. Soc. Faraday Trans.* **1958**, *54*, 1120–1127.
150. Hadži, D. *Pure Appl. Chem.* **1965**, *11*, 435–453.
151. Van Santen, R. A. *Recl. Trav. Chim. Pays-Bas* **1994**, *113*, 423–425.
152. Chakarova, K.; Hadjiivanov, K. *J. Phys. Chem. C* **2011**, *115*, 4806–4817.
153. Chakarova, K.; Hadjiivanov, K. *Chem. Commun.* **2011**, *47*, 1878–1880.
154. Chakarova, K.; Hadjiivanov, K. *Microporous Mesoporous Mater.* **2013**, *177*, 59–65.
155. Chakarova, K.; Hadjiivanov, K. *Microporous Mesoporous Mater.* **2011**, *143*, 180–188.
156. Maache, M.; Janin, A.; Lavalley, J. C.; Joly, J. F.; Benazzi, E. *Zeolites* **1993**, *13*, 419–426.
157. Mirsojew, I.; Ernst, S.; Weitkamp, J.; Knözinger, H. *Catal. Lett.* **1994**, *24*, 235–248.
158. Datka, J.; Gil, B.; Weglarski, J. *Microporous Mesoporous Mater.* **1998**, *21*, 75–79.
159. Ordomsky, V.; Murzin, V.; Monakhova, Y.; Zubavichus, Y. V.; Knyazeva, E. E.; Nesterenko, N. S.; Ivanova, I. I. *Microporous Mesoporous Mater.* **2007**, *105*, 101–110.
160. Nachtigall, P.; Bludsky, O.; Grajciar, L.; Nachtigallova, D.; Delgado, M. R.; Otero Areán, C. *Phys. Chem. Chem. Phys.* **2009**, *11*, 791–802.
161. Odinokov, S. E.; Jogansen, A. V. *Spectrochim. Acta A* **1972**, *28*, 2343–2449.

162. Chakarova, K.; Drenchev, N.; Hadjiivanov, K. *J. Phys. Chem. C* **2012**, *116*, 17101–17109.
163. Hadjiivanov, K.; Vayssilov, G. *Adv. Catal.* **2002**, *47*, 307–511.
164. Bates, S.; Dwyer, J. *J. Phys. Chem.* **1993**, *97*, 5897–5900.
165. Dimitrov, L.; Mihaylov, M.; Hadjiivanov, K.; Mavrodinova, V. *Microporous Mesoporous Mater.* **2011**, *143*, 291–301.
166. Cairon, O.; Chevreau, T. *J. Chem. Soc. Faraday Trans.* **1998**, *94*, 323–330.
167. Thibault-Starzyk, F.; Travert, A.; Saussey, J.; Lavalley, J.-C. *Top. Catal.* **1998**, *6*, 111–118.
168. Catana, G.; Baetens, D.; Mommaerts, T.; Schoonheydt, R. A.; Weckhuysen, B. M. *J. Phys. Chem. B* **2001**, *105*, 4904–4911.
169. Hadjiivanov, K.; Chakarova, K.; Drenchev, N.; Mihaylov, M. *Curr. Phys. Chem.* **2012**, *2*, 151–161.
170. Kotrel, S.; Lunsford, J. H.; Knözinger, H. *J. Phys. Chem. B* **2001**, *105*, 3917–3921.
171. Navarro, U.; Trujillo, C. A.; Oviedo, A.; Lobo, R. *J. Catal.* **2002**, *211*, 64–74.
172. Montanari, T.; Finocchio, E.; Busca, G. *J. Phys. Chem. C* **2011**, *115*, 937–943.
173. Daniell, W.; Topsøe, N.-Y.; Knözinger, H. *Langmuir* **2001**, *17*, 6233–6239.
174. Hensen, E. J. M.; Poduval, D. G.; Degirmenci, V.; Ligthart, D. A. J. M.; Chen, W.; Mauge, F.; Rigutto, M. S.; van Veen, J. A. R. *J. Phys. Chem. C* **2012**, *116*, 21416–21429.
175. Makarova, M. A.; Al-Ghefaili, K. M.; Dwyer, J. *J. Chem. Soc. Faraday Trans.* **1994**, *90*, 383–386.
176. Cairon, O.; Thomas, K.; Chambellan, A.; Chevreau, T. *Appl. Catal. A* **2003**, *238*, 167–183.
177. Cairon, O. *ChemPhysChem* **2013**, *14*, 244–251.
178. Cairon, O.; Chevreau, T.; Lavalley, J.-C. *J. Chem. Soc. Faraday Trans.* **1998**, *94*, 3039–3047.
179. Lavalley, J. C.; Anquetil, R.; Czyzniewska, J.; Ziolek, M. *J. Chem. Soc. Faraday Trans.* **1996**, *92*, 1263–1266.
180. Echoufi, N.; Gelin, P. *J. Chem. Soc. Faraday Trans.* **1992**, *88*, 1067–1073.
181. Góra-Marek, K.; Datka, J. *Appl. Catal. A* **2006**, *302*, 104–109.
182. Gallo, J. M. R.; Bisio, C.; Gatti, G.; Marchese, L.; Pastore, H. O. *Langmuir* **2010**, *26*, 5791–5800.
183. Jiang, M.; Karge, H. C. *J. Chem. Soc. Faraday Trans.* **1996**, *92*, 2641–2649.
184. Maache, M.; Janin, A.; Lavalley, J.-C.; Benazzi, F. *Zeolites* **1995**, *15*, 507–516.
185. Nesterenko, N. S.; Thibault-Starzyk, F.; Montouillout, V.; Yuschenko, V. V.; Fernandez, C.; Gilson, J.-P.; Fajula, F.; Ivanova, I. I. *Microporous Mesoporous Mater.* **2004**, *71*, 157–166.
186. Armandi, M.; Bonelli, N.; Garrone, R.; Ardizzi, M.; Cavani, F.; Dal Pozzo, L.; Maselli, L.; Mezzogori, R.; Calestani, G. *Appl. Catal. B* **2007**, *70*, 585–596.
187. Shigeishi, R. A.; Chiche, B. H.; Fajula, F. *Microporous Mesoporous Mater.* **2001**, *43*, 211–226.
188. Boscoboinik, J. A.; Yu, X.; Emmez, E.; Yang, B.; Shaikhutdinov, S.; Fischer, F. D.; Sauer, J.; Freund, H.-J. *J. Phys. Chem. C* **2013**, *117*, 13547–13556.
189. Gallo, J. M. R.; Bisio, C.; Marchese, L.; Pastore, H. O. *Microporous Mesoporous Mater.* **2011**, *145*, 124–130.
190. Zecchina, A.; Bordiga, S.; Spoto, G.; Scarano, D.; Petrini, G.; Leofanti, G.; Padovan, M. *J. Chem. Soc. Faraday Trans.* **1992**, *88*, 2959–2969.
191. Smirnov, K. S.; Tsyganenko, A. A.; Staudte, B. *Chem. Phys. Lett.* **1991**, *182*, 127–131.
192. Kustov, L. M.; Kazansky, V. B.; Beran, S.; Kubelkova, L.; Jiru, P. *J. Phys. Chem.* **1987**, *91*, 5247–5251.
193. Datka, J.; Gil, B.; Kawazek, M.; Staudte, B. *J. Mol. Struct.* **1999**, *511–512*, 133–139.

194. Bleken, B. T. L.; Mino, L.; Giordanino, F.; Beato, P.; Svelle, S.; Lillerud, K. P.; Bordiga, S. *Phys. Chem. Chem. Phys.* **2013**, *15*, 13363–13370.
195. Gil, B.; Mokrzycki, L.; Sulikowski, B.; Olejniczak, Z.; Walas, S. *Catal. Today* **2010**, *152*, 24–32.
196. Holm, M. S.; Svelle, S.; Joensen, F.; Beato, P.; Christensen, C. H.; Bordiga, S.; Bjørgen, M. *Appl. Catal. A* **2009**, *356*, 23–30.
197. Simon, A.; Delmotte, L.; Chezeau, J.-M.; Janin, A.; Lavalley, J.-C. *Phys. Chem. Chem. Phys.* **1999**, *1*, 1659–1664.
198. Barbera, K.; Bonino, F.; Bordiga, S.; Janssens, T. V. W.; Beato, P. *J. Catal.* **2011**, *280*, 196–205.
199. Neyman, K. M.; Strodel, P.; Ruzankin, S. P.; Schlensog, N.; Knözinger, H.; Rösch, N. *Catal. Lett.* **1995**, *31*, 273–285.
200. Sigl, M.; Ernst, S.; Weitkamp, J.; Knözinger, H. *Catal. Lett.* **1997**, *45*, 27–33.
201. Armaroli, T.; Simon, L. J.; Digne, M.; Montanari, T.; Bevilacqua, M.; Valtchev, V.; Patarin, J.; Busca, G. *Appl. Catal. A* **2006**, *306*, 78–84.
202. Wakabayashi, F.; Kondo, J.; Domen, K.; Hirose, C. *J. Phys. Chem.* **1995**, *99*, 10573–10580.
203. Areán, C. O. *J. Mol. Struct.* **2008**, *880*, 31–37.
204. Areán, C. O.; Turnes Palomino, G.; Escalona Platero, E.; Peñarroya Mentruit, M. *J. Chem. Soc. Dalton Trans.* **1997**, *26*, 873–879.
205. Szanyi, J.; Paffett, M. T. *Microporous Mater.* **1996**, *7*, 201–218.
206. Romotowski, T.; Komorek, J.; Paukshtis, Y. A.; Yurchenko, E. N. *Zeolites* **1991**, *11*, 497–501.
207. Van de Water, L. G. A.; van der Waal, J. C.; Jansen, J. C.; Cadoni, M.; Marchese, L.; Maschmeyer, T. *J. Phys. Chem. B* **2003**, *107*, 10423–10430.
208. Zecchina, A.; Geobaldo, F.; Lamberti, C.; Bordiga, S.; Turnes Palomino, G.; Otero Areán, C. *Catal. Lett.* **1996**, *42*, 25–33.
209. Areán, C. O.; Turnes Palomino, G.; Geobaldo, F.; Zecchina, A. *J. Phys. Chem.* **1996**, *100*, 6678–6690.
210. Martins, G. V. A.; Berlier, G.; Bisio, C.; Coluccia, S.; Pastore, H. O.; Marchese, L. *J. Phys. Chem. C* **2008**, *112*, 7193–7200.
211. Smith, L.; Cheetham, A. K.; Marchese, L.; Thomas, J. M.; Wright, P. A.; Chen, J.; Gianotti, E. *Catal. Lett.* **1996**, *41*, 13–16.
212. Bordiga, S.; Regli, L.; Cocina, D.; Lamberti, C.; Bjørgen, M.; Lillerud, K. P. *J. Phys. Chem. B* **2005**, *109*, 2779–2784.
213. Van Oers, C. J.; Góra-Marek, K.; Sadowska, K.; Mertens, M.; Meynen, V.; Datka, J.; Cool, P. *Chem. Eng. J.* **2014**, *237*, 372–379.
214. Mintova, S.; Valtchev, V.; Onfroy, T.; Marichal, C.; Knözinger, H.; Bein, T. *Microporous Mesoporous Mater.* **2006**, *90*, 237–245.
215. Bisio, C.; Martra, G.; Coluccia, S.; Massiani, P. *J. Phys. Chem. C* **2008**, *112*, 10520–10530.
216. Góra-Marek, K.; Datka, J.; Dzwigaj, S.; Che, M. *J. Phys. Chem. B* **2006**, *110*, 6763–6767.
217. Kefirov, R.; Ivanova, E.; Hadjiivanov, K.; Dzwigaj, S.; Che, M. *Catal. Lett.* **2008**, *125*, 209–214.
218. Kalvachev, Yu.; Jaber, M.; Mavrodinova, V.; Dimitrov, L.; Nihtianova, D.; Valtchev, V. *Microporous Mesoporous Mater.* **2013**, *177*, 127–134.
219. Penkova, A.; Dzwigaj, S.; Kefirov, R.; Hadjiivanov, K.; Che, M. *J. Phys. Chem. C* **2007**, *111*, 8623–8631.
220. Mihaylova, A.; Hadjiivanov, K.; Dzwigaj, S.; Che, M. *J. Phys. Chem. B* **2006**, *110*, 19530–19536.
221. Bisio, C.; Massiani, P.; Fajerwerg, K.; Sordelli, L.; Stievano, L.; Silva, E. R.; Coluccia, S.; Martra, G. *Microporous Mesoporous Mater.* **2006**, *90*, 175–187.

222. Delgado, M. R.; Areán, C. O. *Energy* **2011**, *36*, 5286–5291.
223. Hadjiivanov, K.; Ivanova, E.; Kefirov, R.; Janas, J.; Plesniar, A.; Dzwigaj, S.; Che, M. *Microporous Mesoporous Mater.* **2010**, *131*, 1–12.
224. Manoilova, O. V.; Dakka, J.; Sheldon, R. A.; Tsyganenko, A. A. *Stud. Surf. Sci. Catal.* **1995**, *94*, 163–170.
225. Bisio, C.; Gatti, G.; Boccaleri, E.; Marchese, L.; Bertinetti, L.; Coluccia, S. *Langmuir* **2008**, *24*, 2808–2819.
226. Onida, B.; Borello, L.; Bonelli, B.; Geobaldo, F.; Garrone, E. *J. Catal.* **2003**, *214*, 191–199.
227. Onida, B.; Geobaldo, F.; Testa, F.; Crea, F.; Garrone, E. *Microporous Mesoporous Mater.* **1999**, *30*, 119–127.
228. Delgado, M. R.; Bulánek, R.; Chlubná, P.; Areán, C. O. *Catal. Today* **2014**, *227*, 42–45.
229. Geobaldo, F.; Fiorilli, S.; Onida, B.; Giordano, G.; Katovic, A.; Garrone, E. *J. Phys. Chem. B* **2003**, *107*, 1258–1262.
230. Góra-Marek, K.; Derewiński, M.; Sarv, P.; Datka, J. *Catal. Today* **2005**, *101*, 131–138.
231. Regli, L.; Zecchina, A.; Vitillo, J. G.; Cocina, D.; Spoto, G.; Lamberti, C.; Lillerud, K. P.; Olsbye, U.; Bordiga, S. *Phys. Chem. Chem. Phys.* **2005**, *7*, 3197–3203.
232. Rachwalik, R.; Olejniczak, Z.; Jiao, J.; Huang, J.; Hunger, M.; Sulikowski, B. *J. Catal.* **2007**, *252*, 161–170.
233. Bordiga, S.; Turnes Palomino, G.; Paze, C.; Zecchina, A. *Microporous Mesoporous Mater.* **2000**, *34*, 67–80.
234. Kondo, J.; Yoda, E.; Ishikawa, H.; Wakabayashi, F.; Domen, K. *J. Catal.* **2000**, *191*, 275–281.
235. Chakarova, K.; Hadjiivanov, K. *Microporous Mesoporous Mater.* **2009**, *123*, 123–128.
236. Chakarova, K.; Petrova, G.; Dimitrov, M.; Dimitrovd, L.; Vayssilov, G.; Tsoncheva, T.; Hadjiivanov, K. *Appl. Catal. B* **2011**, *106*, 186–194.
237. Onida, B.; Gabelica, Z.; Lonrenco, J. P.; Ribeiro, M. F.; Garrone, E. *J. Phys. Chem. B* **1997**, *101*, 9244–9249.
238. Ristić, A.; Novak Tušar, N.; Arčon, I.; Logar, N. Z.; Thibault-Starzyk, F.; Czyzniewska, J.; Kaučic, V. *Chem. Mater.* **2003**, *15*, 3643–3649.
239. Beebe, T. P.; Gelin, P.; Yates, J. T., Jr. *Surf. Sci.* **1984**, *148*, 526–550.
240. Rodionova, T. A.; Tsyganenko, A. A.; Filimonov, V. N. *Adsorbtsia i Adsorbenty (in Russian)* **1982**, *10*, 33–46.
241. Daniell, W.; Schubert, U.; Glöckler, R.; Meyer, A.; Noweck, K.; Knözinger, H. *Appl. Catal. A* **2000**, *196*, 247–260.
242. Ghiotti, G.; Garrone, E.; Morterra, C.; Boccurzi, F. *J. Phys. Chem.* **1979**, *83*, 2863–2869.
243. Zecchina, A.; Bordiga, S.; Spoto, G.; Marchese, I.; Petrini, G.; Leofanti, G.; Padovan, M. *J. Phys. Chem.* **1992**, *96*, 4991–4997.
244. Storozhev, P. Y.; Areán, C. O.; Garrone, E.; Ugliengo, P.; Ermoshin, V. A.; Tsyganenko, A. A. *Chem. Phys. Lett.* **2003**, *374*, 439–445.
245. Degirmenci, V.; Uner, D.; Cinlar, B.; Shanks, B. H.; Yilmaz, A.; van Santen, R. A.; Hensen, E. J. M. *Catal. Lett.* **2011**, *141*, 33–42.
246. Camarota, B.; Ugliengo, P.; Garrone, E.; Areán, C. O.; Delgado, M. R.; Inagaki, S.; Onida, B. *J. Phys. Chem. C* **2008**, *112*, 19560–19567.
247. Crépeau, G.; Montouillout, V.; Vimont, A.; Mariey, L.; Cseri, T.; Mauge, F. *J. Phys. Chem. B* **2006**, *110*, 15172–15185.
248. Bonelli, B.; Onida, B.; Chen, J. D.; Galarneau, A.; Di Renzo, F.; Fajula, F.; Garrone, E. *Microporous Mesoporous Mater.* **2004**, *67*, 95–106.
249. Bordiga, S.; Escalona Platero, E.; Otero Areán, C.; Lamberti, C.; Zecchina, A. *J. Catal.* **1992**, *137*, 179–185.

250. Garrone, E.; Onida, B.; Bonelli, B.; Busco, C.; Ugliengo, P. *J. Phys. Chem. B* **2006**, *110*, 19087–19092.

251. Nachtigall, P.; Frolich, K.; Drobná, H.; Bludsky, O.; Nachtigallová, D.; Bulánek, R. *J. Phys. Chem. C* **2007**, *111*, 11353–11362.

252. Venkov, T. V.; Hess, C.; Jentoft, F. C. *Langmuir* **2007**, *23*, 1768–1777.

253. Hadjiivanov, K.; Reddy, B. M.; Knözinger, H. *Appl. Catal. A* **1999**, *188*, 355–360.

254. Morterra, C.; Magnacca, G.; Del Favero, N. *Langmuir* **1993**, *9*, 642–645.

255. Spina, G.; Bonelli, B.; Palmero, P.; Montanaro, L. *Mater. Chem. Phys.* **2013**, *143*, 286–295.

256. Oliviero, L.; Vimont, A.; Lavalley, J.-C.; Romero Sarria, F.; Gaillard, M.; Mauge, M. *Phys. Chem. Chem. Phys.* **2005**, *7*, 1861–1869.

257. Drenchev, N.; Spassova, I.; Ivanova, E.; Khristova, M.; Hadjiivanov, K. *Appl. Catal. B* **2013**, *138–139*, 362–372.

258. Muddada, N. B.; Olsbye, U.; Fuglerud, T.; Vidotto, S.; Marsella, A.; Bordiga, S.; Gianolio, D.; Leofanti, G.; Lamberti, C. *J. Catal.* **2011**, *284*, 236–246.

259. El-Nadjar, W.; Bonne, M.; Trela, E.; Rouleau, L.; Mino, A.; Hocine, S.; Payen, E.; Lancelot, C.; Lamonier, C.; Blanchard, P.; Courtois, X.; Can, F.; Duprez, D.; Royer, S. *Microporous Mesoporous Mater.* **2012**, *158*, 88–98.

260. Montanaria, T.; Castoldi, L.; Lietti, L.; Busca, G. *Appl. Catal. A* **2011**, *400*, 61–69.

261. Smirnova, M. Yu.; Toktarev, A. V.; Ayupov, A. B.; Echevsky, G. V. *Catal. Today* **2010**, *152*, 17–23.

262. Bonelli, B.; Bottero, I.; Ballarini, N.; Passeri, S.; Cavani, F.; Garrone, E. *J. Catal.* **2009**, *264*, 15–30.

263. Ghiotti, G.; Boccuzzi, F.; Scala, R. *J. Catal.* **1985**, *92*, 79–97.

264. Mihaylov, M.; Penkova, A.; Hadjiivanov, K.; Knözinger, H. *J. Phys. Chem. B* **2004**, *108*, 679–688.

265. Vimont, A.; Goupil, J. M.; Lavalley, J.-C.; Daturi, M.; Surblé, S.; Serre, C.; Millange, F.; Férey, G.; Audebrand, N. *J. Am. Chem. Soc.* **2006**, *128*, 3218–3227.

266. Vimont, A.; Leclerc, H.; Maugé, F.; Daturi, M.; Lavalley, J.-C.; Surblé, S.; Serre, C.; Férey, G. *J. Phys. Chem. C* **2007**, *111*, 383–388.

267. Manoilova, O. V.; Podkolzin, S. G.; Tope, B.; Lercher, J.; Stangland, E. E.; Goupil, J. M.; Weckhuysen, B. M. *J. Phys. Chem. B* **2004**, *108*, 15770–15781.

268. Sergent, N.; Gélin, P.; Périer-Camby, L.; Praliaud, H.; Thomas, G. *Phys. Chem. Chem. Phys.* **2002**, *4*, 4802–4808.

269. Hadjiivanov, K.; Lamotte, J.; Lavalley, J.-C. *Langmuir* **1887**, *13*, 3374–3381.

270. Panayotov, D.; Burrows, S.; Mihaylov, M.; Hadjiivanov, K.; Tissue, B. M.; Morris, J. R. *Langmuir* **2010**, *26*, 8106–8112.

271. Mino, L.; Spoto, G.; Bordiga, S.; Zecchina, A. *J. Phys. Chem. C* **2013**, *117*, 11186–11196.

272. Martra, G. *Appl. Catal. A* **2000**, *200*, 275–285.

273. Mino, L.; Spoto, G.; Bordiga, S.; Zecchina, A. *J. Phys. Chem. C* **2012**, *116*, 17008–17018.

274. Minella, M.; Faga, M. G.; Maurino, V.; Minero, C.; Pelizzetti, E.; Coluccia, S.; Martra, G. *Langmuir* **2010**, *26*, 2521–2527.

275. Hadjiivanov, K.; Lavalley, J. C. *Catal. Commun.* **2001**, *2*, 129–133.

276. Spielbauer, D.; Mekhemer, G. A. H.; Zaki, M. I.; Knözinger, H. *Catal. Lett.* **1996**, *40*, 71–79.

277. Spielbauer, D.; Mekhemer, G. A. H.; Riemer, T.; Zaki, M. I.; Knözinger, H. *J. Phys. Chem. B* **1997**, *101*, 4681–4688.

278. Manoilova, O. V.; Olindo, R.; Areán, C. O.; Lercher, J. A. *Catal. Commun.* **2007**, *8*, 865–870.

279. Lebarbier, V.; Houalla, M.; Onfroy, T. *Catal. Today* **2012**, *192*, 123–129.

280. Mihaylov, M.; Hadjiivanov, K.; Knözinger, H. *Phys. Chem. Chem. Phys.* **2006**, *8*, 407–417.

281. Lebarbier, V.; Clet, G.; Houalla, M. *J. Phys. Chem. B* **2006**, *110*, 22608–22617.

282. Piumetti, M.; Armandi, M.; Garrone, E.; Bonelli, B. *Microporous Mesoporous Mater.* **2012**, *164*, 111–119.

283. Ivanova, E.; Hadjiivanov, K.; Dzwigaj, S.; Che, M. *Microporous Mesoporous Mater.* **2006**, *89*, 69–77.

284. Ivanova, E.; Mihaylov, K.; Hadjiivanov, K.; Blasin-Aube, V.; Marie, O.; Plesniar, A.; Daturi, M. *Appl. Catal. B* **2010**, *93*, 325–338.

285. Ivanova Shor, E. A.; Shor, A. M.; Nasluzov, V. A.; Vayssilov, G. N.; Rösch, N. *J. Chem. Theory Comput.* **2005**, *1*, 459–471.

286. Hadjiivanov, K.; Knözinger, H. *J. Phys. Chem. B* **2001**, *105*, 4531–4534.

287. Tsyganenko, A. A.; Storozhev, P. Y.; Areán, C. O. *Kinet. Catal.* **2004**, *45*, 562–573.

288. Hadjiivanov, K.; Penkova, A.; Centeno, M. A. *Catal. Commun.* **2007**, *8*, 1715–1718.

289. Nachtigall, P.; Delgado, M. R.; Nachtigallova, D.; Areán, C. O. *Phys. Chem. Chem. Phys.* **2012**, *14*, 1552–1569.

290. Bulánek, R.; Koudelková, E. *Microporous Mesoporous Mater.* **2012**, *151*, 149–156.

291. Leydier, F.; Chizallet, C.; Costa, D.; Raybaud, P. *Chem. Commun.* **2012**, *48*, 4076–4078.

292. Wakabayashi, F.; Kondo, J.; Domen, K.; Hirose, C. *J. Phys. Chem.* **1996**, *100*, 4154–4159.

293. Wakabayashi, F.; Kondo, J.; Wada, A.; Domen, K.; Hirose, C. *J. Phys. Chem.* **1993**, *97*, 10761–10768.

294. Garrone, E.; Kazansky, V. B.; Kustov, L. M.; Sauer, J.; Senchenya, I. N.; Ugliengo, P. *J. Phys. Chem.* **1992**, *96*, 1040–1045.

295. Bordiga, S.; Garrone, E.; Lamberti, C.; Zecchina, A.; Areán, C. O.; Kazansky, V. B.; Kustov, L. M. *J. Chem. Soc. Faraday Trans.* **1994**, *90*, 3367–3372.

296. Zecchina, A.; Bordiga, S.; Vitillo, J. C.; Ricchiardi, G.; Lamberti, C.; Spoto, G.; Bjørgen, M.; Lill, K. P. *J. Am. Chem. Soc.* **2005**, *127*, 6361–6366.

297. Makarova, M.; Zholobenko, V. L.; Al-Ghefaili, K. M.; Thompson, N. E.; Dewing, J.; Dwyer, J. *J. Chem. Soc. Faraday Trans.* **1994**, *90*, 1047–1054.

298. Ene, A. B.; Bauer, M.; Archipov, T.; Roduner, E. *Phys. Chem. Chem. Phys.* **2010**, *12*, 6520–6531.

299. Hadjiivanov, K. *Catal. Rev. Sci. Eng.* **2000**, *42*, 71–144.

300. Penkova, A.; Hadjiivanov, K.; Mihaylov, M.; Daturi, M.; Saussey, J.; Lavalley, J.-C. *Langmuir* **2004**, *20*, 5425–5431.

301. Ferrari, A. M.; Huber, S.; Knözinger, H.; Neyman, K. M.; Röch, N. *J. Phys. Chem. B* **1998**, *102*, 4548–4555.

302. Wakabayashi, F.; Kondo, J. N.; Domen, K.; Hirose, C. *Microporous Mater.* **1997**, *8*, 29–37.

303. Chen, L.; Lin, L.; Xu, Z.; Zhang, T.; Xin, Q.; Ying, P.; Li, G.; Li, C. *J. Catal.* **1996**, *161*, 107–114.

304. Li, C.; Yan, W.; Xin, Q. *Catal. Lett.* **1994**, *24*, 249–256.

305. Chen, L.; Lin, L.; Xu, Z.; Zhang, T.; Liang, D. *Catal. Lett.* **1995**, *35*, 245–258.

306. Kazanskii, V. B.; Serykh, A. I.; Bell, A. T. *Kinet. Catal.* **2002**, *43*, 453–460.

307. Papp, H.; Hinsen, W.; Do, N. T.; Baerns, M. *Thermochim. Acta* **1984**, *82*, 137–148.

308. McDonald, R. S. *J. Am. Chem. Soc.* **1957**, *79*, 850–854.

309. Yamazaki, T.; Watanabe, M.; Saito, H. *Bull. Chem. Soc. Japan* **2000**, *73*, 1353–1358.

310. Scarano, D.; Bertarione, S.; Spoto, G.; Zecchina, A.; Areán, C. O. *Thin Solid Films* **2001**, *400*, 50–55.

311. Trombetta, M.; Armaroli, T.; Alejandre, A. G.; Solis, J. R.; Busca, G. *Appl. Catal. A* **2000**, *192*, 125–136.

312. Kazansky, V. B.; Kustov, L. M.; Borovkov, V. Y. *Zeolites* **1983**, *3*, 77–81.
313. Datka, J.; Piwowarska, Z. *Zeolites* **1988**, *8*, 30–34.
314. Loeffler, E.; Lohse, U.; Peuker, C.; Oehlmann, G.; Kustov, L. M.; Zholobenko, V. L.; Kazansky, V. B. *Zeolites* **1990**, *10*, 266–271.
315. Spoto, G.; Bordiga, S.; Ricchiardi, G.; Scarano, D.; Zecchina, A.; Borello, E. *J. Chem. Soc. Faraday Trans.* **1994**, *90*, 2827–2835.
316. Bordiga, S.; Civalleri, B.; Spoto, G.; Paze, C.; Lamberti, C.; Ugliengo, P.; Zecchina, A. *J. Chem. Soc. Faraday Trans.* **1997**, *93*, 3893–3898.
317. Kazansky, V. B.; Subbotina, I. R.; Jentoft, F. *J. Catal.* **2006**, *240*, 66–72.
318. Howard, J.; Lux, P. J.; Yarwood, J. *Zeolites* **1988**, *8*, 427–431.
319. Hegde, S. G.; Ratnasamy, P.; Kustov, L. M.; Kazansky, V. B. *Zeolites* **1988**, *8*, 137–141.
320. Datka, J. *Zeolites* **1991**, *11*, 739–741.
321. Bjørgen, M.; Lillerud, K.; Olsbye, U.; Bordiga, S.; Zecchina, A. *J. Phys. Chem. B* **2004**, *108*, 7862–7870.
322. Nivarthy, G. S.; Feller, A.; Seshan, K.; Lercher, J. A. *Microporous Mesoporous Mater.* **2000**, *35–36*, 75–87.
323. Paze, C.; Sazak, B.; Zecchina, D.; Dwyer, J. *J. Phys. Chem. B* **1999**, *103*, 9978–9996.
324. Kondo, J. N.; Liqun, S.; Wakabayashi, F.; Domen, K. *Catal. Lett.* **1997**, *47*, 129–133.
325. Kondo, J. N.; Domen, K.; Wakabayashi, F. *Microporous Mesoporous Mater.* **1998**, *21*, 429–437.
326. Trombetta, M.; Gutierrez Alejandre, A.; Ramirez Solis, J.; Busca, G. *Appl. Catal. A* **2000**, *198*, 81–93.
327. Jacobs, W. P. J. H.; Demuth, D. G.; Schunk, S. A.; Schiith, F. *Microporous Mater.* **1997**, *10*, 95–109.
328. Goyal, R.; Fitch, A. N.; Jobic, H. *J. Phys. Chem. B* **2000**, *104*, 2878–2884.
329. Tzoulakia, D.; Jentys, A.; Pérez-Ramirez, H.; Egeblad, K.; Lercher, J. A. *Catal. Today* **2012**, *198*, 3–11.
330. Datka, J.; Abramowicz, T. *J. Chem. Soc. Faraday Trans.* **1994**, *90*, 2417–2421.
331. Armaroli, T.; Bevilacqua, M.; Trombetta, M.; Gutierrez Alejandre, A.; Ramirez, J.; Busca, G. *Appl. Catal. A* **2001**, *220*, 181–190.
332. Trombetta, M.; Armaroli, T.; Gutierrez Alejandre, A.; Gonzalez, H.; Ramirez Solis, J.; Busca, G. *Catal. Today* **2001**, *65*, 285–292.
333. Armaroli, T.; Trombetta, M.; Gutierrez Alejandre, A.; Ramirez Solis, J.; Busca, G. *Phys. Chem. Chem. Phys.* **2000**, *2*, 3341–3348.
334. Kinger, G.; Lugstein, A.; Swagera, R.; Ebel, M.; Jentys, A.; Vinek, H. *Microporous Mesoporous Mater.* **2000**, *39*, 307–317.
335. Kustov, L. M.; Kazansky, A. B.; Figueras, F.; Tichit, D. *J. Catal.* **1994**, *150*, 143–149.
336. Tripathi, A. K.; Sahasrabudhe, A.; Mitra, S.; Mukhopadhyay, R.; Gupta, N. M.; Kartha, V. B. *Phys. Chem. Chem. Phys.* **2001**, *3*, 4449–4455.
337. Datka, J.; Boczar, M.; Rymarowicz, P. *J. Catal.* **1988**, *114*, 368–376.
338. Datka, J.; Boczar, M. *Zeolites* **1991**, *11*, 397–400.
339. Kiricsi, I.; Flego, C.; Pazzuconi, G.; Parker, W. O., Jr.; Millini, R.; Perego, C.; Bellussi, G. *J. Phys. Chem.* **1994**, *98*, 4627–4634.
340. Hegde, S. G.; Kumar, R.; Bhat, R. N.; Ratnasamy, P. *Zeolites* **1989**, *9*, 231–237.
341. Guisnet, M.; Ayrault, P.; Coutanceau, C.; Alvarez, M. F.; Datka, J. *J. Chem. Soc. Faraday Trans.* **1997**, *93*, 1661–1665.
342. Su, B. L.; Norberg, V. *Zeolites* **1997**, *19*, 65–74.
343. Guisnet, M.; Ayrault, P.; Datka, J. *Microporous Mesoporous Mater.* **1998**, *20*, 283–291.
344. Zholobenko, V. L.; Makarova, M. A.; Dwyer, J. *J. Phys. Chem.* **1993**, *97*, 5962–5964.
345. Su, B. L.; Barthomeuf, D. *Zeolites* **1993**, *13*, 626–633.

346. Archipov, T.; Santra, S.; Ene, A. B.; Stoll, H.; Rauhut, G.; Roduner, E. *J. Phys. Chem. C* **2009**, *113*, 4107–4116.
347. Dzwigaj, S.; Briend, M.; Shikholeslami, A.; Peltre, M. J.; Barthomeuf, D. *Zeolites* **1990**, *10*, 157–162.
348. Höchtl, M.; Jentys, A.; Vinek, H. *Microporous Mesoporous Mater.* **1999**, *31*, 271–285.
349. Datka, J.; Kawalek, M. *J. Chem. Soc. Faraday Trans.* **1993**, *89*, 1829–1831.
350. Muller, G.; Eder-Mirth, G.; Kessler, H.; Lercher, J. A. *J. Phys. Chem.* **1995**, *99*, 12327–12331.
351. Scokart, P. O.; Rouxhet, P. G. *J. Colloid Interface Sci.* **1982**, *86*, 96–97.
352. Sempels, R. E.; Rouxhet, P. G. *J. Colloid Interface Sci.* **1976**, *55*, 263–273.
353. Jentys, A.; Pham, N. H.; Vinek, H. *J. Chem. Soc. Faraday Trans.* **1996**, *92*, 3287–3291.
354. Haaland, D. M. *Surf. Sci.* **1981**, *102*, 405–423.
355. Scokart, P. O.; Selim, S. A.; Damon, J. P.; Rouxhet, P. G. *J. Colloid Interface Sci.* **1979**, *70*, 209–222.
356. Busca, G.; Saussey, H.; Saur, O.; Lavalley, J. C.; Lorenzelli, V. *Appl. Catal.* **1985**, *14*, 245–260.
357. Hadjiivanov, K.; Klissurski, D.; Busca, G.; Lorenzelli, V. *J. Chem. Soc. Faraday Trans.* **1991**, *87*, 175–178.
358. Busca, G.; Ramis, G. *Appl. Surf. Sci.* **1986**, *27*, 114–126.
359. Lonyi, F.; Valyon, J.; Engelhardt, J.; Mizukami, F. *J. Catal.* **1996**, *160*, 279–289.
360. Busca, G.; Zerlia, T.; Lorenzelli, V.; Girelli, A. *J. Catal.* **1984**, *88*, 131–136.
361. Pelmenschikov, A. G.; van Santen, R. A.; Jänchen, J.; Meijer, E. *J. Phys. Chem.* **1993**, *97*, 11071–11074.
362. Paze, C.; Zecchina, A.; Spera, S.; Spano, G.; Rivetti, F. *Phys. Chem. Chem. Phys.* **2000**, *2*, 5756–5760.
363. Tvaruzkova, Z.; Habersberger, K.; Jiru, P. *React. Kinet. Catal. Lett.* **1991**, *44*, 361–365.
364. Morterra, C.; Mentruit, M. P.; Cerrato, G. *Phys. Chem. Chem. Phys.* **2002**, *4*, 676–687.
365. Florian, J.; Kubelkova, L. *J. Phys. Chem.* **1994**, *98*, 8734–8741.
366. Kubelkova, L.; Kotrla, J.; Florian, J. *J. Phys. Chem.* **1995**, *99*, 10285–10293.
367. Pelmenschikov, A. G.; van Wolput, J. H. M. C.; Jänchen, J.; van Santen, R. A. *J. Phys. Chem.* **1995**, *99*, 3612–3617.
368. Pelmenschikov, A. G.; van Santen, R. A. *J. Phys. Chem.* **1993**, *97*, 10678–10680.
369. Anquetil, R.; Saussey, J.; Lavalley, J. C. *Phys. Chem. Chem. Phys.* **1999**, *1*, 555–560.
370. Paze, C.; Bordiga, S.; Civalleri, B.; Zecchina, A. *Phys. Chem. Chem. Phys.* **2001**, *3*, 1345–1347.
371. Marie, O.; Thibault-Starzyk, F.; Lavalley, J.-C. *Phys. Chem. Chem. Phys.* **2000**, *2*, 5341–5349.
372. Wichterlova, B.; Tvaruzkova, Z.; Sobalik, Z.; Sarv, P. *Microporous Mesoporous Mater.* **1998**, *24*, 223–233.
373. Evans, J.; Bernstein, E. *Can. J. Chem.* **1955**, *33*, 1746–1753.
374. Cho, J.-S.; Cho, H.-G. *Bull. Korean Chem. Soc.* **2009**, *30*, 803–809.
375. Ivanov, A. V.; Graham, G. W.; Shelef, M. *Appl. Catal. B* **1999**, *21*, 243–258.
376. Chakarova, K.; Nikolov, P.; Hadjiivanov, K. *Catal. Commun.* **2013**, *41*, 38–40.
377. Jolly, S.; Saussey, J.; Lavalley, J. C. *Catal. Lett.* **1994**, *24*, 141–146.
378. Busca, G.; Montanari, T.; Bevilacqua, M.; Finocchio, E. *Colloids Surf. A* **2008**, *320*, 205–212.
379. Vimont, A.; Lavalley, J. C.; Sahibed-Dine, A.; Areán, C. O.; Rodriguez Delgado, M.; Daturi, M. *J. Phys. Chem. B* **2005**, *109*, 9656–9664.
380. Yim, S.-G.; Son, D. H.; Kim, K. *J. Chem. Soc. Faraday Trans. 1* **1993**, *89*, 837–843.
381. Aboulayt, A.; Binet, C.; Lavalley, J. C. *J. Chem. Soc. Faraday Trans. 1* **1995**, *91*, 2913–2921.
382. Koubowetz, F.; Latzel, J.; Noller, H. *J. Colloid Interface Sci.* **1980**, *74*, 322–330.

383. Lorenzelli, V.; Busca, G.; Sheppard, N. *J. Catal.* **1980**, *66*, 28–35.
384. Nortier, P.; Fourre, P.; Saad, A. B. M.; Saur, O.; Lavalley, J.-C. *Appl. Catal.* **1990**, *61*, 141–160.
385. Krietenbrink, H.; Knözinger, H. *Z. Phys. Chem. (Leipzig)* **1976**, *102*, 43–56.
386. Čejka, J.; Krejči, A.; Zilkova, N.; Kotrla, J.; Ernst, S.; Weber, A. *Microporous Mesoporous Mater.* **2002**, *53*, 121–133.
387. Cerruti, M.; Morterra, C.; Ugliengo, P. *Chem. Mater.* **2005**, *17*, 1416–1423.
388. Rostovshchikova, T.; Smirnov, V.; Kiseleva, O.; Yushcenko, V.; Tzodikov, M.; Maksimov, Y.; Suzdalev, I.; Kustov, L.; Tkachenko, O. *Catal. Today* **2010**, *132*, 48–53.
389. Jänchen, J.; Peeters, M. P. J.; van Wolput, J. H. M. C.; Wolthuizen, J. P.; van Hooff, J. H. C.; Lohse, U. *J. Chem. Soc. Faraday Trans.* **1994**, *90*, 1033–1039.
390. Kotrla, J.; Kubelkova, L.; Lee, C.-C.; Gorte, R. J. *J. Phys. Chem. B* **1998**, *102*, 1437–1443.
391. Paze, C.; Zecchina, A.; Spera, S.; Cosma, A.; Merlo, E.; Spano, G.; Girotti, G. *Phys. Chem. Chem. Phys.* **1999**, *1*, 2627–2629.
392. Mees, F. D. P.; Van Der Voort, P.; Cool, P.; Martens, L. R. M.; Janssen, K. J. G.; Verberckmoes, A. A.; Kennedy, G. J.; Hall, R. B.; Wang, K.; Vansant, E. F. *J. Phys. Chem. B* **2003**, *107*, 3161–3167.
393. Haffad, D.; Chambellan, A.; Lavalley, J. C. *Catal. Lett.* **1998**, *54*, 227–233.
394. Jänchen, J.; van Wolput, J. H. M. C.; van de Ven, L. J. M.; de Haan, J. W.; van Santen, R. A. *Catal. Lett.* **1996**, *39*, 147–152.
395. Kucherov, A. V.; Shigapov, A. N.; Ivanov, A. V.; Shelef, M. *J. Catal.* **1999**, *186*, 334–344.
396. Jolly, S.; Saussey, J.; Lavahey, J. C. *Mol. Catal.* **1994**, *86*, 401–421.
397. Chen, J.; Thomas, J. M.; Sankar, G. *J. Chem. Soc. Faraday Trans.* **1994**, *90*, 3455–3459.
398. Paze, C.; Turnes Palomino, G.; Zecchina, A. *Catal. Lett.* **1999**, *60*, 139–143.
399. Brabec, L.; Jeschke, M.; Klik, R.; Novakova, J.; Kubelkova, L.; Freude, D.; Bosacek, V.; Meusinger, J. *Appl. Catal. A* **1998**, *167*, 309–320.
400. Brabec, L.; Jeschke, M.; Klik, R.; Novakovaa, L.; Kubelkova, L.; Meusinger, J. *Appl. Catal. A* **1998**, *170*, 105–116.
401. Flores, J. H.; da Silva, M. I. P. *Colloids Surf. A* **2008**, *322*, 113–123.
402. Wichterlova, B.; Zilkova, N.; Uvarova, E.; Cejka, J.; Sarv, P.; Paganinic, C.; Lercher, J. A. *Appl. Catal. A* **1999**, *182*, 297–308.
403. Van Donk, S.; Bus, E.; Broersma, A.; Bitter, J. H.; de Jong, K. P. *J. Catal.* **2002**, *212*, 86–93.
404. Penzien, J.; Abraham, A.; van Bokhoven, J. A.; Jentys, A.; Muller, T. E.; Sievers, C.; Lercher, J. A. *J. Phys. Chem. B* **2004**, *108*, 4116–4126.
405. Travert, A.; Vimont, A.; Lavalley, J.-C.; Montouillout, V.; Rodriguez Delgado, M.; Pascual, J. J. C.; Areán, C. O. *J. Phys. Chem. B* **2004**, *108*, 16499–16507.
406. Davit, P.; Martra, G.; Coluccia, S.; Augugliaro, V.; Garcia López, E.; Loddo, V.; Marci, G.; Palmisano, L.; Schiavello, M. *J. Mol. Catal. A* **2003**, *204–205*, 693–701.
407. Augugliaro, V.; Coluccia, S.; Garcia López, E.; Loddo, V.; Marci, G.; Martra, G.; Palmisano, L.; Schiavello, M. *Top. Catal.* **2005**, *35*, 237–244.
408. Al-Abadleh, H. A.; Grassian, V. H. *Langmuir* **2003**, *19*, 341–347.
409. Takeuchi, M.; Bertinetti, L.; Martra, G.; Coluccia, S.; Anpo, M. *Appl. Catal. A* **2006**, *307*, 13–20.
410. Hoffmann, P.; Knözinger, E. *Surf. Sci.* **1987**, *188*, 181–198.
411. Zecchina, A.; Geobaldo, F.; Spoto, G.; Bordiga, S.; Ricchiardi, G.; Buzzoni, R.; Petrini, G. *J. Phys. Chem.* **1996**, *100*, 16584–16599.
412. Müller, G.; Bódis, J.; Kornatowski, J. *Microporous Mesoporous Mater.* **2004**, *69*, 1–7.
413. Sárkány, J. *Appl. Catal. A* **1999**, *188*, 369–379.

414. Parker, L. M.; Bibby, D. M.; Burns, G. R. *Zeolites* **1991**, *11*, 293–297.
415. Ison, A.; Gorte, R. J. *J. Catal.* **1984**, *89*, 150–158.
416. Jentys, A.; Warecka, G.; Lercher, J. A. *Mol. Catal.* **1989**, *51*, 309–327.
417. Jentys, A.; Warecka, G.; Derewinski, M.; Lercher, J. A. *J. Phys. Chem.* **1989**, *93*, 4837–4843.
418. Wakabayashi, F.; Kondo, J. N.; Domen, K.; Hirose, C. *Catal. Lett.* **1996**, *38*, 15–19.
419. Wakabayashi, F.; Kondo, J. N.; Domen, K.; Hirose, C. *J. Phys. Chem.* **1996**, *100*, 1442–1444.
420. Kondo, J. N.; Iizuka, M.; Domen, K.; Wakabayashi, F. *Langmuir* **1997**, *13*, 747–750.
421. Parker, L. M.; Bibby, D. M.; Burns, G. R. *Zeolites* **1993**, *13*, 107–112.
422. Krossner, M.; Sauer, J. *J. Phys. Chem.* **1996**, *100*, 6199–6211.
423. Jobic, H.; Tuel, A.; Krossner, M.; Sauer, J. *J. Phys. Chem.* **1996**, *100*, 19545–19550.
424. Jungsuttiwong, S.; Limtrakul, J.; Truong, T. N. *J. Phys. Chem. B* **2005**, *109*, 13342–13351.
425. Zygmunt, S. A.; Curtiss, L. A.; Iton, L. E.; Erhardt, M. K. *J. Phys. Chem.* **1996**, *100*, 6663–6671.
426. Buzzoni, R.; Bordiga, S.; Ricchiardi, G.; Spoto, G.; Zecchina, A. *J. Phys. Chem.* **1995**, *99*, 11937–11951.
427. Zecchina, A.; Marchese, L.; Bordiga, S.; Paze, C.; Gianotti, E. *J. Phys. Chem.* **1997**, *101*, 10128–10135.
428. Datka, J.; Góra-Marek, K. *Catal. Today* **2006**, *114*, 205–210.
429. Bordiga, S.; Roggero, I.; Ugliengo, P.; Zecchina, A.; Bolis, V.; Artioli, G.; Buzzoni, R.; Marra, G.; Rivetti, F.; Spanò, G.; Lamberti, C. *J. Chem. Soc. Dalton Trans.* **2000**, 3921–3929.
430. Gianotti, E.; Marchese, L.; Guidotti, M.; Ravasio, N.; Psaro, R.; Coluccia, S. *Stud. Surf. Sci. Catal.* **2005**, *155*, 311–320.
431. Niwa, M.; Nishikawa, S.; Katada, N. *Microporous Mesoporous Mater.* **2005**, *82*, 105–112.
432. Lok, B. M.; Marcus, B. K.; Angell, C. L. *Zeolites* **1986**, *6*, 185–194.
433. Sobalik, Z.; Belhekar, A. A.; Tvaružková, Z.; Wichterlová, B. *Appl. Catal. A* **1999**, *188*, 175–186.
434. Datka, J.; Kawałek, M.; Góra-Marek, M. *Appl. Catal. A* **2003**, *243*, 293–299.
435. De Ménorval, B.; Ayrault, P.; Gnep, N. S.; Guisnet, M. *J. Catal.* **2005**, *230*, 38–61.
436. Bourgeat-Lami, E.; Massiani, P.; Di Renzo, F.; Espiau, P.; Fajula, F.; Des Courieres, T. *Appl. Catal.* **1991**, *72*, 139–152.
437. Vedrine, J. C.; Auroux, A.; Dejaifve, P.; Ducarme, V.; Hoser, H.; Zhou, S. *J. Catal.* **1982**, *73*, 147–160.
438. Kosslick, H.; Berndt, H.; Lanh, H. D.; Martin, A.; Miessner, H.; Tuan, V. A. *J. Chem. Soc. Faraday Trans.* **1994**, *90*, 2837–2844.
439. Stockenhuber, M.; Lercher, J. A. *Microporous Mater.* **1995**, *3*, 457–465.
440. Onida, B.; Gabelica, Z.; Lourencüo, J.; Garrone, E. *J. Phys. Chem.* **1996**, *100*, 11072–11079.
441. Dongare, M. K.; Sabde, D. P.; Shaikh, R. A.; Kamble, K. P.; Hegde, S. G. *Catal. Today* **1999**, *49*, 267–276.
442. Bonelli, B.; Zanzottera, C.; Armandi, M.; Esposito, S.; Garrone, E. *Catal. Today* **2013**, *218–219*, 3–9.
443. Hu, Y.; Martra, G.; Zhang, J.; Higashimoto, S.; Coluccia, S.; Anpo, M. *J. Phys. Chem. B* **2006**, *110*, 1680–1685.
444. Weglarski, J.; Datka, J.; He, H.; Klinowski, J. *J. Chem. Soc. Faraday Trans.* **1996**, *92*, 5161–5164.
445. Gianotti, E.; Raimondi, M. E.; Marchese, L.; Martra, G.; Maschmeyer, T.; Seddon, J. M.; Coluccia, S. *Catal. Lett.* **2001**, *76*, 21–26.

446. Wang, Z.-M.; Yamaguchi, M.; Goto, I.; Kumagai, M. *Phys. Chem. Chem. Phys.* **2000**, *2*, 3007–3015.
447. Yamauchi, S.; Mori, T.; Yamamura, H. *Appl. Catal. A* **1995**, *132*, 21–27.
448. Desmartin-Chomel, A.; Flores, J. L.; Bourane, A.; Clacens, J. M.; Figueras, F.; Delahay, G.; Giroir Fendler, A.; Lehaut-Burnouf, C. *J. Phys. Chem. B* **2006**, *110*, 858–863.
449. Khadzhiivanov, K.; Davydov, A. *Kinet. Catal.* **1988**, *29*, 460–465.
450. Ramis, G.; Yi, L.; Busca, G. *Catal. Today* **1966**, *28*, 373–380.
451. Pârvulescu, V. I.; Boghosian, S.; Pârvulescu, V.; Jung, S. M.; Grange, P. *J. Catal.* **2003**, *217*, 172–185.
452. Baertsch, C. D.; Soled, S. L.; Iglesia, E. *J. Phys. Chem. B* **2001**, *105*, 1320–1330.
453. Patron, P.; La Ginestra, A.; Ramis, G.; Busca, G. *Appl. Catal. A* **1994**, *107*, 249–266.
454. Fiorilli, S.; Onida, B.; Bonelli, B.; Garrone, E. *J. Phys. Chem. B* **2005**, *109*, 16725–16729.
455. Primet, M.; Pichat, P.; Mathieu, M. V. *J. Phys. Chem.* **1971**, *75*, 1216–1220.
456. Romero Sarria, F.; Marie, O.; Saussey, J.; Daturi, M. *J. Phys. Chem. B* **2005**, *109*, 1660–1662.
457. Romero Sarria, F.; Blasin-Aube, V.; Saussey, J.; Marie, O.; Daturi, M. *J. Phys. Chem. B* **2006**, *110*, 13130–13137.
458. Morterra, C.; Magnassa, G. *Catal. Today* **1996**, *25*, 497–532.
459. Travert, A.; Vimont, A.; Sahibed-Dine, A.; Daturi, M.; Lavalley, J.-C. *Appl. Catal. A* **2006**, *307*, 98–107.
460. Pieterse, J. A. Z.; Veefkind-Reyes, S.; Seshan, K.; Domokos, L.; Lercher, J. A. *J. Catal.* **1999**, *187*, 518–520.
461. Areán, C. O.; Rodriguez Delgado, M.; Montouillout, V.; Lavalley, J. C.; Fernandez, C.; Pascual, J. J. C.; Parra, J. B. *Microporous Mesoporous Mater.* **2004**, *67*, 259–264.
462. Moliner, M.; Díaz-Cabañas, M. J.; Fornés, V.; Martínez, C.; Corma, A. J. *J. Catal.* **2008**, *254*, 101–109.
463. Vimont, A.; Travert, A.; Binet, C.; Pichon, C.; Mialane, P.; Secheresse, F.; Lavalley, J. C. *J. Catal.* **2006**, *241*, 221–224.
464. Buzzoni, R.; Bordiga, S.; Ricchiardi, G.; Lamberti, C.; Zecchina, A. *Langmuir* **1996**, *12*, 930–940.
465. Borade, R. B.; Adnot, A.; Kaliaguine, S. *J. Chem. Soc. Faraday Trans.* **1990**, *86*, 3949–3956.
466. Castella-Ventura, M.; Akacem, Y.; Kassab, E. *J. Phys. Chem. C* **2008**, *112*, 19045–19054.
467. Escalona Platero, E.; Mentruit, M. P.; Areán, C. O.; Zecchina, A. *J. Catal.* **1996**, *162*, 268–276.
468. Dzwigaj, S.; Massiani, P.; Davidson, A.; Che, M. *J. Mol. Catal. A* **2000**, *155*, 169–182.
469. Meriaudeau, P.; Tuan, V. A.; Hung, L. N.; Nghiem, V. T.; Naccache, C. *J. Chem. Soc. Faraday Trans.* **1998**, *94*, 467–471.
470. Benesi, H. A. *J. Catal.* **1973**, *28*, 176–178.
471. Jacobs, P. A.; Heylen, C. F. *J. Catal.* **1974**, *34*, 267–274.
472. Matulewicz, E. R. A.; Kerkhof, F. P. J. M.; Mouljin, L. A.; Reistma, H. J. *Colloid Int. Sci.* **1980**, *77*, 110–119.
473. Morterra, C.; Cerrato, G.; Meligrana, G. *Langmuir* **2001**, *17*, 7053–7060.
474. Knözinger, H.; Krietenbrink, H.; Ratnasamy, P. *J. Catal.* **1976**, *48*, 436–439.
475. Leydier, F.; Chizallet, C.; Chaumonnot, A.; Digne, M.; Soyer, E.; Quoineaud, A. A.; Costa, D.; Raybaud, P. *J. Catal.* **2011**, *284*, 215–229.
476. Lahousse, C.; Aboulayt, A.; Mauge, F.; Bachelier, J.; Lavalley, J. C. *J. Mol. Catal.* **1993**, *84*, 283–297.

477. Onfroy, T.; Clet, G.; Houalla, M. *Microporous Mesoporous Mater.* **2005**, *82*, 99–104.
478. Corma, A.; Fornés, V.; Forni, L.; Marquez, F.; Martınez-Triguero, J.; Moscottiy, D. *J. Catal.* **1998**, *179*, 451–458.
479. Xue, Z.; Ma, J.; Zheng, J.; Zhang, T.; Kang, Y.; Li, R. *Acta Mater.* **2012**, *60*, 5712–5722.
480. Lavalley, J. C. *Catal. Today* **1996**, *27*, 377–401.
481. Busca, G.; Lorenzelli, V. *Mater. Chem. Phys.* **1982**, *7*, 89–126.
482. Morterra, C.; Zecchina, A.; Coluccia, S.; Chiorino, A. *J. Chem. Soc. Faraday. Trans. 1* **1977**, *73*, 1544–1560.
483. Bensitel, M.; Moravek, V.; Lamotte, J.; Saur, O.; Lavalley, J. C. *Spectrochim. Acta* **1987**, *43A*, 1487–1491.
484. Baltrusaitis, J.; Jensen, J. H.; Grassian, V. H. *J. Phys. Chem. B* **2006**, *110*, 12005–12016.
485. Bachelier, J.; Aboulayt, A.; Lavalley, J. C.; Legendre, O.; Luck, F. *Catal. Today* **1993**, *17*, 55–62.
486. Vimont, A.; Thibault-Starzyk, F.; Lavalley, J.-C. *J. Phys. Chem. B* **2000**, *104*, 286–291.
487. Laane, J.; Ohlsen, J. R. *Prog. Inorg. Chem.* **1980**, *27*, 465–513.
488. Sobczak, I.; Rydz, M.; Ziolek, M. *Mater. Res. Bull.* **2013**, *48*, 795–801.
489. Pan, H.; Wang, X.; Xing, N.; Liu, Z. *Catal. Lett.* **2008**, *125*, 123–129.
490. Ma, X.; Wang, X.; Bi, R.; Zhao, Z.; He, H. *J. Mol. Catal. A* **2009**, *303*, 90–95.
491. Lónyi, F.; Valyon, J.; Gutierrez, L.; Ulla, M.; Lombardo, E. A. *Appl. Catal. B* **2007**, *73*, 1–10.
492. Pietrogiacomi, D.; Campa, M. C.; Indovina, V. *Catal. Today* **2010**, *155*, 192–198.
493. Szanyi, J.; Kwak, J. H.; Zhu, H.; Peden, C. H. F. *Phys. Chem. Chem. Phys.* **2013**, *15*, 2368–2380.
494. Thibault-Starzyk, F.; Marie, O.; Malicki, N.; Vos, A.; Schoonheydt, R.; Geerlings, P.; Henriques, C.; Pommier, C.; Massiani, P. *Stud. Surf. Sci. Catal.* **2005**, *158A*, 663–670.
495. Kantcheva, M.; Bushev, V.; Hadjiivanov, K. *J. Chem. Soc. Faraday Trans.* **1992**, *88*, 3087–3089.
496. Venkov, T.; Hadjiivanov, K.; Klissurski, D. *Phys. Chem. Chem. Phys.* **2002**, *4*, 2443–2448.
497. Hadjiivanov, K.; Avreyska, V.; Klissurski, D.; Marinova, T. *Langmuir* **2002**, *18*, 1619–1625.
498. Apostolescu, N.; Schröder, T.; Kureti, S. *Appl. Catal. B* **2004**, *51*, 43–50.
499. Börensen, C.; Kirchner, U.; Scheer, V.; Vogt, R.; Zellner, R. *J. Phys. Chem. A* **2000**, *104*, 5036–5045.
500. Liu, J.; Yu, Y.; Mu, Y.; He, H. *J. Phys. Chem. B* **2006**, *110*, 3225–3230.
501. Fu, H.; Wang, X.; Wu, H.; Yin, Y.; Chen, J. *J. Phys. Chem. C* **2007**, *111*, 6077–6085.
502. Saur, O.; Bensitel, M.; Mohammed Saad, A. B.; Lavalley, J. C.; Tripp, C. P.; Morrow, B. A. *J. Catal.* **1986**, *99*, 104–110.
503. Kamata, H.; Ohara, H.; Takahashi, K.; Yukimura, A.; Seo, Y. *Catal. Lett.* **2001**, *73*, 79–83.
504. Nanayakkara, C. E.; Pettibone, J.; Grassian, V. H. *Phys. Chem. Chem. Phys.* **2012**, *14*, 6957–6966.
505. Lamotte, J.; Morávek, V.; Bensitel, M.; Lavalley, J. C. *React. Kinet. Catal. Lett.* **1988**, *36*, 113–118.
506. Chauvin, C.; Saussey, J.; Lavalley, J. C.; Idriss, H.; Hindermann, J. P.; Kiennemann, A.; Chaumette, P.; Courty, P. *J. Catal.* **1990**, *121*, 56–69.
507. Bensitel, M.; Saur, O.; Lavalley, J. C. *Mater. Chem. Phys.* **1991**, *28*, 309–320.
508. Daturi, M.; Binet, C.; Lavalley, J. C.; Blanchard, G. *Surf. Interface Anal.* **2000**, *30*, 273–277.
509. Binet, C.; Daturi, M.; Lavalley, J. C. *Catal. Today* **1999**, *50*, 207–225.
510. Finocchio, E.; Daturi, M.; Binet, C.; Lavalley, J. C.; Blanchard, G. *Catal. Today* **1999**, *52*, 53–63.

511. Moulin, B.; Oliviero, L.; Bazin, P.; Daturi, M.; Costentin, G.; Mauge, F. *Phys. Chem. Chem. Phys.* **2010**, *13*, 10797–10807.
512. Busca, G.; Rossi, P. F.; Lorenzelli, V.; Banaissa, M.; Travert, J.; Lavalley, J. C. *J. Phys. Chem.* **1985**, *89*, 5433–5439.
513. Akarmazyan, S. S.; Panagiotopoulou, P.; Kambolis, A.; Papadopoulou, C.; Kondarides, D. I. *Appl. Catal. B* **2014**, *145*, 136–148.
514. Busca, G.; Lorenzelli, V. *J. Catal.* **1980**, *66*, 155–161.
515. Li, C.; Kazunare, D.; Ken-ichi, M. *J. Catal.* **1990**, *125*, 445–449.
516. Ramis, G.; Busca, G.; Lorenzelli, V. *J. Chem. Soc. Faraday Trans. 1* **1987**, *83*, 1591–1599.
517. Bianchi, D.; Chafik, T.; Khalfallah, M.; Teichner, S. *Appl. Catal. A* **1995**, *123*, 89–110.
518. Ouyang, F.; Yao, S. *J. Phys. Chem. B* **2000**, *104*, 11253–11257.
519. Natal-Santiago, M. A.; Dumesic, J. A. *J. Catal.* **1998**, *175*, 252–268.
520. Wovchko, E. A.; Camp, J. C.; Glass, J. A., Jr.; Yates, J. T., Jr. *Langmuir* **1995**, *11*, 2592–2599.
521. Kotrla, J.; Nachtigallova, D.; Kubelkova, L.; Heeribout, L.; Doremieux-Morin, C.; Fraissard, J. *J. Phys. Chem. B* **1998**, *102*, 2454–2463.
522. Basu, P.; Panayotov, D.; Yates, J. T., Jr. *J. Am. Chem. Soc.* **1988**, *110*, 2074–2081.
523. Basu, P.; Panayotov, D.; Yates, J. T., Jr. *J. Phys. Chem.* **1987**, *91*, 3133–3136.
524. Miessner, H.; Burkhardt, I.; Gutschick, D.; Zecchina, A.; Morterra, C.; Spoto, G. *J. Chem. Soc. Faraday Trans. 1* **1989**, *85*, 2113–2126.
525. Wong, T. T. T.; Stakheev, A. Y.; Sachtler, W. M. H. *J. Phys. Chem.* **1992**, *96*, 7733–7740.
526. Zaki, M.; Kunzmann, G.; Gates, B. C.; Knözinger, H. *J. Phys. Chem.* **1987**, *91*, 1486–1493.
527. Vayssilov, G. N.; Petrova, G. P.; Ivanova Shor, E. A.; Nasluzov, V. A.; Shor, A. M.; St. Petkov, P.; Rösch, N. *Phys. Chem. Chem. Phys.* **2012**, *14*, 5879–5890.
528. Vayssilov, G. N.; Gates, B. C.; Rösch, N. *Angew. Chem. Int. Ed.* **2003**, *42*, 1391–1394.
529. Vayssilov, G. N.; Rösch, N. *J. Phys. Chem. B* **2004**, *108*, 180–197.
530. Yokomizo, G. H.; Louis, C.; Bell, A. T. *J. Catal.* **1989**, *120*, 1–14.
531. Zanderighi, G. M.; Dossi, C.; Ugo, R.; Psaro, R.; Theolier, A.; Choplin, A.; D'Ornelas, L.; Basset, J. M. *J. Organomet. Chem.* **1985**, *296*, 127–146.
532. Sárkány, J. *Phys. Chem. Chem. Phys.* **1999**, *1*, 5251–5257.
533. Kefirov, R.; Penkova, A.; Hadjiivanov, K.; Dzwigaj, S.; Che, M. *Microporous Mesoporous Mater.* **2008**, *116*, 180–187.
534. Brown, M. A.; Fujimori, Y.; Ringleb, F.; Shao, X.; Stavale, F.; Nilius, N.; Sterrer, M.; Freund, H. J. *J. Am. Chem. Soc.* **2011**, *133*, 10668–10676.
535. Frache, A.; Gianotti, E.; Marchese, L. *Catal. Today* **2003**, 77, 371–384.
536. Nobukawa, T.; Yoshida, M.; Kameoka, S.; Ito, S.; Tomishige, K.; Kunimori, K. *J. Phys. Chem. B* **2004**, *108*, 4071–4079.
537. Mihaylov, M.; Ivanova, E.; Chakarova, K.; Novachka, P.; Hadjiivanov, K. *Appl. Catal. A* **2011**, *391*, 3–10.
538. Lamberti, C.; Bordiga, S.; Salvalaggio, M.; Spoto, G.; Zecchina, A.; Geobaldo, F.; Vlaic, C.; Bellatreccia, M. *J. Phys. Chem. B* **1997**, *101*, 344–360.
539. Karge, H. G.; Wichterlova, B.; Beyer, H. K. *J. Chem. Soc. Faraday Trans.* **1992**, *88*, 1345–1351.
540. Wittayakun, J.; Grisdanurak, N.; Kinger, G.; Vinek, H. *Korean J. Chem. Eng.* **2004**, *21*, 950–955.
541. Jentys, A.; Lugstein, A.; Vinek, H. *Zeolites* **1997**, *18*, 391–397.
542. Mihalyi, R. M.; Beyer, H. K.; Mavrodinova, V.; Minchev, C.; Neinska, Y. *Microporous Mesoporous Mater.* **1998**, *24*, 143–151.
543. Jia, C.; Massiani, P.; Barthomeuf, D. *J. Chem. Soc. Faraday Trans.* **1993**, *89*, 3659–3665.

544. Jia, C.; Beaunier, P.; Massiani, P. *Microporpos Mesoporous Mater.* **1998**, *24*, 69–82.
545. Yang, C.; Xu, Q. *J. Chem. Soc. Faraday Trans.* **1987**, *98*, 1675–1680.
546. Yang, C.; Xu, Q. *Zeolites* **1997**, *19*, 404–410.
547. Jin, Y. S.; Auroux, A.; Vedrine, J. C. *Appl. Catal.* **1998**, *37*, 1–19.
548. Dzwigaj, S.; Peltre, M.-J.; Massiani, P.; Davidson, A.; Che, M.; Sen, T.; Sivasanker, S. *Chem. Commun.* **1998**, *34*, 87–88.
549. Atoguchi, T.; Kanougi, T. *J. Mol. Catal. A* **2004**, *222*, 253–257.
550. Mauvezin, M.; Delahay, G.; Coq, B.; Kieger, S.; Jumas, J. C.; Olivier-Fourcade, J. *J. Phys. Chem. B* **2001**, *105*, 928–935.
551. Pieterse, J. A. Z.; Pirngruber, G. D.; van Bokhoven, J. A.; Booneveld, S. *Appl. Catal. B* **2007**, *71*, 16–22.
552. Zhu, C. Y.; Lee, C. W.; Chong, P. J. *Zeolites* **1996**, *17*, 483–488.
553. Mhamdi, M.; Khaddar-Zine, S.; Ghorbel, A. *Appl. Catal. A* **2009**, *357*, 42–50.
554. Nares, R.; Ramirez, J.; Gutierrez-Alejandre, A.; Louis, C.; Klimova, T. *J. Phys. Chem. B* **2002**, *106*, 13287–13293.
555. Marques, J. P.; Gener, I.; Ayrault, P.; Bordado, J. C.; Lopes, J. M.; Ribeiro, F. R.; Guisnet, M. *C. R. Chim.* **2005**, *8*, 399–410.
556. Kameoka, S.; Nobukawa, T.; Tanaka, S.; Ito, S.; Tomishige, K.; Kunimori, K. *Phys. Chem. Chem. Phys.* **2003**, *5*, 3328–3333.
557. Occelli, M. L.; Eckert, H.; Wolker, A.; Auroux, A. *Microporous Mesoporous Mater.* **1999**, *30*, 219–232.
558. Chester, A. W.; Chu, Y. F.; Dessau, R. M.; Kerr, G. T.; Kresge, C. T. *J. Chem. Soc. Chem. Commun.* **1985**, *21*, 289–290.
559. Woolery, G. L.; Alemany, L. B.; Dessau, R. M.; Chester, A. W. *Zeolites* **1986**, *6*, 14–16.
560. Góra-Marek, K.; Gil, B.; Datka, J. *Appl. Catal. A* **2009**, *353*, 117–122.
561. Dessau, R. M.; Schmitt, K. D.; Kerr, G. T.; Woolery, G. L.; Alemany, L. B. *J. Catal.* **1987**, *104*, 484–489.
562. Omegna, A.; Vasic, M.; van Bokhoven, J. A.; Pirngruber, G.; Prins, R. *Phys. Chem. Chem. Phys.* **2004**, *6*, 447–452.
563. Hadjiivanov, K.; Klissurski, D.; Kantcheva, M.; Davydov, A. *J. Chem. Soc. Faraday Trans.* **1991**, *87*, 907–911.
564. Trombetta, M.; Busca, G. *J. Catal.* **1999**, *187*, 521–523.
565. Thibault-Starzyk, F.; Stan, I.; Abelly, S.; Bonilla, A.; Thomas, K.; Fernandez, C.; Gilson, J.-P.; Pérez-Ramirez, L. *J. Catal.* **2009**, *264*, 11–14.
566. Sadowska, K.; Góra-Marek, K.; Datka, J. *J. Phys. Chem. C* **2013**, *117*, 9237–9244.
567. Domokos, L.; Lefferts, L.; Seshan, K.; Lercher, J. A. *J. Mol. Catal. A* **2000**, *162*, 147–157.
568. Zecchina, A.; Bordiga, S.; Spoto, G.; Scarano, D.; Spano, G.; Geobaldo, F. *J. Chem. Soc. Faraday Trans.* **1996**, *92*, 4863–4875.
569. Sigma-Aldrich. http://www.sigmaaldrich.com/chemistry/chemical-synthesis/learning-ce.
570. Zeynali, M. E. *Diffusion-fundamentals.org* **2010**, *13*, 1–18.
571. Ertl, G., Knözinger, H., Schüth, F., Weitkamp, J. Eds. *Handbook of Heterogeneous Catalysis*; Wiley VCH: Weinheim, Germany, 2008.
572. Jae, J.; Tompsett, G. A.; Foster, A. J.; Hammonda, K. D.; Auerbach, S. M.; Lobo, R. F.; Huber, G. W. *J. Catal.* **2011**, *279*, 257–268.
573. Sebastian, V.; Bosque, J.; Kumakiri, I.; Bredesen, R.; Anson, A.; Macia-Agullo, J. A.; Linares-Solano, A.; Tellez, C.; Coronas, J. *Microporous Mesoporous Mater.* **2011**, *142*, 649–654.
574. Bevilacqua, M.; Gutierrez Alejandre, A.; Resini, C.; Casagrande, M.; Ramirez, J.; Busca, G. *Phys. Chem. Chem. Phys.* **2002**, *4*, 4575–4583.
575. Ding, X.; Geng, S.; Li, C.; Yang, C.; Wang, G. *J. Nat. Gas Chem.* **2009**, *18*, 156–160.

576. Cosseron, A.-F.; Daou, T. J.; Tzanis, L.; Nouali, H.; Deroche, I.; Coasne, B.; Tchamber, V. *Microporous Mesoporous Mater.* **2013**, *173*, 147–154.
577. Lashaki, M. J.; Fayaz, M.; Niknaddaf, S.; Hashisho, Z. *J. Hazard. Mater.* **2012**, *241–242*, 154–163.
578. Bevilacqua, M.; Busca, G. *Catal. Commun.* **2002**, *3*, 497–502.
579. Corma, A.; Fornés, V.; Rey, F. *Zeolites* **1993**, *13*, 56–59.
580. Traa, Y.; Sealy, S.; Weitkamp, J. *Molecular Sieves—Science and Technology*; Springer-Verlag: Berlin, Heidelberg, 2006; pp 103–154.
581. Busca, G. *Curr. Phys. Chem.* **2012**, *2*, 136–150.
582. Knözinger, H. *Adv. Catal.* **1976**, *25*, 184–271.
583. Žilková, N.; Bejblová, M.; Gil, B.; Zones, S. I.; Burton, A. W.; Chen, C.-Y.; Musilová-Pavlačková, Z.; Košová, G.; Čejka, J. *J. Catal.* **2009**, *266*, 79–91.
584. Xue, Z.; Zhang, T.; Ma, J.; Miao, H.; Fan, W.; Zhang, Y.; Li, R. *Microporous Mesoporous Mater.* **2012**, *151*, 271–276.
585. Armaroli, T.; Bevilacqua, M.; Trombetta, M.; Milella, F.; Gutiérrez Alejandre, A.; Ramirez, J.; Notari, B.; Willey, R. J.; Busca, G. *Appl. Catal. A* **2001**, *216*, 59–71.
586. Montanari, T.; Bevilacqua, M.; Busca, G. *Appl. Catal. A* **2006**, *307*, 21–29.
587. Trombetta, M.; Busca, G.; Storaro, L.; Lenarda, M.; Casagrande, M.; Zambon, A. *Phys. Chem. Chem. Phys.* **2000**, *2*, 3529–3537.
588. Bjørgen, M.; Bonino, F.; Kolboe, S.; Lillerud, K.; Zecchina, A.; Bordiga, S. *J. Am. Chem. Soc.* **2003**, *125*, 15863–15868.
589. Nunan, J.; Cronin, J.; Cunningham, J. *J. Catal.* **1984**, *87*, 77–85.
590. Areán, C. O.; Escalona Platero, E.; Penarroya Mentruit, M.; Rodriguez Delgado, M.; Llabres i Xamena, F. X.; Garcia-Raso, A.; Morterra, C. *Microporous Mesoporous Mater.* **2000**, *34*, 55–60.
591. Qin, G.; Zheng, L.; Xie, Y.; Wu, C. *J. Catal.* **1985**, *95*, 609–612.
592. Zholobenko, V. L.; Kustov, L. M.; Borovkov, V. Y.; Kazansky, V. B. *Zeolites* **1988**, *8*, 175–178.
593. Trombetta, M.; Busca, G.; Lenarda, M.; Storaro, L.; Pavan, M. *Appl. Catal. A* **1999**, *182*, 225–235.
594. Zholobenko, V. L.; Lukyanov, D. B.; Dwyer, J.; Smith, W. J. *J. Phys. Chem. B* **1998**, *102*, 2715–2721.
595. Lukyanov, D. B.; Zholobenko, V. L.; Dwyer, J.; Barri, S. A. I.; Smith, W. J. *J. Phys. Chem. B* **1999**, *103*, 197–202.
596. Yoda, E.; Kondo, J. N.; Wakabayashi, F.; Domen, K. *Phys. Chem. Chem. Phys.* **2003**, *5*, 3306–3310.
597. Alberti, A. *Zeolites* **1997**, *19*, 411–415.
598. Moreau, F.; Ayrault, P.; Gnep, N. S.; Lacombe, S.; Merlen, E.; Guisnet, M. *Microporous Mesoporous Mater.* **2002**, *51*, 211–221.
599. Trombetta, M.; Busca, G.; Lenarda, B.; Storaro, L.; Ganzerla, R.; Piovesan, L.; Lopez, A. J.; Alcantara-Rodriguez, M.; Rodriguez-Castellón, E. *Appl. Catal. A* **2000**, *193*, 55–69.
600. Duncan, J. F.; Cook, G. B. *Isotopes in Chemistry*; Clarendon Press: Oxford, 1968; pp 124–145.
601. Jencks, W. P. *Catalysis in Chemistry and Enzymology*; McGraw-Hill: New York, 1969; pp 243–281.
602. Chang, R. *Physical Chemistry for Chemical and Biological Sciences*; University Science Books: Sausalito, CA, 2000; pp 445–493.
603. Katz, J. *J. Am. Sci.* **1960**, *48*, 544–580.
604. Soper, A. K.; Benmore, C. *J. Phys. Rev. Lett.* **2008**, *101*, 065502.
605. Wiberg, K. B. *Chem. Rev.* **1955**, *55*, 713–743.

606. German, E. D.; Kuznetsov, A. M.; Dogonadze, R. R. *J. Chem. Soc. Faraday Trans. 2* **1980**, *76*, 1129–1146.
607. Duprez, D. *Isotopes in Heterogeneous Catalysis*; In: Hargreaves, J. S. J., Lackson, S. D., Webb, G. Eds.; Imperial College Press: London, UK, 2006.
608. Westheimer, F. H. *Chem. Rev.* **1961**, *61*, 265–273.
609. Paabo, M.; Bates, R. G. *J. Phys. Chem.* **1970**, *74*, 706–710.
610. McConnell, R.; Godwin, W.; Stanley, B.; Shane Green, M. *J. Ark. Acad. Sci.* **1997**, *57*, 135–140.
611. Marx, D.; Tuckerman, M. E.; Hutter, J.; Parrinello, M. *Nature* **1999**, *397*, 601–604.
612. Hadjiivanov, K. *Ordered Porous Solids—Recent Advances and Prospects*; In: Valtchev, V., Mintova, S., Tsapatis, M. Eds.; Elsevier: Amsterdam, 2008; pp 263–281.
613. Burneau, A.; Carteret, C. *Phys. Chem. Chem. Phys.* **2000**, *2*, 3217–3226.
614. Gallas, J.-P.; Goupil, J.-M.; Vimont, A.; Lavalley, J.-C.; Gil, B.; Gilson, J.-P.; Miserque, O. *Langmuir* **2009**, *15*, 5825–5834.
615. Kondo, J. N.; Nishitani, R.; Yoda, E.; Yokoi, T.; Tatsumi, T.; Domen, K. *Phys. Chem. Chem. Phys.* **2010**, *12*, 11576–11586.
616. Poduval, D. G.; van Veen, J. A. R.; Rigutto, M. S.; Hensen, E. J. M. *Chem. Commun.* **2010**, *46*, 3466–3468.
617. Hohmeyer, J.; Kondratenko, E. V.; Bron, M.; Kröhnert, J.; Jentoft, F. C.; Schlögl, R.; Claus, P. *J. Catal.* **2010**, *269*, 5–14.
618. Stolle, S.; Summchen, L.; Roland, U.; Herzog, K.; Salzer, R. *J. Mol. Struct.* **1995**, *349*, 93–96.
619. Dwyer, J.; Dewing, J.; Thompson, N. E.; O'Malley, P. J.; Karim, K. *J. Chem. Soc. Chem. Commun.* **1989**, *25*, 843–844.
620. Aronson, M. T.; Gorte, R. J.; Farneth, W. E. *J. Catal.* **1987**, *105*, 455–468.
621. Kolyagin, Y. G.; Ordomsky, V. V.; Khimyak, Y. Z.; Rebrov, A. I.; Fajula, F.; Ivanova, I. I. *J. Catal.* **2006**, *238*, 122–133.
622. Lee, B.; Kondo, J. N.; Wakabayashi, F.; Domen, K. *Catal. Lett.* **1999**, *59*, 51–54.
623. Roessner, F.; Steinberg, K.-H.; Rechenburg, S. *Zeolites* **1987**, *7*, 488–489.
624. Jacobs, W. P. J. H.; van Wolput, J. H. M. C.; van Santen, R. A.; Jobic, H. *Zeolites* **1994**, *14*, 117–125.
625. Scarano, D.; Bordiga, S.; Lamberti, C.; Ricchiardi, G.; Bertarione, S.; Spoto, G. *Appl. Catal. A* **2006**, *307*, 3–12.
626. Ueda, R.; Kusakari, T.; Tomishige, K.; Fujimoto, K. *J. Catal.* **2000**, *194*, 14–22.
627. Hensen, E. J. M.; Poduval, D. G.; Ligthart, D. A. J. M.; van Veen, J. A. R.; Rigutto, M. S. *J. Phys. Chem. C* **2010**, *114*, 8363–8374.
628. Jia, M.; Lechert, H.; Förster, H. *Zeolites* **1992**, *12*, 32–36.
629. Wang, P.; Yang, S.; Kondo, J. N.; Domen, K.; Babay, T. *Chem. Lett.* **2003**, *32*, 792–793.
630. Datka, J.; Gil, B.; Baran, P. *Microporous Mesoporous Mater.* **2003**, *58*, 291–294.
631. Zholobenko, V. *Mendeleev Commun.* **1993**, *3*, 67–68.
632. Yoda, E.; Kondo, J. N.; Wakabayashi, F.; Domen, K. *Appl. Catal. A* **2000**, *194–195*, 275–283.
633. Strunk, J.; Kähler, K.; Xia, X.; Muhler, M. *Surf. Sci.* **2009**, *603*, 1776–1783.
634. Chang, C. C.; Dixon, L. T.; Kokes, R. J. *J. Phys. Chem.* **1973**, *77*, 2634–2639.
635. Guglielminotti, E. *Langmuir* **1990**, *6*, 1455–1460.
636. Davies, L. E.; Bonini, N. A.; Locatelli, S.; Gonzo, E. E. *Latin Am. Appl. Res.* **2005**, *35*, 23–28.
637. Kandiah, M.; Usseglio, S.; Svelle, S.; Olsbye, U.; Lillerud, K. P.; Tilset, M. *Mater. Chem.* **2010**, *20*, 9848–9851.
638. Paulidou, A.; Nix, R. M. *Surf. Sci.* **2000**, *470*, L104–L108.

639. Vayssilov, G. N.; Mihaylov, M.; St. Petkov, P.; Hadjiivanov, K. I.; Neyman, K. J. Phys. Chem. C 2011, 115, 23435–23454.
640. Jacobs, G.; Patterson, P. M.; Graham, U. M.; Sparks, D. E.; Davis, B. H. Appl. Catal. A 2004, 269, 63–73.
641. Tabakova, T.; Boccuzzi, F.; Manzoli, M.; Sobczak, J. W.; Idakiev, V.; Andreeva, D. J. Appl. Catal. A 2006, 298, 127–143.
642. Valentin, R.; Bonelli, B.; Garrone, E.; Di Renzo, F.; Quignard, F. Biomacromolecules 2007, 8, 3646–3650.
643. Mielke, Z.; Sobczyk, L. Isotope Effects in Chemistry and Biology; In: Kohen, A., Limbach, H. H. Eds.; Taylor & Francis: Boca Raton, 2006; pp 281–304.
644. Ghalla, H.; Rekik, N.; Baazaoui, M.; Oujia, B.; Wójcik, J. M. J. Mol. Struct. THEOCHEM 2008, 855, 102–110.
645. Coutinho, J.; Torres, V. J. B.; Pereira, R. N.; Bech Nielsen, B.; Jones, R.; Briddon, P. R. Phys. B 2006, 376–377, 126–129.
646. Zhdanov, S. P.; Kosheleva, L. S.; Titova, T. I. Langmuir 1987, 3, 960–967.
647. Detoni, S.; Hadzi, D.; Juranji, M. Spectrochim. Acta 1974, 30A, 249–253.
648. Pinchas, S.; Laulicht, N. Infrared Spectra of Labelled Compounds; Academic Press: London & New York, 1971.
649. Morrow, B. A.; Devi, A. Can. J. Chem. 1970, 48, 2454–2456.
650. Geidel, E.; Lechert, H.; Döbler, J.; Jobic, H.; Calzaferri, G.; Bauer, F. Microporous Mesoporous Mater. 2003, 65, 31–42.
651. Hadjiivanov, K.; Saur, O.; Lamotte, J.; Lavalley, J.-C. Z. Phys. Chem. 1994, 187, 281–300.
652. Kim, S.; Sorescu, D. C.; Yates, J. T., Jr. J. Phys. Chem. C 2007, 111, 18226–18235.
653. Galhotra, P.; Navea, J. G.; Larsena, S. C.; Grassian, V. H. Energy Environ. Sci. 2009, 2, 401–409.
654. Badri, A.; Binet, C.; Lavalley, J. C. J. Chem. Soc. Faraday Trans. 1996, 92, 4669–4673.
655. Takeuchi, M.; Martra, G.; Coluccia, S.; Anpo, M. J. Phys. Chem. B 2005, 109, 7387–7391.
656. Gallas, J.-P.; Lavalley, J.-C.; Burneau, A.; Barres, O. Langmuir 1991, 7, 1235–1240.
657. Emeis, C. A. J. Catal. 1993, 141, 347–354.
658. Jacobs, P. A.; Uytterhoeven, J. B. J. Catal. 1972, 26, 175–190.
659. Khabtou, S.; Chevreau, T.; Lavalley, J. C. Microporous Mater. 1994, 3, 133–148.
660. Hughes, T. R.; White, H. M. J. Phys. Chem. 1967, 71, 2192–2201.
661. Jacobs, P. A.; Theng, B. K.; Uytterhoeven, J. B. J. Catal. 1972, 26, 191–201.
662. Bortnovsky, O.; Melichar, Z.; Sobalik, Z.; Wichterlova, B. Microporous Mesoporous Mater. 2001, 42, 97–102.
663. Taouli, A.; Klemt, A.; Breede, M.; Reschetilowski, W. Stud. Surf. Sci. Catal. 1999, 125, 307–314.
664. Datka, J.; Turek, A. M.; Jehng, J. M.; Wachs, I. E. J. Catal. 1992, 135, 186–199.
665. Take, J.; Yamaguchi, T.; Miyamoto, K.; Ohyama, H.; Misono, N. Stud. Surf. Sci. Catal. 1986, 28, 495–501.
666. Coluccia, S.; Marchese, L.; Lavagnino, S.; Anpo, M. Spectrochim. Acta 1987, 43A, 1573–1576.
667. Diwald, O.; Sterrer, M.; Knözinger, E. Phys. Chem. Chem. Phys. 2002, 4, 2811–2817.
668. Knözinger, E.; Jacob, K.-H.; Singh, S.; Hofmann, P. Surf. Sci. 1993, 290, 388–402.
669. Kappers, M.; Dossi, C.; Psaro, R.; Recchia, S.; Fusi, A. Catal. Lett. 1996, 39, 183–189.
670. Anderson, P. J.; Horlock, R. F.; Olivier, J. F. Trans. Faraday Soc. 1965, 61, 2754–2762.
671. Chizallet, C.; Costentin, G.; Che, M.; Delbecq, F.; Sautet, P. J. Am. Chem. Soc. 2007, 129, 6442–6452.
672. Schiek, M.; Al-Shamery, K.; Kunat, M.; Traeger, F.; Wöll, C. Phys. Chem. Chem. Phys. 2006, 8, 1505–1512.

673. Busca, G. *Catal. Today* **2014**, *226*, 2–13.
674. Peri, J. B. *J. Phys. Chem.* **1965**, *69*, 220–221.
675. Busca, G.; Lorenzelli, V.; Ramis, G.; Willey, R. J. *Langmuir* **1993**, *9*, 1492–1499.
676. Busca, G.; Lorenzelli, V.; Sanchez Escribano, V.; Guidetti, R. *J. Catal.* **1991**, *131*, 167–177.
677. Heitmann, G. P.; Dahlhoff, G.; Hölderich, W. F. *J. Catal.* **1999**, *186*, 12–19.
678. Primet, M.; Pichat, P.; Mathieu, M. V. *J. Phys. Chem.* **1971**, *75*, 1216–1220.
679. Khadzhiivanov, K.; Davydov, A.; Klissurski, D. *Kinet. Catal.* **1988**, *29*, 161–167.
680. Dzwigaj, S.; Arrouvel, C.; Breysse, M.; Geantet, C.; Inoue, S.; Toulhoat, H.; Raybaud, P. *J. Catal.* **2005**, *236*, 245–250.
681. Arrouvel, C.; Digne, M.; Breysse, M.; Toulhoat, H.; Raybaud, P. *J. Catal.* **2004**, *222*, 152–166.
682. Jones, P.; Hockey, J. A. *Trans. Faraday Soc.* **1972**, *68*, 907–1001.
683. Hadjiivanov, K. *Appl. Surf. Sci.* **1998**, *135*, 331–338.
684. Morterra, C.; Cerrato, S.; Di Ciero, S. *Appl. Surf. Sci.* **1998**, *126*, 107–128.
685. Azambre, B.; Zenboury, L.; Koch, A.; Weber, J. W. *J. Phys. Chem. C* **2009**, *113*, 13287–13299.
686. Korhonen, S. T.; Calatayud, M.; Krause, A. O. I. *J. Phys. Chem. C* **2008**, *112*, 6469–6476.
687. Yamaguchi, T.; Nakano, Y.; Tanabe, K. *Bull. Chem. Soc. Jpn.* **1978**, *51*, 2482–2487.
688. Klose, B. S.; Jentoft, F. C.; Schlögl, R. *J. Catal.* **2005**, *233*, 68–80.
689. Miller, T. M.; Grassian, V. H. *Catal. Lett.* **1997**, *46*, 213–221.
690. Ma, Z. Y.; Yang, C.; Wei, W.; Li, W. H.; Sun, Y. H. *J. Mol. Catal. A* **2005**, *227*, 119–124.
691. Paier, J.; Penschke, C.; Sauer, J. *Chem. Rev.* **2013**, *113*, 3949–3985.
692. Lykhach, Y.; Johánek, V.; Aleksandrov, H. A.; Kozlov, S. M.; Happel, M.; Skála, T.; St. Petkov, P.; Tsud, N.; Vayssilov, G. N.; Prince, K. C.; Neyman, K. M.; Matolín, V.; Libuda, J. *J. Phys. Chem. C* **2012**, *116*, 12103–12113.
693. Binet, C.; Badri, A.; Lavalley, J. C. *J. Phys. Chem.* **1994**, *98*, 6392–6398.
694. Daturi, M.; Finocchio, E.; Binet, C.; Lavalley, J. C.; Fally, F.; Perrichon, V. *J. Phys. Chem. B* **1999**, *103*, 4884–4891.
695. Agarwal, S.; Lefferts, L.; Mojet, B. L. *ChemCatChem* **2013**, *5*, 479–489.
696. Pozdnyakova, O.; Teschner, D.; Wootsch, A.; Kröhnert, J.; Steinhauer, B.; Sauer, H.; Toth, L.; Jentoft, F. C.; Knop-Gericke, A.; Paál, Z.; Schlögl, R. *J. Catal.* **2006**, *237*, 1–16.
697. Farra, R.; Wrabetz, S.; Schuster, M. E.; Stotz, E.; Hamilton, N. G.; Amrute, A. P.; Pérez-Ramírez, J.; López, N.; Teschner, D. *Phys. Chem. Chem. Phys.* **2013**, *15*, 3454–3465.
698. Holmgren, A.; Andersson, B.; Duprez, D. *Appl. Catal. B* **1999**, *22*, 215–230.
699. Daturi, M.; Bion, N.; Saussey, K.; Lavalley, J. C.; Hedouin, C.; Seguelong, T.; Blanchard, G. *Phys. Chem. Chem. Phys.* **2001**, *3*, 252–255.
700. Timofeeva, M. N.; Jhung, S. H.; Hwang, Y. K.; Kim, D. K.; Panchenko, V. N.; Melgunov, M. S.; Chesalov, Y. A.; Chang, J.-S. *Appl. Catal. A* **2007**, *317*, 1–10.
701. Areán, C. O.; López Bellan, A.; Peñarroya Mentruit, M.; Rodriguez Delgado, M.; Turnes Palomino, G. *Microporous Mesoporous Mater.* **2000**, *40*, 35–42.
702. Pan,, Y.-x; Mei, D.; Liu,, C.-j; Ge, Q. *J. Phys. Chem. C* **2011**, *115*, 10140–10146.
703. Pohle, R.; Fleischer, M.; Meixner, H. *Sens. Actuators B* **2000**, *68*, 151–156.
704. Klingenberg, B.; Vannice, M. A. *Chem. Mater.* **1996**, *8*, 2755–2768.
705. Tsyganenko, A. A.; Lamotte, J.; Gallas, J. P.; Lavalley, J. C. *J. Phys. Chem.* **1989**, *93*, 4179–4183.
706. Emiroglu, S.; Barsan, N.; Weimar, U.; Hoffmann, V. *Thin Solid Films* **2001**, *391*, 176–185.
707. Borade, R. B.; Adnot, A.; Kaliaguine, S. *Zeolites* **1991**, *11*, 710–719.

708. Sasidharan, M.; Hegde, S. G.; Kumar, R. *Microporous Mesoporous Mater.* **1998**, *24*, 59–67.
709. Montanari, T.; Busca, G. *Vib. Spectrosc.* **2008**, *46*, 45–51.
710. Giordanino, F.; Vennestrøm, P. N. R.; Lundegaard, L. F.; Stappen, F. N.; Mossin, S.; Beato, P.; Bordiga, S.; Lamberti, C. *Dalton Trans.* **2013**, *42*, 12741–12761.
711. Lónyi, F.; Solt, H. E.; Paszti, Z.; Valyon, J. *Appl. Catal. B* **2014**, *150–151*, 218–229.
712. Bonino, F.; Lamberti, C.; Chavan, S.; Vitillo, J. G.; Bordiga, S. *Metal Organic Frameworks as Heterogeneous Catalysts*; In: Llabres i Xamena, F. X., Gascon, J. Eds.; Thomas Graham House, Science Park: Cambridge, UK, 2013; pp 76–142.
713. Volkringer, C.; Loiseau, T.; Guillou, N.; Ferey, G.; Elkaim, E.; Vimont, A. *Dalton Trans.* **2009**, 2241–2249.
714. Ravon, U.; Chaplais, G.; Chizallet, C.; Seyyedi, B.; Bonino, F.; Bordiga, S.; Bats, N.; Farrusseng, D. *ChemCatChem* **2010**, *2*, 1235–1238.
715. Vimont, A.; Travert, A.; Bazin, P.; Lavalley, J.-C.; Daturi, M.; Serre, C.; Férey, G.; Bourrelly, S.; Llewellyn, P. L. *Chem. Commun.* **2007**, *43*, 3291–3293.
716. Gascon, J.; Aktay, U.; Hernandez-Alonso, M. D.; van Klink, G. P. M.; Kapteijn, F. *J. Catal.* **2009**, *261*, 75–87.
717. Serra-Crespo, P.; Gobechiya, E.; Ramos-Fernandez, E. V.; Juan-Alcañiz, J.; Martinez-Joaristi, A.; Stavitski, E.; Kirschhock, C. E. A.; Martens, J. A.; Kapteijn, F.; Gascon, J. *Langmuir* **2012**, *28*, 12916–12922.
718. Zornoza, B.; Martinez-Joaristi, A.; Serra-Crespo, P.; Tellez, C.; Coronas, J.; Gascon, J.; Kapteijn, F. *Chem. Commun.* **2011**, *47*, 9522–9524.
719. Goesten, M. G.; Juan-Alcañiz, J.; Ramos-Fernandez, E. V.; Sai Sankar Gupta, K. B.; Stavitski, E.; van Bekkum, C. H.; Gascon, J.; Kapteijn, F. *J. Catal.* **2011**, *281*, 177–187.
720. Cavka, J. H.; Jakobsen, S.; Olsbye, U.; Guillou, N.; Lamberti, C.; Bordiga, S.; Lillerud, K. P. *J. Am. Chem. Soc.* **2008**, *130*, 13850–13851.
721. Wiersum, A. D.; Soubeyrand-Lenoir, E.; Yang, Q.; Moulin, B.; Guillerm, V.; Yahia, M. B.; Bourrelly, S.; Vimont, A.; Miller, S.; Vagner, C.; Daturi, M.; Clet, G.; Serre, C.; Maurin, G.; Llewellyn, P. L. *Chem. Asian J.* **2011**, *6*, 3270–3280.
722. Miller, S. R.; Pearce, G. M.; Wright, P. A.; Bonino, F.; Chavan, S.; Bordiga, S.; Margiolaki, I.; Guillou, N.; Férey, G.; Bourrelly, S.; Llewellyn, P. L. *J. Am. Chem. Soc.* **2008**, *130*, 15967–15981.
723. Masciocchi, N.; Galli, S.; Colombo, V.; Maspero, A.; Palmisano, G.; Seyyedi, B.; Lamberti, C.; Bordiga, S. *J. Am. Chem. Soc.* **2010**, *132*, 7902–7904.
724. Mino, L.; Colombo, V.; Vitillo, J. G.; Lamberti, C.; Bordiga, S.; Gallo, E.; Glatzel, P.; Maspero, A.; Galli, S. *Dalton Trans.* **2012**, *41*, 4012–4019.

CHAPTER THREE

Structural, Surface, and Catalytic Properties of Aluminas

Guido Busca

Dipartimento di Ingegneria Civile, Chimica e Industriale, Università di Genova, Genova, Italia

Contents

Abstract

The published data concerning the structural, surface, and catalytic properties of aluminas are reviewed, and these properties are related to the preparation procedures. The experimental and computational investigations of the structural characteristics of the polymorphs most useful for applications in catalysis, which are γ-, η-, δ-, and θ-Al$_2$O$_3$, are critically analyzed. The thermodynamics of the various polymorphs and the kinetics of the phase transitions are considered. The available information on Brønsted sites (i.e., hydroxyl groups), Lewis acid sites, and acid–base pairs on the surface of aluminas is discussed. Data regarding the application of aluminas as a catalyst and as a catalyst support are summarized. Suggestions for future research are proposed.

ABBREVIATIONS

ccp cubic close packed
DME dimethylether
FCC fluid catalytic cracking
hcp hexagonal close packing
HRTEM high-resolution transmission electron microscopy
MAS magic angle spinning
sh shoulder
TEM transmission electron microscopy
XP(S) X-ray photoelectron (spectroscopy)

1. INTRODUCTION

Metal oxides constitute an important family of solid materials because of their vast number of applications, many of which are large scale operations or have technological significance *(1)*. Several of these applications are associated with the surface activity that many metal oxides exhibit, examples are adsorption *(2)*, sensing, and catalysis *(3)*. Among the surface-active metal oxides, aluminum(III) oxide, more often referred to as "alumina," is particularly relevant. There is, however, no single alumina, because Al_2O_3 is characterized by pronounced polymorphism, and "alumina" can refer to any of a number of materials that are distinguished only by their structures. Aluminas enjoy wide-spread application in catalytic processes and adsorption technologies. Additionally, aluminas have become reference materials in the solid-state chemistry and the surface science of oxides: much of the present knowledge goes back to pioneering work on aluminas. The findings were later extrapolated to other oxides, or the concepts were expanded to include other materials. Notwithstanding the substantial amount of knowledge, there are many open questions and also significant disagreement in the literature. In the present review, we will attempt to summarize the state of the art regarding the characterization of the solid-state, surface, and catalytic properties of aluminas and will seek to relate these properties to the preparation procedure.

2. PREPARATION AND SOLID-STATE CHEMISTRY OF ALUMINAS

2.1. Methods of alumina preparation

Most of the preparations of aluminas are performed by thermal decomposition of alumina "hydrates" (i.e., hydroxides, oxyhydroxides) previously precipitated from solutions containing Al^{3+} ions or organometallic compounds. The characteristics of the obtained aluminas are closely related to those of the hydrate precursor. Thus, a short review of the structural characteristics of these compounds and of the conditions for their precipitation will be given in Sections 2.1.1 to 2.1.3.

2.1.1 Crystal structures of aluminum hydroxides and oxyhydroxides

At least seven different crystalline phases of alumina hydrates are known (Table 3.1) (4–6). Among them are four polymorphs of the trihydroxide $Al(OH)_3$ and two polymorphs of the oxyhydroxide AlOOH. The four trihydroxides are bayerite (7,8), usually denoted as α-$Al(OH)_3$, gibbsite (8,9), usually denoted as γ-$Al(OH)_3$, and the less common doyleite (10,11), and nordstrandite (12). Their structures are closely related. While characterization of bayerite and gibbsite has been the subject of many investigations, details on the structure of nordstrandite and doyleite are scarce. The availability of refined theoretical methods and computational power and the combination of the respective results with diffraction and vibrational spectroscopic data recently led to reanalysis of the structure of all of these

Table 3.1 Crystal data of aluminum hydroxides and oxyhydroxides

Mineral name	Formula	Space group	Z
Bayerite	α-$Al(OH)_3$	$P2_1/n$	8
Gibbsite	γ-$Al(OH)_3$	$P2_1/n$	8
Nordstrandite	$Al(OH)_3$	$P\bar{1}$	4
Doyleite	$Al(OH)_3$	$P\bar{1}$ or $P1$	2
Diaspore	α-AlOOH	$Pbnm$	4
Boehmite	γ-AlOOH	$P2_1/c$ or $Cmc2_1$	4
Akdalaite, Tohdite	$5Al_2O_3 \cdot H_2O$ or $2Al_5O_7(OH)$	$P6_3mc$, $P3\,1c$, or $Cmc2_1$	2

phases, whereby particular attention was paid to the positions of the hydrogen atoms and the geometries of the hydrogen bonds. Modern thermodynamics calculations agree that gibbsite is the most stable $Al(OH)_3$ polymorph at a temperature of 25 °C *(13,14)*.

The four $Al(OH)_3$ polymorphs result from different stacking sequences of a single kind of layer. Each layer in the sequence is comprised of $Al(OH)_6$ edge-sharing octahedra that are arranged in a planar pseudohexagonal pattern (Figure 3.1). All hydroxyl groups are of the bridging type, and each of them connects two octahedrally coordinated aluminum atoms. The hydroxyl groups cover the external surface of both sides of a layer. The different $Al(OH)_3$ polymorphs are thus also distinguished by different geometries of both interlayer (Figure 3.2 *(6)*) and intralayer hydrogen bonds. Bayerite is the most dense phase, with an Al–Al interlayer distance of 4.79 Å, in comparison with gibbsite, which has two different interlayer distances of 4.84 and 4.94 Å, and nordstrandite and doyleite with distances of 4.98 and 5.08 Å, respectively.

The peak positions of octahedral aluminum in the ^{27}Al magic angle spinning (MAS) NMR spectra of gibbsite and bayerite *(15)* were found to be different. More recent measurements of gibbsite *(16)* and use of the ^1H-decoupled ^{27}Al 3Q-MAS NMR technique allowed the distinction between two nonequivalent octahedral aluminum ions in both structures *(17)*. The diversity of the aluminum sites was explained by the different character of the hydrogen bonds in which the hydroxyls forming the octahedron around each aluminum site are involved. With the help of high-resolution solid-state ^1H CRAMPS (combination of rotation and

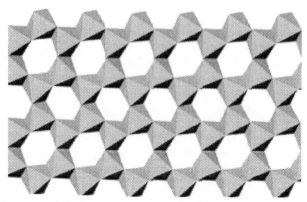

Figure 3.1 Structure of the layers common to $Al(OH)_3$ polymorphs.

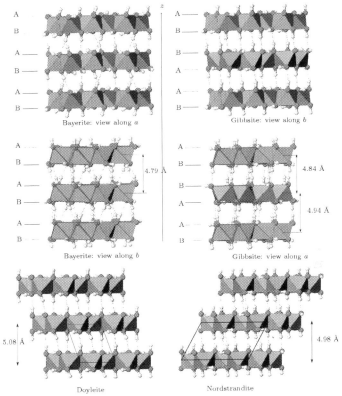

Figure 3.2 Layer stacking and hydrogen bonds in aluminum trihydroxides Al(OH)₃. Red spheres, oxygen atoms; white spheres, hydrogen atoms. *Reprinted with permission from Ref.* (6). (See the color plate.)

multiple-pulse spectroscopy) measurements, four signals with an intensity ratio of 1:2:2:1 were resolved. On the basis of these data, six nonequivalent hydrogen sites were inferred in the gibbsite crystal structure and ascribed to two types of structural OH groups, associated with intralayer and interlayer hydrogen bonds *(16)*. Evidence for the different arrangements of the interlayer hydrogen bonds also comes from the different Raman *(10,18,19)* and IR spectra *(6,20–22)* in the region of OH stretching and deformation vibrations.

In contrast to the closely related structures of the four trihydroxides, the structures of the two oxyhydroxides are significantly different from each other. The structure of boehmite *(23,24)*, depicted in Figure 3.3 *(25)*, is characterized by double layers of oxygen octahedra partially filled with aluminum cations. The stacking arrangement of the three oxygen layers in each

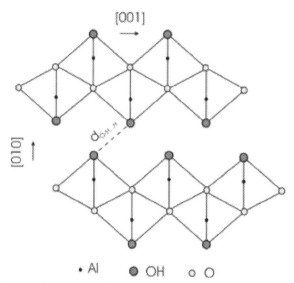

Figure 3.3 The crystal structure of boehmite projected on a plane perpendicular to the [100] direction. Double layers of octahedra are held together through hydrogen bonds. *Reprinted with permission from Ref. (25).*

double layer is such that the oxygen atoms form a cubic close packing. Oxide ions in the middle of the double layer are shared by four octahedra (i.e., they are coordinated to four Al^{3+} ions), whereas the hydroxyl groups on the outside are only shared by two octahedra (i.e., they form a bridge between two Al^{3+} ions). The stacking of the layers is such that the hydroxyl groups of one layer are located over the depression between the hydroxyl groups in the adjacent layer. Zig-zag chains of hydrogen bonds, whose exact geometry has not been completely identified, are formed between the layers. Poorly crystallized boehmite is often prepared as a precursor of γ-Al_2O_3, and is usually referred to as "pseudoboehmite."

In the structure of diaspore *(26,27)*, the oxygen atoms are in hexagonal close packing (hcp) with two thirds of the octahedral sites filled with aluminum cations (Figure 3.4 *(28)*). The occupied octahedra are linked together by edge-sharing, forming double chains in the *c*-axis direction. Diaspore differs from boehmite in the coordination of the oxygen atoms: both oxide ions and hydroxyl groups bridge three aluminum ions. Hydroxyl groups are H bonded to oxide ions. This, in comparison with boehmite, compact arrangement explains the greater density of diaspore and the low frequency of the OH stretching vibration shown by infrared absorption.

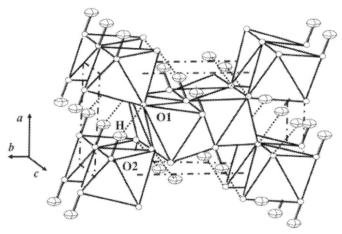

Figure 3.4 The crystal structure of diaspore. *Reprinted with permission from Ref.* (28).

Al 2p XP spectra of aluminum hydroxides and oxyhydroxides show a single peak at a binding energy of 73.9–74.9 eV that has not been found useful for the distinction between phases. In contrast, the O 2p XP spectra can be used to discriminate the different phases, including even the distinction between boehmite and pseudoboehmite. The O 2p peak of the hydroxyl groups is centered at 531.7–531.9 eV, whereas the peak of oxide species is found near 530.6 eV for oxyhydroxides and also for aluminas *(29,30)*. Water present on the surface is characterized by binding energies around 533.0 eV. A distinction can be made between boehmite and pseudoboehmite based on the slightly lower ratio of oxygen to hydroxyl groups and the presence of water in pseudoboehmite.

A seventh phase exists, denoted as akdalaite, and has been found to have the same structure as the phase denoted as tohdite. The composition corresponds to the formula $5Al_2O_3 \cdot H_2O$. Three different space groups for the unit cell have been proposed *(5,31)* which were found to be nearly equivalent in terms of both geometry and energy. Of the 10 aluminum atoms in the unit cell, eight are octahedrally coordinated in edge-sharing oxygen octahedra (resulting in two different crystallographic positions for Al^{3+}) and two are tetrahedrally coordinated in corner-sharing oxygen tetrahedra. The structure is actually that of an oxyhydroxide with a terminal H-bonded hydroxyl group on top of an octahedrally coordinated aluminum cation. Thus, akdalaite is an oxyhydroxide and its composition would be better represented by the formula $Al_5O_7(OH)$.

2.1.2 Precipitation of aluminum hydroxides and oxyhydroxides

The mineral bauxite, which is a mixed hydroxide of iron and aluminum, is the raw material at the origin of the industrial chemistry of aluminum compounds. Bauxite is transformed into sodium-contaminated gibbsite in the so-called Bayer process *(32,33)*. Industrial Bayer-type gibbsite can be redissolved in acids or in strongly basic solutions, and from these solutions all other aluminum compounds are produced including aluminum hydroxides and oxides *(34,35)*.

According to their amphoteric nature, aluminum hydroxides have their minimum solubility in water at a close-to-neutral pH value, that is, in the pH range of 6–7. At a pH larger than 8.5, the solubility increases, and the $[Al(OH)_4]^-$ anion is formed. At a pH smaller than 4, the aquo cation $[Al(OH_2)_6]^{3+}$ is reported to be predominant, whereas at a pH between 4 and 6, its dissociated form $[Al(OH_2)_5(OH)]^{2+}$ prevails *(36,37)*, likely both with less tightly bound water molecules in the secondary hydration shell. However, the speciation of aluminum in aqueous environments is very complicated. Depending on the conditions, such as pH, concentration (or hydrolysis ratio), and anions present, a variety of polynuclear species can be found in solution. In pure water, polynuclear species include dimers $[Al_2(OH)_4(H_2O)_4]^{4+}$, trimers $[Al_3(OH)_4(H_2O)_9]^{5+}$, tridecamers $[Al_{13}O_4(OH)_{24}(H_2O)_{12}]^{7+}$ (also termed Al13-mers), and the recently characterized $[Al_{30}O_8(OH)_{56}(H_2O)_{26}]^{18+}$ species (Al30-mers) *(38,39)*. Al13 and Al30 species are reported to predominate at pH values in the range of approximately 4.0–6.0, depending on the aluminum ion concentration.

The chemistry of the precipitation of aluminum hydroxides and oxides is very complex *(40)*. When the solubility is exceeded, "gelatinous" precipitates, which are found to be amorphous by X-ray diffraction, usually form initially. ^{27}Al MAS NMR shows the predominance of octahedrally coordinated Al^{3+} ions in these amorphous hydroxides, as are present in the crystalline trihydroxides and oxyhydroxides. However, in the amorphous materials some pentacoordinated and tetracoordinated Al^{3+} ions are also found *(15)*. As discussed above, there are many different crystalline hydroxides or oxyhydroxides, and which of them will be formed depends on the conditions *(41)*. Primary factors are temperature and pH, as well as aging time; however, the nature of the anions present and the possible presence of organic components *(42,43)* also play a role. At low temperature in an excess of water, the hydroxides are preferentially formed, specifically bayerite at pH values between 5.8 and 9 or gibbsite for pH values smaller than 5.8 or larger than 9.

As stated above, gibbsite, or γ-Al(OH)$_3$, is an intermediate phase in the commercial production of alumina by the Bayer process. The industrial process involves a leaching step using hot, caustic aluminate solution (referred to as "digestion") which, after removal of insoluble residue, is followed by the seeded precipitation of purified gibbsite, at a slightly lower temperature of about 90 °C. The product contains smaller crystals that have the shape of plates and prisms and larger particles that are agglomerates of tabular and prismatic crystals. The basic crystal habit is pseudohexagonal tabular. Gibbsite from the Bayer process usually contains several tenths of a percent of alkali metal cations; the technical product, precipitated from a sodium aluminate solution, contains up to 0.5 mass% Na$_2$O *(32)* which cannot be washed out even by using dilute HCl.

Gibbsite of high purity is needed to prepare aluminas for catalytic applications. A number of synthesis protocols are reported, all characterized by very long aging times for the crystallization to avoid high sodium concentrations, and with the final steps sometimes at acidic pH *(44)*. Nanometer-sized crystals of gibbsite samples can be prepared by neutralization of sodium aluminate solutions *(45)* or at acidic pH *(46)*.

Bayerite can be produced by partial neutralization of acidic solutions of Al^{3+} with ammonia *(47)*, or by neutralization of basic aluminate [Al(OH)$_4$]$^-$ solutions with CO$_2$ *(48)* or protonic acids such as nitric acid *(49)*. In most procedures, there is little control of the precipitation process, and a mixture of different particles shapes and sizes is obtained. Only by following particular protocols may uniform particles be produced. Uniform spherical bayerite particles can be prepared from sodium aluminate and carbonate mixtures at a temperature of 50 °C *(50)*, whereas nanorods are obtained at constant pH *(49)*. Pure bayerite that contains not alkali metal ions can be prepared, and there is in fact evidence that bayerite converts irreversibly to gibbsite in the presence of alkali metal cations (i.e., Na$^+$ and K$^+$).

At conditions that normally favor the production of bayerite, the addition of chelating agents (e.g., ethylenediamine, ethylene glycol) leads to the formation of nordstrandite *(51)*. A number of different synthesis protocols are reported for nordstrandite preparation *(51–53)*.

At temperatures higher than about 80 °C, the oxyhydroxides become thermodynamically more stable than the trihydroxides *(54)* and thus tend to form. Poorly crystallized solids designated as "pseudoboehmite" form at lower temperatures, whereas well-crystallized boehmite forms at higher temperatures or after longer aging. The typical conditions for the formation of poorly crystallized, gelatinous pseudoboehmite are pH values of about

6–8 and temperatures of 25–80 °C (300–350 K) (55). The simplest method to create the precipitation medium is to use soluble aluminum salts as starting compounds. The most common industrial source of aluminum is sodium aluminate, which is inexpensive but also has some disadvantages. Specifically, the precipitate may contain nonnegligible amounts of alkali metal cations, which must be washed off. This procedure seems to be applied, for example, in the preparation of the Versal alumina from UOP (previously from LaRoche Chemicals). In this case, precipitation from a sodium aluminate solution (56) leads to the production of pseudoboehmite with a fibrous plate-like structure, which is evident from transmission electron micrographs (57). Precipitation from $NaAlO_2$ solutions with aluminum sulfate produces well-defined fibrous boehmite, which can be converted into well-defined fibrous aluminas by calcination (58). Electron diffractions studies of fibrous boehmite show a diffuse halo indicating a highly disordered structure caused by random intercalation of water molecules into the octahedral layers (59).

Hydrothermal transformation of gibbsite at temperatures above 150 °C is a common method for the synthesis of well-crystallized boehmite. Higher temperatures and the presence of alkali metal cations increase the rate of transformation. Boehmite crystals with a diameter of 5–10 mm are produced by this method. Hydrothermally produced boehmite usually contains excess water, that is, about 1–2 wt.% more than the stoichiometric amount of 15 wt.%. Boehmites obtained from pseudoboehmite or gibbsite are generally "lamellar" or "plate-like," whereby a number of different habits have been reported such as platy ellipse-like, rhombic, hexagonal, and lath-like with hexagonal profile (as shown Figure 3.5 (60)). A fibrous habit is possible, too (58). Hydrothermal and solvothermal methods involving water–organic solvent mixtures have been developed to control the morphology (61), and boehmite nanoparticles with the following shapes have been prepared: fibers or rods (62), sheets (63), plates (64), cubes (65), and hierarchical nanostructures (66).

Diaspore, which is thermodynamically more stable than boehmite, can also be obtained by hydrothermal transformation of gibbsite or boehmite, but its formation is slow. Higher temperatures (i.e., >200 °C) and pressures (>15 MPa) are required for the synthesis, and the presence of diaspore seed crystals helps to avoid boehmite formation. Methods to produce well-crystallized diaspore have been reported; these include hydrothermal synthesis at 300 °C and 3.45×10^7 Pa (5000 psi) over a 72-h period (67) or high-pressure calcination of boehmite (68).

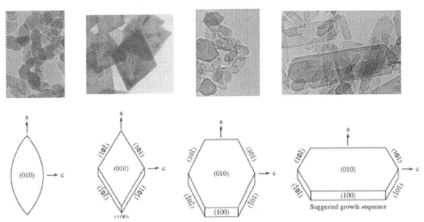

Figure 3.5 Proposed sequence for the growth of plate-like well-crystallized boehmite crystals in acetic acid solution. Growth starts from gibbsite or pseudoboehmite precursors. Top: Micrographs showing the crystal shapes of well-crystallized synthetic boehmite prepared by the method described in Ref. *(60)* (full size images with scale available in original paper). Bottom: proposed growth sequence with crystallographic relationships. *Reprinted from Ref. (60).*

Tohdite (akdalaite) can be manufactured by hydrothermal synthesis in a relatively narrow range of conditions *(69,70)*, that is, at temperatures above those used for the other hydroxides and oxyhydroxides but below the temperature for direct formation of corundum (α-Al_2O_3), which is reported to occur above 385 °C *(70,71)*.

2.1.3 Preparation of aluminum hydroxides from alkoxides

Several preparation methods of aluminas are based on the conversion of aluminum alkoxides. These precursor compounds allow the preparation of pure aluminas that do not contain significant amounts of alkali metals. The most frequently applied procedures can be grouped under the generic term "sol–gel" methods. Characteristic of these methods is the production of a gel through hydrolysis of an alkoxide; this gel is aged and later thermally dried. A similar commercial method to prepare microporous boehmite, which is applied by Sasol (and formerly by Condea), takes advantage of the chemistry of the Ziegler process and is known in industry as the ALFOL process. The original goal of the ALFOL process is to produce linear fatty alcohols, and the process starts with the synthesis of trialkyl aluminum by an oligomerization reaction of ethylene involving aluminum metal centers. Subsequent oxidation of the aluminum trialkyls yields aluminum trialkoxides, which can be hydrolyzed to alcohols and boehmite.

A modification of this process allows the production of aluminum trialkoxides and H_2 from alcohols and aluminum metal. The aluminum alkoxides are hydrolyzed, and the obtained suspension is aged in an autoclave, preferably, at a steam pressure of 1–30 bar, corresponding to a temperature of 100–235 °C, for between 0.5 and 20 h; the aging step is preferably carried out with stirring at a peripheral speed of 1–6 m s^{-1} (72). The boehmite product (Pural from Sasol) is reported to consist of a mixture of crystals with different habits including rhombic plates and blocks exposing mainly (020) and (120) faces (73). These materials are largely free of alkali metals but may contain some titanium and iron impurities. Using a modified process, the production of hyperpure aluminas (with a total impurity content of only a few ppm) have been commercialized (73).

Aerogels are materials prepared by supercritical drying of gels produced by hydrolysis of alkoxides (74). High-surface-area boehmite (75, 76) or pseudoboehmite (77) can be prepared by this method, and these materials can subsequently be transformed into high-surface-area aluminas.

2.1.4 Thermal decomposition of aluminum hydroxides

A scheme outlining the most relevant paths of phase transformations between various alumina polymorphs is presented in Figure 3.6. As is well known (78), thermal transformations of aluminas always ultimately lead to α-Al$_2$O$_3$ (corundum). This phase is the thermodynamically stable polymorph if the free energy of the bulk is considered. In this polymorph, the oxide ions are arranged in hcp and all aluminum cations are octahedrally coordinated. This polymorph can be produced by various methods: (i) by high temperature (>350 °C) solvothermal precipitation and aging; (ii) by

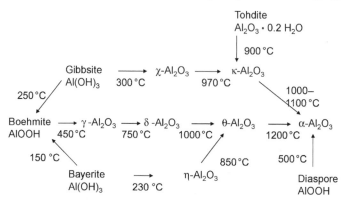

Figure 3.6 Most common evolution paths for phase transformations of aluminum hydroxides and oxides upon heat treatment.

low temperature (≈ 500 °C) thermal decomposition of the oxyhydroxide diaspore (usually denoted as α-AlOOH); or (iii) by calcination of any other aluminum oxide or hydroxide at sufficiently high temperature (i.e., higher than 800–1200 °C). Obviously, the powders produced by these three methods differ significantly in regard to surface area, porosity, and morphology.

The phases that are intermediates in the pathways shown in Figure 3.6 are referred to as "transition" aluminas. Of these polymorphs several are applied in the catalysis field, namely those that may be prepared by thermal decomposition of boehmite or bayerite in air ("calcination"). The structure of these alumina polymorphs differs from that of corundum: they are characterized by cubic close packing of the oxide ions. The product obtained by calcination of boehmite in air is designated as γ-Al_2O_3, whereas the product obtained by calcination of bayerite in air is designated as η-Al_2O_3. Both transformations result in a significant increase of the surface area of the material (48,79). Spray pyrolysis without previous drying of the sol can also be performed (80). The thermal decompositions of (pseudo)boehmite and of bayerite are topotactical transformations. Thus, the original morphology of the hydroxide is retained during conversion into transition aluminas. For example, lamellar and fibrous γ-Al_2O_3 can be prepared by decomposing lamellar and fibrous (pseudo)boehmite, respectively.

γ-Al_2O_3 and η-Al_2O_3, whose closely related structures have not yet been exactly determined, are converted into θ-Al_2O_3 upon calcination in air at higher temperatures. The structure of θ-Al_2O_3 has been identified to be of the β-Ga_2O_3-type. η-Al_2O_3 usually converts directly into θ-Al_2O_3, whereas intermediate phases (viz., δ-Al_2O_3 or γ'-Al_2O_3) are often (but not always) observed during conversion of γ-Al_2O_3 into θ-Al_2O_3. Bayerite is normally converted to η-Al_2O_3; except when coarse bayerite is heated slowly in air, then the particles may lose water and transform into boehmite (γ-AlOOH), which subsequently can pass through the usual sequence via γ-Al_2O_3 (47).

In Figures 3.7 and 3.8, various data are shown that illustrate the evolution of the phase composition with increasing calcination temperature, with (pseudo)boehmite as the starting compound. In Figure 3.7, the evolution of the surface area upon heat treatment of pseudoboehmite is reported in parallel to the differential thermal analysis (DTA) curve; and the phase composition in each temperature range is indicated (81). In Figure 3.8 (82), the X-ray diffraction (XRD) patterns and the ^{27}Al MAS NMR spectra of the products of boehmite calcination are shown. The XRD pattern of boehmite

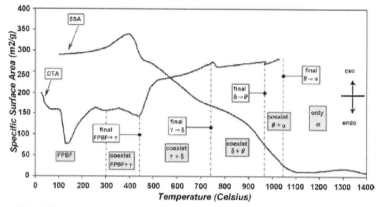

Figure 3.7 Differential thermal analysis trace, surface area, and phase evolution of pseudoboehmite as a function of increasing temperature. *Reprinted with permission from Ref. (81).*

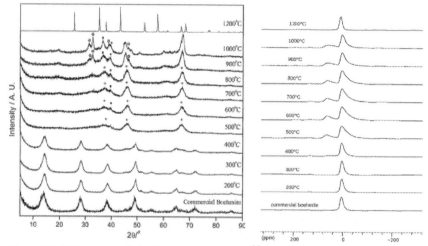

Figure 3.8 XRD patterns (left, Cu K_α excitation) and ^{27}Al NMR spectra (right) of the products of calcination of a commercial boehmite sample. Calcination temperatures are indicated in the graph. *Reprinted with permission from Ref. (82).*

is substituted by that of γ-Al_2O_3 between 400 and 500 °C. The diffractograms then show a progressively higher fraction of δ-Al_2O_3, until the pattern of θ-Al_2O_3 emerges at 1000 °C. At 1200 °C, narrow reflections indicate well-crystallized α-Al_2O_3. Consistently, the ^{27}Al NMR spectra show first the single peak characterizing the octahedrally coordinated aluminum ions of boehmite, at about 5 ppm, and then the formation of a broader peak at 63–67 ppm that can be ascribed to the tetrahedrally coordinated aluminum ions typical of transition aluminas. This peak becomes more complex at higher calcination temperatures and finally disappears when α-Al_2O_3 is formed, which exclusively contains aluminum ions in octahedral coordination.

Thermal treatment of small crystals of gibbsite at 300 °C produces χ-Al_2O_3, which transforms into κ-Al_2O_3 at 650–750 °C, which in turn transforms into α-Al_2O_3 above 1000 °C *(83,84)*. An intermediate phase is sometimes observed in the γ-$Al(OH)_3 \rightarrow \chi$-$Al_2O_3$ transformation *(85)*. Very small gibbsite plates are reported to produce high-surface-area χ-Al_2O_3 (400 $m^2\,g^{-1}$), which transforms into κ-Al_2O_3 at 1000 °C and into α-Al_2O_3 above 1100 °C *(86)*. In contrast, large particles of gibbsite are converted into boehmite, and then the normal boehmite decomposition path (see Figure 3.6) is followed *(60)*.

Grinding or heating in vacuum prior to the thermal decomposition can lead to a different phase sequence during calcination. Grinding of the hydroxide precursors favors the intermediate formation of amorphous materials, from which amorphous alumina (see Section 2.2.1), sometimes denoted as ρ-Al_2O_3, may form *(87)*, or reduces the phase transformation temperatures *(88)*. The thermal decomposition of previously ground bayerite may proceed according to the following path α-$Al(OH)_3 \rightarrow \gamma$-$AlOOH \rightarrow \gamma$-$Al_2O_3 \rightarrow \alpha$-$Al_2O_3$, that is, without the formation of the θ-Al_2O_3 transition phase *(89)* Grinding of gibbsite before thermal treatment may decrease the temperature for the transformation into corundum *(90)*. Other investigations *(91)* indicate, however, that mild heating of gibbsite, that is heating to a temperature of 350 °C, yields a largely amorphous solid containing only traces of boehmite (γ-$AlOOH$). This solid is designated as ρ-Al_2O_3 and can be converted into γ-Al_2O_3 by calcination and, subsequently, into θ-Al_2O_3, and, finally, into α-Al_2O_3. The decomposition of both bayerite and gibbsite in vacuum is reported to follow the sequence $Al(OH)_3 \rightarrow \rho$-$Al_2O_3 \rightarrow \eta$-$Al_2O_3 \rightarrow \theta$-$Al_2O_3 \rightarrow \alpha$-$Al_2O_3$ *(92)*.

The calcination of nordstrandite is reported to give cubic spinels assumed to be either γ-Al_2O_3 *(53)* or η-Al_2O_3 *(51,93)*. The decomposition of tohdite leads to κ-Al_2O_3 and κ'-Al_2O_3 *(94,95)*.

2.1.5 Flame hydrolysis of aluminum chloride

The so-called "fumed aluminas" are obtained by flame hydrolysis of $AlCl_3$. The high temperature reaction of $AlCl_3$ vapor in a H_2/O_2 flame yields very small, nonporous alumina particles *(96)* that are characterized by a low packing density and thus become easily airborne. The reaction conditions influence the properties of the product. The obtained powder may be comprised of γ-Al_2O_3 crystals with an approximate size of 10 nm and 130 $m^2 g^{-1}$ surface area. Alternatively, the powder may contain a mixture of phases, with δ-Al_2O_3 or θ-Al_2O_3 predominating, and have a surface area as low as 65 $m^2 g^{-1}$ along with a slightly larger particle size. Materials produced by flame hydrolysis are largely free of alkali metals, but may contain significant amounts (i.e., 0.5–1 wt%) of chlorine (97,98), silicon, or titanium. Commercial powders belonging to this family of materials are Aeroxide from Evonik (previously Degussa) and SpectrAl® from Cabot. These materials are commonly used in academic catalysis research, whereas we suspect that their application as catalysts in industry is not so widespread. In a recent patent, an improvement of the method was reported that made possible the production of aluminum oxide with a BET surface area of 135–190 $m^2 g^{-1}$ (99).

2.1.6 Other preparation methods for aluminas

Hydrothermal precipitation at relatively high temperatures (400 °C) and under pressure allows the direct synthesis of α-Al_2O_3 powders; by adjusting the reaction conditions, nanosized particles of corundum can be obtained (40).

Aluminas can also be prepared by thermal decomposition of aluminum salts. Aluminum nitrate decahydrate decomposes at a temperature of 180 °C producing amorphous alumina (100); the spray pyrolysis of aluminum nitrate solutions yields γ-Al_2O_3 (80).

The thermal decomposition of crystalline ammonium aluminum carbonate hydroxide (AACH), $NH_4[Al(OOH)HCO_3]$, at 500 °C results in high–surface-area γ-Al_2O_3 (101). The decomposition of crystalline aluminum hydroxy acetate $Al(OH)(CH_3COO)_2$ at 300 °C also produces γ-Al_2O_3 (102).

The decomposition of aluminum hydroxychloride $AlCl_3 \cdot 2Al(OH)_3 \cdot 6H_2O$ at low temperatures initially gives an amorphous product, and subsequently, transition aluminas are formed. However, if the evolving gas is rapidly removed, the amorphous decomposition product directly transforms into a fine corundum (α-Al_2O_3) powder (103) at 530 °C.

The decomposition of $Al_2(SO_4)_3 \cdot 18H_2O$ in four major stages also leads to Al_2O_3; at temperatures below 350 °C, the compound progressively loses water molecules, and near 850 °C, the remaining aluminum sulfate starts decomposing (104). The calcination of ammonium alum, $NH_4Al(SO_4)_2$, at 800 °C yields aluminum sulfate, $Al_2(SO_4)_3$, whereas at higher temperatures (900–1000 °C), γ-Al_2O_3 is produced, which transforms into corundum at even higher temperatures *(105,106)*. By adding ammonia to aluminum sulfate solutions, basic aluminum sulfates can be prepared, which are good starting materials for aluminum hydroxides and oxides *(107)*.

Combustion synthesis *(108)*, which is a method suitable for creating complex oxide materials, can also be applied to prepare aluminas. By using the glycine–nitrate process, amorphous materials are produced that can be converted into γ-Al_2O_3 by further heating *(109)*. The combination of tartaric acid and nitrate produces amorphous alumina that will convert to pseudoboehmite under hydrothermal conditions *(110)*. Medium surface area γ-Al_2O_3 can be prepared by combustion synthesis using aluminum alkoxides as starting compounds *(111)*.

Several techniques have been developed to deposit alumina films on different surfaces such as those of semiconductors or metals. These films find application in various areas. In most cases, amorphous alumina films are desired. Depending on the deposition techniques, various precursors may be used; the following combinations have been reported: plasma-enhanced atomic layer deposition using trimethylaluminum *(112)*, metal-organic chemical vapor deposition using aluminum tri-iso-propoxide *(113)*, and condensation from the gas phase using laser-evaporated alumina *(114)*. Similar "evaporation" techniques can also be applied to prepare γ-Al_2O_3 powders *(115,116)*.

2.2. Crystal structures and morphologies of aluminas

A number of different polymorphs of alumina have been reported. Here, we will summarize data on the most common polymorphs (Table 3.2), while also reporting the intermediate phases that appear during solid-state transformations.

2.2.1 The nature of amorphous aluminas

As stated above, a variety of deposition techniques have been developed in recent years to produce amorphous alumina coatings *(112–115)*, among them the anodization of aluminum. Amorphous aluminas, denoted as a-Al_2O_3 or sometimes as ρ-Al_2O_3, can also be prepared by conventional liquid-phase reactions *(134)*, by the decomposition of some hydroxides, or by sol-gel procedures aimed at the preparation of mesoporous materials. In several instances, IR *(110)* and Raman *(135)* spectra have demonstrated that amorphous aluminas may contain impurities originating from the

Table 3.2 Structural data of aluminas

	SG	$Z_{Al_2O_3}$	a (Å) b (Å)	c (Å)	Structure type	Al_{IV}%	Al_{VI}%	Al_{IV}%	Density (g cm^{-3})	Reference
γ-Al$_2$O$_3$	$Fd\bar{3}m$	10,66	7.90	7.90	Cubic disordered non-stoichiometric spinel	30–60	70–40	1–2	3.65	(117)
γ-Al$_2$O$_3$	$I4_1/amd$	5,33	5.616	7.836	Tetragonal disordered non-stoichiometric spinel	35.6	64.4	0		(118)
γ-Al$_2$O$_3$	$P2_1/m$	8	5.587 8.413	8.068 β 90.59°	Orthorhombic nonspinel structure	25	75			(119)
γ'-Al$_2$O$_3$	$P\bar{4}m2$	16	5.611	24.450	Tetragonal spinel superstructure	34.4	65.6	0		(120)
δ-Al$_2$O$_3$	$P\bar{4}m2$	16	7.93	23.50	Tetragonal spinel superstructure	30–45	55–70			(78)
δ-Al$_2$O$_3$	$P2_12_12_1$	32	15.89 7.94	11.75	Orthorhombic spinel superstructure					(78)
θ-Al$_2$O$_3$	$C2/m$	4	5.62 2.91	11.79 β 103.8°	Monoclinic β-Ga$_2$O$_3$ structure, ordered spinel-related structure	50	50			(14)
η-Al$_2$O$_3$	$Fd\bar{3}m$	8	7.91	7.91	Cubic disordered non-stoichiometric spinel	20–50	80–50			(117,14)
κ-Al$_2$O$_3$	$Pna2_1$	8	4.844 8.330	8.955	Orthorhombic	25	75		3.98	(121–127)
χ-Al$_2$O$_3$	$P6/mm$ or $P6_3/mcm$	8?	5.56	13.44	Hexagonal	20	73	7		(128)
χ-Al$_2$O$_3$	Cubic									(129)
ρ-Al$_2$O$_3$					Amorphous	25–15	30–55	20–50		(130,131)
α-Al$_2$O$_3$	$R\text{-}3c$	6	4.76	12.99	Hexagonal–rhombohedral, corundum structure	0	100		3.99	(132,133)

precursors; for example, oxalate ions have been found in anodic amorphous alumina *(135)*. IR spectra have also indicated the presence of hydroxyl groups on the surface of amorphous aluminas *(110,135,136)*. Adsorption of basic probe molecules such as pyridine revealed the presence of Lewis acid sites of medium strength, while the strong Lewis acid sites typical of transition aluminas are lacking *(110,137)*. Analysis of the region of skeletal vibrations in spectra of a sample prepared by combustion synthesis suggests that the predominating coordination of Al^{3+} ions is higher than fourfold *(138)*. A number of recent investigations of differently prepared samples agree that in addition to tetra- and hexacoordinated ions, pentacoordinated Al^{3+} ions can often be present and even be predominant *(139,140)*. Amorphous alumina samples may exhibit varying degrees of disorder, depending on the preparation conditions *(112)*.

Amorphous aluminas have moderate thermal stability in dry atmospheres, that is, up to temperatures of less than 700 °C. However, the thermal stability of amorphous alumina is reported to significantly increase with decreasing particle size *(141)*. Upon hydrothermal treatment, they tend to convert easily into bayerite, γ-Al_2O_3, or η-Al_2O_3 *(110,131,142)*. Amorphous aluminas appear to have only moderate activity as acid–base catalysts *(110,130)*. Surprisingly, they have been reported to have redox properties *(143)*. In summary, amorphous aluminas have so far not shown high activity for any catalytic reaction, and because of their limited structural stability, they are also expected to have poor stability as catalysts. However, they can serve as reference materials in the catalysis field.

A number of mesoporous aluminas are reported to have amorphous walls *(144)*, but the wall material may crystallize under appropriate conditions.

2.2.2 Crystal structure and morphologies of γ-Al_2O_3

2.2.2.1 On the bulk structure of γ-Al_2O_3

As stated above, γ-Al_2O_3 is, by definition, the solid product of the decomposition of boehmite, the oxyhydroxide denoted as γ-AlOOH. Since the early investigation by Verwey *(145)* and subsequent work by Lippens and de Boer *(146)*, the structure of γ-alumina has been conventionally described as a cubic defective (or non-stoichiometric) spinel. The cubic spinel structure (Figure 3.9) is typical of compounds with AB_2O_4 stoichiometry and belongs to the space group $Fd\bar{3}m$ and the number of formula units in the unit cell, Z, is 8. The structure is characterized by a cubic close packed (ccp) sublattice of oxide anions, with A^{2+} and B^{3+} ions occupying tetrahedral (T_d) and octahedral (O_h) voids, with Wyckoff positions 8a and 16d,

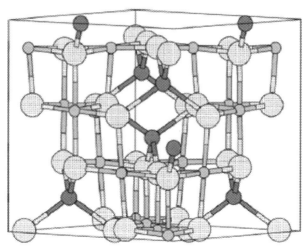

Figure 3.9 The unit cell of the spinel structure. Large spheres are oxide ions, small dark spheres are tetrahedrally coordinated cations, and small light-colored spheres are octahedrally coordinated cations.

respectively. In normal spinels, A^{2+} ions occupy the T_d (8a) positions and B^{3+} ions occupy the O_h (16d) positions. In inverted spinels, A^{2+} ions occupy half of the O_h (16d) positions and B^{3+} ions occupy the T_d (8a) positions and the other half of the O_h (16d) positions. In so-called random spinels, the two types of cation are distributed randomly among the two positions. In fact, even in normal and inverted spinels, the distribution of the cations does not completely correspond to the perfect structure model, that is, there is always some disorder in the occupancy.

For γ-Al_2O_3, the formula $Al_{21+1/3} \square_{2+2/3}O_{32}$ is often used to describe the unit cell, where $\square$ denotes a cation vacancy. This formula implies that the cubic spinel oxygen sublattice is fully occupied, whereas the bivalent cations are essentially replaced by trivalent cations and vacancies to achieve charge balance, resulting in a cation deficiency with respect to the spinel stoichiometry. The Al^{3+} ions partially occupy the T_d (8a) and O_h (16d) sites, and, because some of the cation positions remain vacant, γ-Al_2O_3 is considered a defective spinel. The percentages of cations in T_d and O_h positions can theoretically vary over a significant range. Extreme values that can be reached are 25% T_d and 75% O_h positions if exclusively T_d sites are vacant, or 37.5% T_d and 62.5% O_h positions if exclusively O_h sites are vacant. Investigations of the distribution of cations and vacancies in spinel-type models of γ-Al_2O_3 suggest that spinel structures with vacant octahedral sites should be more stable than structures with vacant tetrahedral sites *(147,148)*.

The position of aluminum cations can be determined experimentally by ^{27}Al NMR spectra (Figure 3.8) and by Rietveld analysis of X-ray and neutron scattering patterns *(117)*. The results indicate that in γ-Al_2O_3, 25–33% of all Al^{3+} ions are in tetrahedral positions, whereas the rest are almost exclusively in octahedral positions, with a few Al^{3+} ions in fivefold *(149,150)* or highly distorted tetrahedral coordination. It has been reported that milling can increase the fraction of pentacoordinated aluminum up to a value of 20% *(151)*; the fivefold coordinations arise through an increased disorder similar to that discussed above for amorphous alumina.

In principle, the deviation of the cation stoichiometry from that of the spinel structure could be compensated by the presence of protons and (a smaller number of) Al^{3+}, as an alternative solution to the presence of Al^{3+} and vacancies. The protons could form surface hydroxyl groups. Indeed, as noted by Soled *(152)*, γ-Al_2O_3 surfaces are always largely hydroxylated, and dehydroxylation is only achieved at so high a temperature that phase transformation to θ-Al_2O_3 or α-Al_2O_3 is already occurring. The so-called "hydrogen spinel" structure is another possibility that was proposed on the basis of spectroscopic measurements *(153)* and supported by computational chemistry results *(154,155)*. In the hydrogen spinel, protons located in the bulk balance the cation deficiency of γ-Al_2O_3 with respect to the spinel stoichiometry.

Additional structures have been proposed in the literature. The results of several analyses, including Rietveld refinement of XRD patterns, suggest that a fraction of the aluminum cations may occupy "nonspinel positions" *(156)*, specifically the sites of the ccp oxide lattice that are unoccupied in stoichiometric spinels. These nonspinel sites are other sites with T_d (8b and 48f) or O_h (16c) coordination. The structure of most γ-Al_2O_3 samples actually appears to be tetragonally distorted *(78)*, whereby the distortion decreases with increasing temperature *(120)*. Some reports classify the structure as tetragonal, as a result of a contraction of the cubic lattice along one of the cubic crystallographic directions. The space group $I4_1/amd$ was proposed and occupancy of nonspinel sites was inferred *(118)*, with overall 32.8% of the aluminum in tetrahedral and 67.2% in octahedral positions. This model, by Paglia *et al. (118)*, was further supported by the results of Rozita *et al. (156b)*, Peitinger *et al. (14)* and Samain *et al. (157)*.

In recent years, modern methods of theoretical chemistry were used to address the problem of the structure of γ-Al_2O_3. Digne *et al. (158,159)* and Krokidis *et al. (119)* proposed a structure model characterized by an orthorhombically distorted ccp oxide lattice and nonspinel cationic positions (space group $P2_1/m$, $Z=8$), with 25% of the Al^{3+} ions in tetrahedral positions and no structural vacancies. According to the calculations of these authors, this nonspinel structure should be more stable than possible spinel-type structures. Additionally, most authors nowadays tend to explicitly exclude the "hydrogen spinel" structure, which is, according to calculations, less stable than other structures. However, the free energy difference may be very small (0–0.2 eV/Al_2O_3 formula unit) (148).

The variations in the thermodynamic stability of the bulk structure are not necessarily significant enough to determine which of the polymorphs exists under particular conditions. It recently became evident that the contribution of the surface to the overall energy and thus stability is larger than that of the bulk, provided the crystallites are small (i.e., in the case of fine powders) (160). Moreover, as will be discussed in Section 2.2.10, the phase transformations may proceed so slowly that kinetics can have a more prominent role than thermodynamic stability in determining which of the alumina polymorphs actually exist.

Ferreira et al. (161) performed DFT calculations to consider a spinel-type model with 37.5% tetrahedrally coordinated and 62% octahedrally coordinated Al^{3+} ions. The authors compared this model with the nonspinel-type model of Krokidis et al. (119) and concluded that the best agreement with experimental data was obtained for the spinel-type model. The same authors performed ab initio calculations of ^{27}Al MAS NMR spectra for different structural models of γ-Al_2O_3 (162) and concluded that the best fit is obtained using a spinel-type structure with space group $Fd\overline{3}m$ and Z equal to 32, and with the aluminum ions distributed as follows: 64% octahedral coordination, 34.4% tetrahedral coordination, and 1.6% pentacoordinated.

Most authors, including Smrčok et al. (156), agree that the oxygen sublattice is fairly well ordered. This interpretation is consistent with the sharpness of the (222) reflection in powder X-ray diffractograms; this reflection is known to primarily arise from scattering by the oxygen sublattice, as already noted by Lippens and DeBoer (146). However, according to Streitz et al. (163), the cubic arrangement of oxide ions may be disordered. In fact, if the cation vacancies (in a spinel-type model) concern octahedral positions, some oxygen atoms may be missing two or even three Al^{3+} neighbors, and would thus be in an unstable situation. The incomplete coordination will provoke a slight changes in the positions of the affected oxide ions and also in the charges of the surrounding cations and anions, thus significant disorder is created in a defective spinel structure. The presence of large numbers of oxygen vacancies in defective spinels was confirmed in investigations of $Mg_{1-x}Al_2O_{4-x}$ ($0 \leq x \leq 0.35$) defective spinels (164,165). Disorder in the oxygen sublattice can also result from stacking faults that are introduced during the formation of alumina from boehmite *(166)*.

Data from a variety of analytical techniques indicate that disorder is indeed present in the bulk and/or on the surface of γ-Al_2O_3. According to factor group analysis, four IR active modes are expected for cubic spinels, including defective spinels. *Ab initio* calculations by Ishii *et al.* *(167)* suggest that the four modes include stretching motions of octahedral relative to tetrahedral aluminum (350–380 cm^{-1}), bending motions of oxygen relative to aluminum ions (450–600 cm^{-1}), stretching motions of anions relative to octahedral cations (730–780 cm^{-1}), and stretching motions of anions relative to octahedral and tetrahedral cations (780–830 cm^{-1}). All these modes are also expected to undergo transverse optical–longitudinal optical (TO/LO) splitting; thus, there might be up to eight absorption maxima in the wavenumber region between 1000 and 300 cm^{-1}. The experimentally observed spectrum of γ-Al_2O_3 is typically characterized by two very broad, poorly resolved bands in the mid-IR region at wavenumbers of 580 and 880 cm^{-1}, which may show some substructure *(168–171)*. The IR spectra of the different structural models have been simulated with the help of various calculation methods, but the results have so far been largely inconclusive *(161,172)*.

Raman spectroscopic measurements also provide evidence of structural disorder in γ-Al_2O_3. In fact, strong fluorescence is observed that extends over almost the entire range, and the fundamental vibrational modes (expected between 1000 and 300 cm^{-1}) cannot be detected. As recently reviewed and discussed by Kim *et al.* *(173)*, this behavior, which is observed when analyzing transition aluminas with visible range laser radiation, is still not well understood. The fluorescence has repeatedly been associated with structural defects and has thus been called "defect fluorescence." However, other phenomena cannot be excluded, such as the fluorescence activity of impurities or anharmonic coupling between high-frequency OH stretching vibrations of surface hydroxyl groups and low frequency lattice vibrations. Fluorescence has also been reported for η-Al_2O_3 *(174)*. In contrast, the fundamental Raman vibrations of nondefective structures, which include θ-Al_2O_3 *(175)*, α-Al_2O_3 *(176)*, and aluminum hydroxides *(19)*, can be detected; these results demonstrate that fluorescence is weak or negligible for nondefective structures.

More experimental evidence that supports the idea of disorder in the structure of γ-Al_2O_3 has been collected: Pentacoordinated Al^{3+} cations have been observed in small amounts *(177)*, mainly located at the surface, and

these species become fewer or even disappear during phase transformation to
θ-Al₂O₃ *(150)*.

The disorder in the structure is evident in the unusual broadening of
most of the reflections in X-ray diffractograms of γ-Al₂O₃ (Figure 3.10
(178)). The broadening has been explained by strong faulting in several sets
of planes, viz. {111}, {110}, and {100}, in the spinel-like structure of these
materials. Individual and/or interconnecting defects in the different plane
families *(178)* are responsible for the faulting.

In summary, the disorder in the bulk is generally assumed to be mainly
associated with the random distribution of cations and cationic vacancies in
spinels, concerning spinel-type sites (i.e., those typically occupied in spinels)
and possibly also nonspinel-type sites. In addition, electronic and atomic
structure calculations by Dyan *et al. (179)* have indicated that much disorder
is found at the surface of γ-Al₂O₃ when surface reconstruction is permitted
within the model.

It was recognized years ago *(153)* that in the structures of transition aluminas, strongly H-bonded OH groups are present, which are very likely
located in the subsurface layers or in the bulk. In the region of the O–H
stretching vibrations of surface hydroxyl groups, the IR spectra of
γ-Al₂O₃ are characterized by well-defined bands between 3800 and

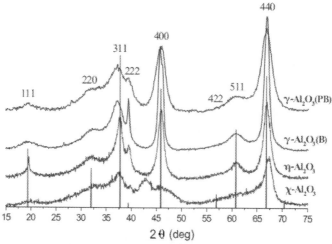

Figure 3.10 Experimental X-ray powder diffraction patterns of the low temperature
Al₂O₃ polymorphs. Vertical lines indicates the positions of the characteristic spinel
reflections, label with the appropriate indices. *Reprinted with permission from Tsybulya,
S. V.; Kryukova, G. N. Phys. Rev. B 77, 024112, 2008. Copyright (2008) by the American
Physical Society* (178).

3500 cm^{-1} and a broad absorption that extends over the range of $3500-2500 \text{ cm}^{-1}$ and is present even after outgassing at high temperature (Figure 3.11, upper spectrum). After H/D-exchange with D_2O, the bands due to stretching vibrations of the surface OH groups are shifted to the corresponding positions of stretching vibrations of surface OD groups ($2800-2600 \text{ cm}^{-1}$). Moreover, the broad absorption is shifted accordingly to $2600-1800 \text{ cm}^{-1}$ (Figure 3.11, lower spectrum), confirming the assignment to OH (OD) stretching vibrations. The OH (OD) groups responsible for this absorption are obviously H bonded and are apparently not located in the outermost surface layers: the broad absorption is not affected by adsorption of molecular probes and only gradually decreases upon transformation of $\gamma\text{-Al}_2\text{O}_3$ and $\delta\text{-Al}_2\text{O}_3$ into $\theta\text{-Al}_2\text{O}_3$. Hence, IR spectroscopy provides clear evidence for the presence of protons in the alumina bulk. It cannot be excluded that these species contribute to the "defect fluorescence" in Raman spectra that was described above.

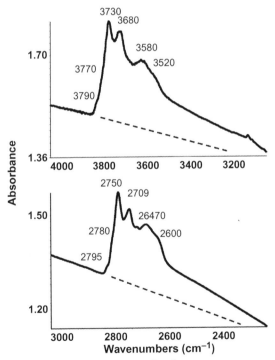

Figure 3.11 IR spectra of $\gamma\text{-Al}_2\text{O}_3$ (Akzo, 185 $\text{m}^2 \text{ g}^{-1}$) after out gassing at a temperature of 400 °C (upper spectrum) and after deuteration with D_2O and outgassing at 400 °C (lower spectrum).

In conclusion, the real structure of γ-Al$_2$O$_3$ has still not been completely solved. The spinel model with occupancy of nonspinel sites and significant disorder, including the possible presence of protons and formation of surface OH groups, is currently the best interpretation of available data.

2.2.2.2 On the morphology of γ-Al$_2$O$_3$ particles

To comprehensively interpret the results on the surface and catalytic properties of γ-Al$_2$O$_3$, not only the bulk structure must be carefully analyzed but also precise information on the particle morphology is also needed.

As described above, there are several different preparation procedures for the commercial production of γ-aluminas. Some of the procedures start with the preparation of boehmite, which is the layered oxyhydroxide denoted as γ-AlOOH, and this intermediate phase is then decomposed to the product. The morphological properties of the resulting γ-Al$_2$O$_3$ are strictly determined by those of the starting boehmite material *(119)*. Nanocrystalline hydrous oxyhydroxide, sometimes denoted as "pseudoboehmite," is the best starting material for high-surface-area γ-aluminas for catalytic applications *(180)*. Calcination of nanocrystalline (pseudo)boehmite at a temperature of about 450 °C produces highly microporous, high-surface-area γ-Al$_2$O$_3$ (with a BET surface area of ≈ 500 m^2 g^{-1}). If well-crystallized boehmite with a lower surface area is calcined, medium-surface-area powders (≈ 100 m^2 g^{-1}) with lamellar structure are obtained.

As already mentioned, transmission electron microscopy (TEM) images show that precipitated (pseudo)boehmites usually exhibit some predominant habits: rhombic, lamellar, fibrous *(60, 181)*, or blocks (73). These morphologies are retained during the transformation into γ-Al$_2$O$_3$. Correspondingly, γ-Al$_2$O$_3$ particles obtained by topotactic decomposition *(157)* are reported to expose predominantly spinel-like (110) faces, which correspond to the basal (101) face of boehmite. Other faces are also present; mainly (100) faces are found if the particles are lamellar *(136)*, and (111) faces if the crystals have rhombic shape *(115, 178, 182, 183)*. High-resolution images show that the (110) face is not atomically flat but significantly reconstructed, and exhibits nanoscale (111) facets, as shown in Figure 3.12 *(183)*. Heavily stepped (111) terminations can be found and are assumed to be associated to excess cation termination *(156b)*. Thus, the (111) face may actually be more abundant than the (110) face, which is usually assumed to be predominant. Moreover, there is a very high density of edges. Other materials such as "Versal" alumina *(57)* are evidently microfibrous with a poorly defined habit, as can be seen in Figure 3.13.

In contrast, boehmites obtained from Ziegler-type alkoxides, such as Pural from Sasol (formerly Condea), seem to have a less pronounced lamellar

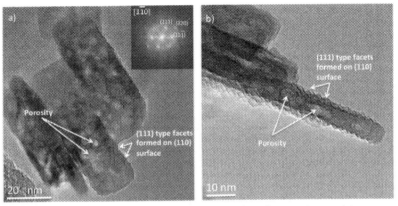

Figure 3.12 HRTEM images of lamellar γ-Al_2O_3 particles. The inset in panel A represents an indexed diffractogram (Fourier spectrum) obtained from the upper particle. *Reprinted with permission from Ref. (183).*

Figure 3.13 TEM image of Versal γ-Al_2O_3.

or layered structure than precipitated boehmites *(184–186)* and show rather bulky prismatic particles. Through calcination, these particles are converted into largely globular γ-Al_2O_3 crystals, implying that basal planes are not predominant.

Fumed aluminas have a peculiar morphology that is characterized by the absence of significant porosity. The phase composition of fumed aluminas has been found to vary, and different samples have been reported to consist of pure γ-Al_2O_3 *(188–190)*, pure δ-Al_2O_3 *(186–191)*, or a mixture of γ-Al_2O_3 and δ-Al_2O_3 *(186–192)*. TEM images of fumed aluminas

(186,191,193) have been interpreted as showing (110) cubic planes as prevailing terminations and, to a lesser extent, (100) and (111) planes. Wischert *et al. (192)* concluded that fumed and precipitated aluminas present the same faces in similar ratios, which, considering the difference in pore shapes and sizes between the two types of material, seems to be an oversimplification.

Other commercial γ-aluminas prepared by methods different than those already cited in this paragraph have an evident spherical or globular habit *(115,156b,194,195)* as shown in Figure 3.14 *(194)*. This morphology is, for example, found for aluminas produced by spray pyrolysis *(80,196)*. TEM analyses of these materials show cuboctahedral particles, implying that (111) and (100) faces should predominate, as illustrated in Figure 3.15 *(115)*.

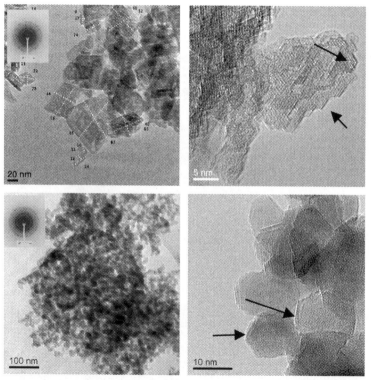

Figure 3.14 Bright field TEM images of γ-Al$_2$O$_3$ samples from Alfa Aesar (top row) and Johnson Matthey (bottom row). Black arrows indicated facetted particles. Insets: selected area diffraction patterns. *Reprinted with permission from Ref. (194).* (See the color plate.)

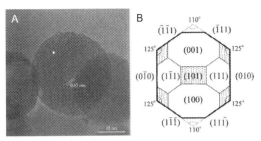

Figure 3.15 TEM image of a "spherical," that is, a cuboctahedral γ-Al$_2$O$_3$ particle (left) and image interpretation (right). *Reprinted with permission from Ref. (115).*

It is obvious in light of these observations that the properties of γ-alumina cannot be ascribed to one or a few predominant faces of a single structure.

2.2.3 Crystal structure and morphologies of δ-Al$_2$O$_3$

As mentioned in Section 2.1.4, only a few authors report the direct conversion of γ-Al$_2$O$_3$ into θ-Al$_2$O$_3$ during calcination at temperatures near 650 °C, whereas in most investigations, the formation of intermediate phases is observed *(78)*. According to Lippens and de Boer *(146)* and Wilson and McConnel *(197)*, a phase with a structure similar to that of spinels and γ-Al$_2$O$_3$, denoted as δ-Al$_2$O$_3$, evolves in the range 500–650 °C. δ-Al$_2$O$_3$ has been determined to be either a tetragonal or an orthorhombic *(78)* spinel superstructure, characterized by a unit cell consisting of three spinel unit blocks, likely with a partially ordered distribution of aluminum ions at octahedral sites *(198)*. Different space groups have been proposed for this phase (or these phases). According to Levin and Brandon *(78)*, the space group is $P2_12_12_1$ or $P4_1$ with 64 cations per unit cell. According to Paglia *et al. (120)*, a slightly different spinel superstructure is formed, denoted as γ′-Al$_2$O$_3$, with space group $P\overline{4}m2$, $Z = 16$, and 66% octahedral and 34% tetrahedral aluminum cation coordination. However, data on the location of the aluminum ions in δ-Al$_2$O$_3$ are still incomplete *(199)* because of the difficulty to prepare phase-pure samples for ^{27}Al NMR experiments. Evidence for the existence of δ- or γ′-Al$_2$O$_3$ comes from XRD patterns, which are distinctly different from those of both γ-Al$_2$O$_3$ and θ-Al$_2$O$_3$ *(120,200)*. IR spectra of the skeletal vibrations of samples rich in δ-Al$_2$O$_3$ show a number of sharp features superimposed to a broad absorption *(169,170,201)*. To our knowledge, detailed Raman spectroscopic investigations of δ-Al$_2$O$_3$ have not appeared in the literature up to now. Roy and Sood *(202)* reported that a Raman spectrum of δ-Al$_2$O$_3$ is not detectable, as is the case for γ-Al$_2$O$_3$, whereas Xiong *et al. (135)* attributed Raman peaks at 740 and 830 cm^{-1}, observed when analyzing anodic layers on aluminum, to δ-Al$_2$O$_3$. By combining results from high-angle annular dark field electron microscopy imaging, X-ray diffractogram refinement, and density functional theory calculations it has been recently concluded that the structure of δ-Al$_2$O$_3$ is built through complex structural intergrowth of two slightly different crystallographic variants (203).

Several surface investigations of δ-Al$_2$O$_3$ or δ-Al$_2$O$_3$-rich samples have been published. Fumed alumina "Aeroxide-C" from Evonik (formerly Degussa) is considered by most authors to predominantly consist of δ-Al$_2$O$_3$. High-resolution TEM (HRTEM) images of a sample of this material show hexagonal particles of very regular shape, and further analysis suggests that cubic (110) faces are the most abundant terminations (191). The surface of Aeroxide-C has been the focus of a number of investigations (97,98,185). It must to be taken into account, though, that this material is prepared by a less commonly applied procedure and contains different impurities (specifically chlorine instead of alkali) than "normal" δ-Al$_2$O$_3$ samples prepared from boehmites. Moreover, samples of this material frequently consist of a mixture of phases. Thus, it is misleading to consider this material as representative of δ-Al$_2$O$_3$. Métivier et al. (204) reported IR spectra of the hydroxyl groups of a sample assumed to be pure δ-Al$_2$O$_3$ of the "Ziegler-type" boehmite. The surface acidity of phase mixtures of δ-Al$_2$O$_3$ and θ-Al$_2$O$_3$ has been investigated by adsorption of basic probe molecules (205). Actually, δ-Al$_2$O$_3$ or δ-Al$_2$O$_3$-rich powders show a behavior similar to that of other transition aluminas; some additional peculiarities will be discussed below. For example, the behavior as an acid catalyst is similar to that of η- and γ-Al$_2$O$_3$ (206). However, according to Sung et al. (207), δ-Al$_2$O$_3$ is a much less active catalyst for methanol dehydration than η-, γ-, and θ-Al$_2$O$_3$ are and has also weaker Lewis acid sites than these three phases, whereas it is more acidic and more active than α- and κ-Al$_2$O$_3$.

TEM shows fumed alumina rich in δ-Al$_2$O$_3$ to have a pronounced globular morphology, whereas δ-Al$_2$O$_3$ produced by calcination of boehmite essentially retains the morphology of the starting oxyhydroxide, such as lamellar, fibrous, or nanorod like (57,208).

2.2.4 Crystal structure and morphologies of η-Al$_2$O$_3$

η-Al$_2$O$_3$ is obtained by decomposing bayerite (α-Al(OH)$_3$) at temperatures of 200–300 °C,. The differences between the structures of η-Al$_2$O$_3$ and γ-Al$_2$O$_3$ are minor, as can be deduced from the very similar X-ray (Figure 3.10) and neutron diffraction patterns (117), which differ only in the relative intensity and perhaps in the width of some reflections. The likeness of the patterns implies that the difference between these phases lies in the occupancy of the sites by cations and vacancies more than in the locations of the crystallographic positions. The XRD pattern of η-Al_2O_3 shows, like that of γ-Al_2O_3, unusually broad features (Figure 3.10) as a result of structural disorder. The IR spectrum of the skeletal vibrations of η-Al_2O_3 (209,210) is very similar to that of γ-Al_2O_3. The Raman spectrum of the fundamental skeletal modes of η-Al_2O_3 (174) is not observed, likely because of strong defect fluorescence.

Given these data, most authors suppose that η-Al_2O_3 and γ-Al_2O_3 are essentially the same phase, namely, a non-stoichiometric spinel structure. Only, η-Al_2O_3 is characterized by more tetrahedrally coordinated (35%) and fewer octahedrally coordinated aluminum ions (209,211) than γ-Al_2O_3, according to ^{27}Al NMR spectra and Rietveld analysis of X-ray and neutron diffraction patterns. The diffraction patterns of η-Al_2O_3 have been refined with a model characterized by a "quasi-trihedral" coordination of 13% of all aluminum ions; these ions are exposed at the (111) face (117). ^{27}Al NMR spectra show more pentacoordinated aluminum ions in γ-Al_2O_3 than η-Al_2O_3 (209). To our understanding, the "nonspinel," nondefective model of γ-Al_2O_3 (119) does not provide means to capture the slight differences between the two closely related polymorphs η-Al_2O_3 and γ-Al_2O_3 and the variation of the ratio of octahedrally to tetrahedrally coordinated ions. By applying selected area diffraction, Lippens and deBoer showed that η-Al_2O_3 is also tetragonally distorted, but less so than γ-Al_2O_3 (146). According to Peintinger et al. (14) η-Al_2O_3 (assumed to be cubic) is slightly less stable than γ-Al_2O_3 (assumed to be tetragonal).

Notwithstanding their structural similarity, η-Al_2O_3 and γ-Al_2O_3 behave differently in several respects. Calcination of η-Al_2O_3 usually leads directly to θ-Al_2O_3 (198) at temperatures lower than those needed to convert γ-Al_2O_3 to θ-Al_2O_3, and without the formation of intermediate phases. Moreover, most reports suggest that η-Al_2O_3 has a higher population of stronger Lewis acid sites (212–214) and a slightly higher activity than γ-Al_2O_3 as an acid catalyst (206,207,215,216). On the basis of first-principles calculations, the difference in catalytic and surface activity has been attributed to different surface reconstruction phenomena (217). Moreover, a different habit has been reported for η-Al_2O_3 produced by calcination of bayerite, in comparison with γ-Al_2O_3 produced by calcination of (pseudo)boehmite. η-Al_2O_3 is reported to expose preferentially the (111) spinel planes (178,216). However, the termination may depend on the synthesis route, and η-Al_2O_3 produced from aluminum isopropoxide is reported to expose (100), (110), and superimposed (111)/(211) faces (193).

2.2.5 Crystal structure of χ-Al_2O_3

χ-Al_2O_3 is the first phase formed in the thermal decomposition of pure gibbsite (γ-$Al(OH)_3$) and is thus part of the following sequence of phase transitions: γ-$Al(OH)_3 \rightarrow \chi$-$Al_2O_3 \rightarrow \kappa$-$Al_2O_3 \rightarrow \alpha$-$Al_2O_3$ *(83,218,219)*. It has been reported that there is a critical particle size of χ-Al_2O_3 that determines the phase transformation behavior, and this size is around 40 nm. For crystals larger than this size, κ-Al_2O_3 is formed as an intermediate phase during conversion to α-Al_2O_3 *(220)*, whereas smaller particles grow and then transform directly to α-Al_2O_3. Calcination of gibbsite with 25 m^2 g^{-1} surface area at a temperature of 300 °C produces χ-Al_2O_3 with a surface area of 350–370 m^2 g^{-1}. χ-Al_2O_3 has also been prepared by solvothermal reaction of aluminum isopropoxide in mineral oil at 250–300 °C; samples prepared by this route directly transform to α-Al_2O_3 upon further calcination, whereby the phase transition starts once a critical crystallite size around 15 nm is attained *(221)*.

The structure of χ-Al_2O_3 is still not defined. It is either cubic *(129)* or hexagonal *(128)*. According to Tsybulya and Kryukov *(178)*, the structure is also of the spinel-type and closely related to those of γ-Al_2O_3 and η-Al_2O_3; though defects of a different kind cause the structural disorder. This conclusion was made on the basis of the similarity of the respective XRD patterns, which are all characterized by broad reflections (Figure 3.10). ^{27}Al NMR spectra show aluminum ions in tetrahedral and in octahedral coordination and only a negligible fraction of pentacoordinated species *(222)*. According to Favaro *et al.* *(223)*, IR spectra of χ-Al_2O_3 in the region of skeletal vibrations are congruent with that of γ-Al_2O_3, thus suggesting both materials have a similar ratio of tetrahedral to octahedral aluminum. This result is in partial contradiction with the slightly different XRD patterns of the two phases.

Little information is available on the surface and catalytic properties of χ-Al_2O_3. Calorimetric data taken during ammonia adsorption suggest that the Lewis acid site distribution on χ-Al_2O_3 is similar to that on γ-Al_2O_3 and η-Al_2O_3 *(213)*. It has been reported that mixed γ,χ-Al_2O_3 catalysts are more active than pure γ-Al_2O_3 catalysts in dimethylether (DME) synthesis from methanol *(224)*; however, these results are in contradiction with those of another investigation *(207)*. A sample of χ-Al_2O_3 containing 0.3% Na$^+$ as an impurity was found to be a very active catalyst for the Claus reaction *(225)*.

2.2.6 Crystal structure of κ-Al_2O_3

κ-Al_2O_3 is reported to form by thermal decomposition of gibbsite or tohdite. In the latter case, an intermediate phase is observed that is slightly

different from κ-Al$_2$O$_3$, and is thus denoted as κ'-Al$_2$O$_3$ *(94,95)*. The structure of κ-Al$_2$O$_3$ *(121–127)* is essentially nondefective, and well defined,
sharp peaks are observed in the XRD pattern (Figure 3.16). The structure
is orthorhombic (space group *Pna2$_1$*) with pseudoclose-packed planes of
oxygen in an ABAC stacking sequence along the *c*-axis. One quarter of
the aluminum ions occupy tetrahedral positions and three quarters occupy
octahedral positions. The oxygen tetrahedra are corner sharing and the octahedra are edge sharing such that zigzag ribbons are formed, as illustrated in
Figure 3.17 *(127)*.

Only few investigations can be found in the literature concerning the
surface and catalytic properties of κ-Al$_2$O$_3$. IR spectra of adsorbed basic

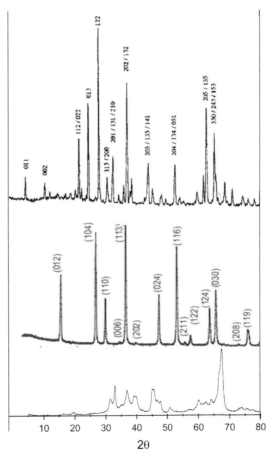

Figure 3.16 XRD patterns (Cu K$_\alpha$ excitation) of θ-Al$_2$O$_3$ (bottom, Ref. *(124)*), α-Al$_2$O$_3$
(middle), and κ-Al$_2$O$_3$ (top).

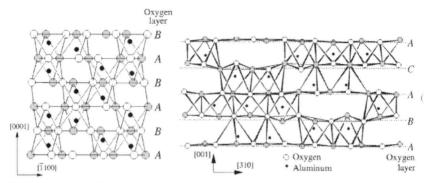

Figure 3.17 Crystal structure of α-Al₂O₃, left, and κ-Al₂O₃, right. *Reprinted with permission from Ruberto, C.; Yourdshahyan, Y.; Lundqvist, B. I. Phys. Rev. 67, 195412, 2003. Copyright (2003) by the American Physical Society (127).*

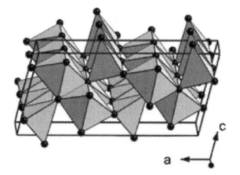

Figure 3.18 Crystal structure of θ-Al₂O₃. Solid black circles are oxygen atoms.

probes indicate medium–strong Lewis acidity similar to that of other transition aluminas *(205)*. However, other authors reported a small amount of Lewis acid sites and very low catalytic activity for κ-Al₂O₃ *(207)*.

2.2.7 Crystal structure and morphologies of θ-Al₂O₃

θ-Al₂O₃ is customarily obtained by calcination of boehmite, bayerite, γ-Al₂O₃, or η-Al₂O₃ at temperatures above 650–800 °C, and its formation is associated with a decrease of the surface area to 70–120 m² g⁻¹ or less. Applications of θ-Al₂O₃ as a catalyst support are mainly motivated by its stability at high temperatures (500–800 °C); namely, it retains a medium-to-high surface area of about 100 m² g⁻¹.

Its monoclinic structure (Figure 3.18), which is the same as that of β-gallia, can be derived from that of a spinel, through deformation and change to a completely ordered distribution of aluminum ions, with 50%

in tetrahedral and 50% in octahedral positions *(232,233)*. The monoclinic structure belongs to the *C2/m* (or C_{2h}^3) space group (No. 12). There are four Al_2O_3 units per crystallographic unit cell (which comprises two primitive unit cells) with two inequivalent cation sites and three inequivalent oxygen sites. Half of the aluminum cations are in the so-called Al(I) sites, which form slightly distorted tetrahedra with four oxide ions, and the other half are in Al(II) sites, which form highly distorted octahedra with six oxide ions. These coordinations are in agreement with ^{27}Al NMR spectra *(234)*. Each oxide ion of the O(I) type is threefold coordinated and is located at the intersection of two octahedra and one tetrahedron. Each oxide ion of the O(II) type is also threefold coordinated and is shared between one octahedron and two tetrahedra. Each O(III) ion is fourfold coordinated and sits at the corner of three octahedra and one tetrahedron. The structure of θ-Al_2O_3 is closely related to that of γ-Al_2O_3, that is, it is also assumed to be a defective spinel-like phase: during the transformation of γ-Al_2O_3 to θ-Al_2O_3 (which definitely requires several steps), the face-centered cubic (fcc) oxygen sublattice is practically unaffected, while ordering of the aluminum ions located at the cation positions occurs *(235,236)*. Although there are no cation vacancies in the structure of θ-Al_2O_3, some extent of disorder, maybe associated with an incomplete phase transformation, is found in typical samples *(209)*, as will be discussed below. The X-ray diffractogram is indicative of a small crystal size and a nondefective structure (Figure 3.16). The orientation relationships (corresponding planes in the two related structures) between γ-Al_2O_3 and θ-Al_2O_3 are $[110]_\gamma$//$[010]_\theta$ and $(001)_\gamma$//$(100)_\theta$ *(197)*. Based on these relationships, it is usually assumed that θ-Al_2O_3 preferentially exposes the (010) face *(226,227)*, which corresponds to the (110) face of the γ-Al_2O_3 spinel structure. The latter has been reported to be preferentially exposed in lamellar particles. Several investigations show that the surface chemistry of θ-Al_2O_3 has much similarity, qualitatively, to that of the γ-Al_2O_3 and η-Al_2O_3 polymorphs *(187,214,228,229)*, while the surface area of θ-Al_2O_3 is lower, implying that the concentration of active sites may also be lower. Correspondingly, the catalytic activity of θ-Al_2O_3 for acid-catalyzed reactions such as methanol dehydration *(207)* and ethanol dehydration *(187)* is reported to be slightly lower per unit mass than that of the other two polymorphs.

2.2.8 Crystal structure and morphologies of corundum (α-Al_2O_3)

All alumina polymorphs are converted into α-Al_2O_3 (corundum) upon calcination at 800–1200 °C. The corundum structure (Figure 3.17) of α-Al_2O_3

belongs to the space group $R\bar{3}c$ (No. 167) with Z equal to 2 and is characterized by hcp of oxide ions with aluminum ions occupying two thirds of the
octahedral sites. Thus, in bulk corundum all aluminum ions are in sixfold
coordination, and all oxide ions are in fourfold coordination. The last step
in the thermal transformation of aluminum oxides and hydroxides is usually
the phase transition from θ-Al_2O_3 to α-Al_2O_3, which occurs abruptly
within a narrow temperature range as indicated by a well-defined exother-
mic DTA signal between 1200 and 1300 °C (230). This transition is accompanied by a loss of porosity and surface area, as well as by some increase in
crystal size. α-Al_2O_3 powders produced by high temperature calcination are
usually characterized by a BET surface area of less than 5 m^2 g^{-1} and a pore
volume smaller than 0.5 cm^3 g^{-1}. By controlling reaction temperature and
time, higher surface areas may be obtained, but the values remain below
50 m^2 g^{-1}. If the particles are nanosized, the transition of θ-Al_2O_3 to
α-Al_2O_3 reportedly occurs without particle growth, and one θ-Al_2O_3 crystallite transforms to one α-Al_2O_3 crystallite at temperatures as low as 800 °C.
It appears that a crystallite size limit of 20–25 nm $(231,237)$ exists; above this
size the θ-Al_2O_3 to α-Al_2O_3 conversion occurs, below this size, θ-Al_2O_3
may be stable. This phenomenon will be further discussed in Section 2.2.10.

α-Al_2O_3 powders can also be prepared by directly calcining the hydroxides or even amorphous precipitates, provided the temperature is sufficiently
high. The behavior of precipitated gels during calcination depends on the
precursor salt and the presence of additives, which can reduce the temperature for formation of α-Al_2O_3 (132).

A common route to obtain α-Al_2O_3 powders is the calcination, respectively decomposition, of diaspore, α-$Al(OH)_3$, at 400–500 °C. By using this
method, a surface area of the order of 65 m^2 g^{-1} may be obtained $(133,238)$.
It is frequently stated that the nearly close-packed hexagonal oxygen sub-
lattices in both diaspore and corundum permit a topotactic transformation
with little atomic rearrangement. The actual process, however, requires that
one quarter of the oxygen atoms is removed and that the hexagonal sublattice
is restored afterward. Powder X-ray diffraction patterns recorded during
reaction demonstrated the presence of an intermediate phase (67), designated
as α'-Al_2O_3. This phase could be indexed with a monoclinic (P2) cell and the
following lattice parameters $a = 9.566(5)$ Å, $b = 5.124(2)$ Å, $c = 9.156(4)$ Å,
and $\beta = 91.76(3)°$. NMR spectra of α'-Al_2O_3 indicated that 15–20% of
the aluminum atoms are at tetrahedral sites, and the rest are in octahedral
coordination. More recent investigations (239) confirm the existence of
an intermediate phase in the diaspore-to-corundum transformation, and

explain the formation of thin dense corundum regions as the result of a misfit between the close-packed oxygen planes of the intermediate and the final corundum phase. These thin corundum regions were separated by nanometer-sized pores, thus forming lamellae with a periodicity of 3.7 nm.

Corundum powders can also be directly precipitated under hydrothermal conditions. This method produces nanoparticles with remarkable surface area; values as high as $160 \, \mathrm{m^2 \, g^{-1}}$ have been reported (240). α-Alumina has been synthesized in 1,4-butanediol solutions, and it was found that multiple variables including reaction time, stirring speed, amount of organic compounds, and solid loading have a strong effect on the size and shape of the obtained particles. By variation of the conditions, α-alumina crystals with different habits could be prepared, including circular platelets, hexagonal platelets, hexagonal platelets with habit modifications on the edges, dodecahedra, polyhedra with 14 faces, polyhedra with 20 faces, hexagonal prisms, and hexagonal prisms with habit modifications on the edges (241). Hexagonally shaped α-Al_2O_3 crystals have been synthesized starting from a gibbsite precursor at a pH value of 9 and a temperature of 300 °C; this product was characterized by a specific surface area of $26 \, \mathrm{m^2 \, g^{-1}}$ (242).

Corundum powders are very relevant ceramic materials with high hardness, a high melting point (2053 °C), and a high specific electrical resistance (about $10^{12} \, \Omega\mathrm{m}$ at 20 °C).

Single crystals of α-Al_2O_3 have been used to investigate the surface of this phase (243). Vibrational sum frequency generation spectroscopy of an α-Al_2O_3 (0001) surface after adsorption of water revealed a sharp band at $3780 \, \mathrm{cm^{-1}}$ and a broad band at $3450 \, \mathrm{cm^{-1}}$ (244). In an IR spectroscopic investigation, a band at $3710 \, \mathrm{cm^{-1}}$ was found, which was assigned to isolated OH groups on the α-Al_2O_3 (0001) surface (245). The medium–strong base piperidine was reported to prefer adsorption on hydroxyl groups over adsorption on Lewis acid sites (244).

The spectra of the surface hydroxyl groups of α-Al_2O_3 powders vary significantly from sample to sample (246–249). The adsorption of basic probe molecules such as pyridines or CO typically reveals Lewis acid sites of medium strength, whereas very strongly acidic sites, which are observed for transition aluminas, are absent (186,247–251). However, the observed sites depend on the calcination conditions, the dehydration conditions applied prior to adsorption, and on the surface area of the sample; some of the surface properties of the original transition alumina may be retained (252). The catalytic activity of α-Al_2O_3 for acid-catalyzed reactions is usually low (207); in fact, even a basic character has been documented (206).

2.2.9 Mesoporous aluminas

To obtain the same variety of materials that exists for silicas and silica–aluminas, many attempts have lately been devoted to the synthesis of transition aluminas with very large surface area and mesoporosity. Such syntheses require the use of templating agents. Although many new materials have been prepared, the properties are often not as desired. Some of the materials appear to have amorphous walls, which explains their low acidity and low catalytic activity (144). Recently, high–surface-area mesoporous forms of crystalline aluminas have been prepared; the structures include γ-Al_2O_3 (200–400 m^2 g^{-1}) (253), η-Al_2O_3 (430 m^2 g^{-1}), and χ-Al_2O_3 (325 m^2 g^{-1}) (254). These aluminas were shaped through assembly of mesostructured surfactant-encapsulated mesophases of boehmite, bayerite, and gibbsite, respectively, which in turn had been made from amorphous aluminum hydroxide by using an amine surfactant as the structure-templating porogen. In conclusion, the recently developed mesostructuration routes have made possible the preparation of highly porous aluminas. However, preparation procedures that lead to surfaces as highly acidic as that of γ-Al_2O_3 generally do not yield improved textural properties; rather, the textural properties remain close to those observed when using classical precipitation and sol-gel procedures (255).

2.2.10 Thermodynamics, polymorphism, and surface reactivity of aluminas

It became evident only in the past two decades that the thermodynamic stability of the surface may even have a larger effect than that of the bulk in determining the structure of the particles in fine powders. It has been reported that, although the α-Al_2O_3 bulk is more stable than the γ-Al_2O_3 bulk, the reverse is true for the respective surfaces (160). Thus, fine powders of γ-Al_2O_3 with surface areas greater than 125–130 m^2 g^{-1} should be, according to the authors of Ref. (160), thermodynamically more stable than fine powders of α-Al_2O_3. In a more recent report, the same group concluded that the surface of amorphous Al_2O_3 is even more stable than that of γ-Al_2O_3. These findings imply that for fine powders with surface areas greater than 370 m^2 g^{-1}, an amorphous state should be more stable than any crystalline phase (114).

It is worthwhile to place these data in context with the observation of a maximum size for θ-Al_2O_3 crystallites, which is about 20–25 nm (231). Larger crystallites tend to convert into corundum. One interpretation of this result is that the surface of θ-Al_2O_3 must be more stable than that of

α-Al$_2$O$_3$, which would become relevant for smaller particles; however, some investigations suggest the opposite *(230)*. Yang *et al. (256)* reported that α-Al$_2$O$_3$ crystals near this critical size may undergo a reverse phase transformation into θ-Al$_2$O$_3$ above temperatures of 800 °C. Theoretical and experimental evidence has been presented suggesting that θ-Al$_2$O$_3$ has a very stable hydroxylated surface *(257)* and that θ-Al$_2$O$_3$ nanocrystals smaller than the above specified critical size are thermodynamically more stable than crystals of α-Al$_2$O$_3$ of the same size, whereas for crystal sizes of 30 nm or larger, the opposite is true *(258)*. The difference in thermodynamic stability of α-Al$_2$O$_3$ and θ-Al$_2$O$_3$ is likely very small, considering the fact that for the corresponding isostructural compounds of gallia, the bulk stability is reversed: β-Ga$_2$O$_3$, isostructural with θ-Al$_2$O$_3$, is more stable than α-Ga$_2$O$_3$, also a corundum structure.

The overall picture that emerges is the following: for the bulk structures, the thermodynamic stability trend is α-Al$_2$O$_3 > \theta$-Al$_2$O$_3 > \gamma$-Al$_2$O$_3 >$ a-Al$_2$O$_3$, whereas for the respective surfaces, the inverse trend seems to be true. These opposing trends satisfactorily explain the solid–state chemistry of alumina powders. The high surface stability of amorphous aluminas also agrees with their poor reactivity and poor catalytic activity compared with those of crystalline "transition" aluminas.

The conclusion that the corundum surface is basically instable, or at least significantly less stable than those of transition aluminas, is surprising in light of the known behavior of α-Al$_2$O$_3$. In fact, the surfaces of α-Al$_2$O$_3$ powders are usually considered to be catalytically inactive in comparison to those of transition aluminas, which are very active. α-Al$_2$O$_3$ powders are also assumed to be less suitable as adsorbents than transition aluminas, which are widely applied in adsorption technologies simply because of their ability to strongly adsorb many species, among them water. The supposition of a high instability of the α-Al$_2$O$_3$ surface seems to conflict with the low activity in catalysis and adsorption (see Section 3.2). However, it must be considered that in most cases, transition aluminas have high surface areas whereas corundum powders have very low surface areas. However, as discussed above, corundum powders with moderately high surface area can be prepared and do not seem to reveal high surface reactivity and adsorption capacity. This point deserves further investigation.

It needs to be emphasized that the above picture of the solid-state chemistry of aluminas is solely based on thermodynamics and the role of solid-state kinetics is neglected. If the transformation into a more stable species is kinetically hindered, metastable species can exist, as is well established. Moreover,

structural relationships are believed to promote certain phase transformations. For example, the relationships of both boehmite and bayerite with
the spinel structure and of diaspore with the corundum structure have been
considered to be the main reason for the preferential transformations
γ-AlOOH $\rightarrow$ γ-Al$_2$O$_3$ *(119,178)*, α-Al(OH)$_3$ $\rightarrow$ η-Al$_2$O$_3$ *(178)*, and
α-AlOOH $\rightarrow$ α-Al$_2$O$_3$ *(67)*. In other words, it has been claimed that,
starting from boehmite or bayerite, the formation of spinel-type aluminas
is faster than that of corundum, whereas the conversion of diaspore into
corundum is both faster and thermodynamically more favored than its transformation into spinels. Also, the transformations of spinel-type aluminas into
θ-Al$_2$O$_3$ *(236)* or α-Al$_2$O$_3$ *(259)* have been reported to involve complex
diffusion processes and displacements of atoms and vacancies: thus, these
transformation are considered to be kinetically hindered. Even the existence
of the two closely related spinel-type structures of γ-Al$_2$O$_3$ and η-Al$_2$O$_3$ was
supposed to be the result of kinetically hindered diffusion of aluminum cat-
ions in the ccp array of oxygen anions. It appears obvious that this complex
reaction network is, at least in part, ruled by kinetics.

2.2.11 Stabilized aluminas

As discussed above, transition aluminas are metastable if only bulk effects are
considered. With increasing temperature, first the low temperature varieties
of the transition phases are transformed into the high temperature varieties
and finally, α-Al$_2$O$_3$ (corundum) is formed. This sequence of transformations is associated with a decrease of the surface area because of intraparticle
porosity loss and particle growth. The solid-state reactions, which may start
by surface diffusion processes, are inhibited by the presence of foreign ionic
species on the surface. Silica deposited on the surface of alumina or
coprecipitated with alumina inhibits these phase transformations, thus providing bulk and morphological stability to transition aluminas. These materials are known as silica-stabilized aluminas and are used as supports for
catalysts that are applied at high temperatures. Examples are the rhodium
catalyst used for the decomposition of N$_2$O *(260)* and the chromia catalysts
used industrially for paraffin dehydrogenation *(81,261)*. The stabilization
can be achieved without significantly affecting the surface properties of
the alumina if the amount of silica added is in the range of a few weight percent. Indeed, small amounts of silica do not induce relevant Brønsted acidity,
which is known to be pronounced in silica-rich silica–aluminas. The Lewis
acidity of alumina is in some cases found to be unaltered by silica, in others it
is reported to be slightly enhanced *(262)*. However, silica may poison some

basic surface sites. Silica is sometimes added to alumina to modify the state of supported species: surface silicate species can occupy anchoring sites and thus shift other species to less active surface sites.

Another way to induce thermal and morphological stability to aluminas is to deposit metal cations on the surface these cations must be so large that they are unable to enter the close packing of oxide ions. Suitable ions are large monovalent, bivalent, and trivalent cations such as Na^+, K^+, Cs^+, Ca^{2+}, Sr^{2+}, Ba^{2+}, and La^{3+}. These ions do not enter spinel structures but tend to form, when combined with Al^{3+}, beta-alumina-type structures, which are characterized by spinel-type alumina layers separated by layers containing these large cations together with ionically conducting oxide species. When deposited on the surface of transition aluminas, these cations inhibit surface diffusion and thus also phase transformation. Alumina stabilized in this way is suitable for high temperature applications, for example as support of manganese catalysts for combustion reactions (263). However, the cations significantly modify the surface chemistry of alumina, by neutralizing acid sites and, at increasing coverage, by enhancing basicity. In many cases, the chemical effects of cation addition are actually beneficial; for example, the activity of alumina for coke formation from organic compounds under reducing conditions may be decreased. The stabilizing effect of such cations has been attributed to a lowering of the surface free energy, which can result in an overall free energy of doped γ-Al_2O_3 nanoparticles that is lower than that of undoped γ-Al_2O_3 nanoparticles and of α-Al_2O_3 particles of the same size (264). However, as already stated, another effect of the large cations is a hindered surface diffusion; that is, the stabilization may also be explained by kinetic inhibition.

2.2.12 Alumina films on metal single crystal surfaces

Thin films of alumina have been prepared on the surface of selected single crystal faces and are commonly used as a model of alumina to investigate its properties as a catalyst and as a support of catalysts. These materials are attractive because they can be analyzed by typical surface science techniques such as low-energy electron diffraction, high-resolution electron energy loss spectroscopy (HREELS), and infrared reflection absorption spectroscopy, together with scanning tunneling microscopy and XRD. Thus, these "model catalysts" can be very well characterized.

Similar alumina layers can be produced on a variety of surfaces including NiAl(110), NiAl(100), Ni$_3$Al(100) (265), Ni$_3$Al(111) (266), Ni(111) (267), CoAl(110) (268), FeAl(110) (269), and Cu–9% Al(111) (270). It has been demonstrated that these layers contain a number of defects (271), and it still has to be clarified to what extent their surface reconstruction resembles that of metastable alumina polymorphs actually used in catalysis.

Model alumina layers grown on a NiAl(110) surface have been reported to exhibit a γ-Al$_2$O$_3$-like structure *(272)*, and HREELS revealed an OH stretching band at 3756 cm^{-1} *(273)*. In another investigation, the structure of a similar layer on NiAl(110) was identified as that of κ-Al$_2$O$_3$ *(274)*. An alumina layer grown on NiAl(100) was reported to also have γ-Al$_2$O$_3$ structure, and to be characterized by an OH stretching band at 3690 cm^{-1} (recorded by HREELS). This sample, however, did not adsorb CO at conditions where adsorption typically occurs *(275)*. In other investigations, the formation of either an amorphous alumina layer *(276)* or a θ-Al$_2$O$_3$ layer *(277)* was observed on the (100) face of NiAl. An alumina layer grown on CoAl(110) was reported to be amorphous; at a temperature of 1000 K it was converted into θ-Al$_2$O$_3$ *(268)*. These model layers present an interesting and still developing research field, but the usefulness of these layers to understand the surface properties of aluminas and of alumina-supported catalysts has yet to be demonstrated.

3. SURFACE STRUCTURES AND CHEMISTRY OF ALUMINAS

3.1. Surface hydroxyl groups on alumina surfaces

Several investigations concerning the surfaces of transition aluminas have focused on the identification of the structure of the surface hydroxyl groups. Spectroscopic measurements, for example by IR or NMR spectroscopy, reveal the presence of hydroxyl groups on the surface of transition aluminas, even after outgassing at relatively high temperature. Significant dehydroxylation of transition aluminas cannot be achieved before conversion of these Al$_2$O$_3$ polymorphs to α-Al$_2$O$_3$ occurs. The density of hydroxyl groups still on the surface of γ-Al$_2$O$_3$ or δ-Al$_2$O$_3$, respectively, after various treatments has been assessed and is estimated to be in the range of 12–16 OH nm^{-2} after outgassing at 100 °C and to be reduced to 2.6 OH nm^{-2} *(160)* or 0.7 OH nm^{-2} *(192)* after outgassing at 700–750 °C.

The presence of hydroxyl groups on alumina-based materials is also evident from the splitting of the XPS O1s signal, which is usually characterized by a contribution near 530 eV that is assigned to oxide species and a second contribution at higher energy that is assigned to hydroxide species *(278)*.

^{1}H NMR spectra of dehydrated γ-Al$_2$O$_3$ show complex spectra that simplify as dehydroxylation progresses *(279)*. However, such spectra are mostly recorded without previous outgassing of the sample, and thus show the partially superimposed peaks of free and H-bonded OH groups and of adsorbed water. De Canio *et al.* observed ^{1}H chemical shifts of −0.3, 0.3, 1.5, 2.4, 4.0, 5.0, 6.5, 7.1, and 7.8 ppm for a γ-Al$_2$O$_3$ sample described as to be "with no impurities" *(280)*. Deng *et al. (281)* assigned a peak at 0 ppm to terminal OH groups attached to octahedrally coordinated aluminum ions, a peak at 2 ppm to bridging OH groups simultaneously connected to octahedrally and tetrahedrally coordinated aluminum ions, and a small peak at 4.3 ppm to triply bridging OH groups. The authors also reported that the peak at 2 ppm is the first affected by introduction of Na$^+$ ions to the sample. The ^{1}H broad signal at 4.5 ppm that is observed when Na$^+$ ions are present disappears in a ^{1}H {^{23}Na} double resonance NMR experiment, indicating that the signal arises from OH groups in interaction with Na$^+$ ions. According to Qu *et al. (282)*, two main peaks exist, a sharper one at −0.2 ppm and a broader one at 1.9 ppm, which were attributed to two different families of OH groups designated as "basic" and "acidic," respectively. Huittinen *et al. (283)* reported distinct features at 7.7 ppm (only observed after drying at higher temperature, i.e., 150 °C), and at 5.6, 3.9, 1.3, 0.9, and 0.3 ppm. The peaks at 1.3 and 0.9 ppm were assigned to adsorbed water. Delgado *et al. (189)* found spectra similar to those described by Huittinen *et al.* and attempted to make assignments with the help of simulations. Specifically, they assigned the peak at 0.3 ppm to isolated terminal OH groups attached to tetrahedrally coordinated Al^{3+} ions, and the peak at about 3 ppm to either terminal or bridging OH groups attached to either penta- or hexacoordinated Al^{3+} ions. The other peaks were interpreted as H-bonded species and adsorbed water. Ferreira *et al. (284)* performed *ab initio* calculations assuming the nonspinel model of γ-Al$_2$O$_3$ and concluded that terminal OH groups on the (100) surface should have a chemical shift in the region of 0.8−5.3 ppm, free triply bridging OH groups (bonded to three cations) should have a chemical shift of 4.9−7.0 ppm, whereas species bridging two aluminum cations should be H bonded. In contrast, bridging OH species on the (110) face are expected to have a chemical shift of 0.5−2.7 ppm. The variations in the reported spectra and the, in part, contradictory interpretations let us conclude that it so far has not been possible to formulate a complete and reliable picture of the OH groups on aluminas on the basis of ^{1}H NMR spectra.

The IR spectra of amorphous aluminas show the presence of hydroxyl groups with bands at approximately 3780−3795, 3730−3750, and

3680 cm^{-1} *(110,134,135,255).* Most IR spectroscopic investigations have been devoted to the surface hydroxyl groups of γ-Al$_2$O$_3$. At least five bands are usually discernable in the hydroxyl stretching vibration region of IR spectra of γ-Al$_2$O$_3$. These bands are typically located at about 3790, 3770, 3740–3720, 3700–3660, and 3580 cm^{-1}. The entire pattern is complex, and some bands evidently have a substructure indicating more than one component, as evident in Figure 3.11. Interestingly, the spectra of different samples of γ-Al$_2$O$_3$ tend to be comparable, notwithstanding different preparation procedures and morphologies: for example, fibrous particles *(187,285)* and spherical particles *(195)* show spectra that are very similar to those of the more common lamellar or prismatic particles. Essentially the same band positions and intensities are also found for highly porous or mesoporous γ-Al$_2$O$_3$ materials *(255)* as well as for alumina aerogels *(75,76)* although in these cases, the bands are perhaps broader and less well resolved. In our opinion, the lack of definition makes the assignments of these bands to species located on particular surface planes unreliable.

Partial dehydroxylation of the γ-Al$_2$O$_3$ phase is possible; however, the temperature must remain in the range in which the phase is stable, that is, it must not exceed 750 °C. As the outgassing temperature is increased, the absorption bands at wavenumbers below 3700 cm^{-1} decrease progressively in intensity until they almost disappear, whereas the bands at higher frequencies appear to be stable *(189,262,286),* as shown in Figure 3.19 *(262).* The bands above 3700 cm^{-1} are also relatively sharp, definitely sharper than those at lower frequency (cf. Figures 3.11 and 3.19).

The bands located at approximately 3790 and 3730 cm^{-1} are both always present in IR spectra of transition aluminas. The band at 3790 cm^{-1} is definitely weaker in intensity than the band at 3730 cm^{-1}, which is the strongest band after outgassing at high temperature. However, the exact position of this band is not always the same. For some samples, this band is observed at 3724 cm^{-1}; for others, it is observed at 3737 cm^{-1}. In some cases, two superimposed components are clearly present.

The band at about 3770 cm^{-1} is characterized by the most complex behavior, and is found to be associated with the most reactive sites. It has been reported to be particularly intense in the spectra of highly reactive aluminas *(98).* This band is not present in the spectra of fumed aluminas that were activated in vacuum *(97,98,286),* but it appears if the activation of the fumed alumina includes a step in wet air *(97,189),* as shown in Figure 3.20 *(97).* On the basis of these results, the lack of a band at

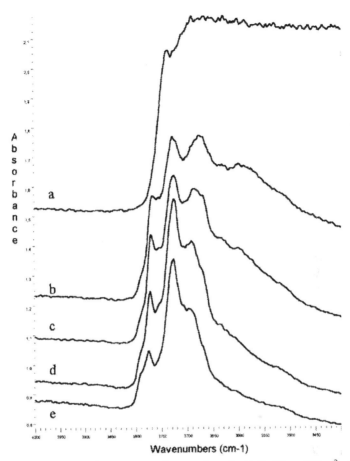

Figure 3.19 FTIR spectra of the OH stretching region of γ-Al$_2$O$_3$ (Akzo, 185 m^2 g^{-1}) after outgassing at 30 °C (a), 450 °C (b), 550 °C (c), 650 °C (d), and 730 °C (e). *Reprinted with permission from Ref. (262).*

3770 cm^{-1} was explained with the presence of chlorine impurities that selectively replace some of the OH groups in these samples *(97)*.

This band is also reduced in intensity after doping with potassium *(287)* or in the presence of sodium impurities *(169,288)*. For samples doped with high concentrations of Na$^+$, the bands at 3790 and 3770 cm^{-1} are both absent, and the most intense band is observed at about 3742 cm^{-1} *(289,290)*. The deposition of phosphate species on the surface also seems to selectively eliminate the band at 3770 cm^{-1} *(291)*, whereas the deposition of silicate species does not *(286)*. The band also seems to disappear, at least partially, upon phase transition to δ-Al$_2$O$_3$ *(204)* or to θ-Al$_2$O$_3$ *(257)* and

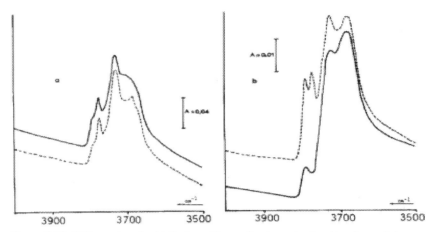

Figure 3.20 FTIR spectra of γ-Al$_2$O$_3$ from Rhone–Poulenc obtained by thermal decomposition of boehmite (left) and of a fumed γδ-Al$_2$O$_3$ from Degussa (right) after activation by outgassing at a temperature of 500 °C with previous heating in dry O$_2$ to 500 °C (solid lines) or in wet air at 500 °C (dashed lines). *Reprinted from Ref. (97).*

gradually disappears as non-stoichiometric spinels such as Mg$_x$Al$_2$O$_{3+x}$ and Ni$_x$Al$_2$O$_{3+x}$ approach spinel stoichiometry *(247)*. The band at 3770 cm^{-1} is also always altered when basic molecules are adsorbed and appears to be the most strongly affected band during adsorption of CS$_2$ *(292)* or CO$_2$ *(293)*. In the latter case, bicarbonates are formed from the hydroxyl group and CO$_2$; however, bicarbonate formation has also been reported for samples not showing this band *(285)*.

All hydroxyl groups characterized by bands above 3700 cm^{-1}, which are expected to be basic in nature according to the model by Knözinger and Ratnasamy (see Ref. *(294)* and discussion below), are involved in the adsorption of weakly and strongly basic molecules (such as CO, nitriles, or pyridine). IR spectra of a γ-Al$_2$O$_3$ sample (Versal from UOP) after activation and after contact with CO at low temperature demonstrate this behavior; the region of OH stretching vibrations is shown in Figure 3.21. It is evident that the three bands at about 3790, 3770, and 3730 cm^{-1} are all perturbed by CO adsorption *(295)*. This perturbation is reversible, that is, the bands can be restored by outgassing of the sample. The extent of the perturbation that weak bases exert on the most intense band (at 3740–3720 cm^{-1}) can be easily evaluated. In the experiment presented in Figure 3.21, the shift Δ𝜈 of the main OH band is approximately −140 cm^{-1} and the stretching frequency of the associated CO species is 2157 cm^{-1}, corresponding to a shift of +17 cm^{-1} relative to the CO gas

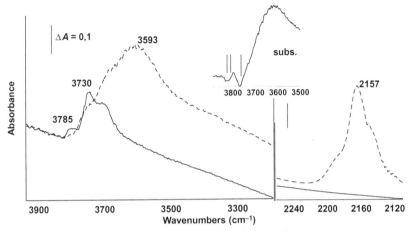

Figure 3.21 FTIR spectra of γ-Al$_2$O$_3$ obtained by thermal decomposition of boehmite after activation by outgassing at 500 °C (solid line) and in contact with CO at a partial pressure of 10 torr at −123 °C (dotted line). Original spectra are shown in the region of OH stretching vibrations and the CO stretching vibrations. Inset shows the difference spectrum in the OH stretching region (dotted spectrum–solid spectrum).

phase vibration. The blue shift of the CO vibration is consistent with a significant bond order increase. The analysis of the acidity of the individual hydroxyl species is difficult, as pointed out by Liu and Truitt *(296)*, because most probes simultaneously adsorb on Lewis sites, which, like the direct interaction with OH groups, also causes frequency shifts, provided the OH groups are in the proximity of the Lewis site. It can be seen in Figure 3.21 that in this experiment, the spectrum of the free OH groups is almost completely restored after outgassing of the sample at low temperature while the bands at 2185–2240 cm^{-1}, which are indicative of CO adsorbed on Lewis sites, are still present. It follows that the shift observed for the OH stretching vibrations is the result of a direct interaction of the hydroxyl groups with CO. The apparent shifts of the most intense OH band upon adsorption of bases have been measured for various alumina samples. For CO, the shift ranges from −120 to −140 cm^{-1}; for acetonitrile, the shift is about −190 cm^{-1} *(297)*. According to the criteria of the so-called "hydrogen bonding method" *(298)* for the evaluation of the Brønsted acid strength of surface hydroxyl groups, these values indicate a medium–high acidity of the perturbed alumina hydroxyls. The results of the "base protonation method" also suggest medium Brønsted acidity *(298)*. This method is a simple test of the ability of a surface to protonate bases, which are generally stronger than those used for the hydrogen bonding method. When pyridine (proton affinity $= 912$ kcal mol^{-1}, p$K_a = 5.2$) is adsorbed on alumina, it is not protonated, whereas protonation of the more basic molecules *n*-butylamine *(299)* and piperidine is observed. In Figure 3.22, the spectrum of piperidine (proton affinity $= 933$ kcal mol^{-1}, p$K_a = 11.1$) adsorbed on γ-Al$_2$O$_3$ is reported. The complex of absorption bands at 1650–1550 cm^{-1}, which is absent in the spectrum of pure piperidine, can be assigned to the $=NH_2^+$ deformation of the piperidinium cation, in agreement with the presence of a band in this region in the spectrum of piperidinium chloride *(300)*. Further confirmation of the medium Brønsted acid strength of hydroxyls on alumina comes from the "olefin polymerization method" *(298)*: 1,3-butadiene polymerizes on the surface of alumina *(301)*, whereas simple olefins do not. Still, among the predominantly ionic simple oxides, alumina is one of the strongest Brønsted acids *(302)*. Accordingly, the good performance of pure γ-Al$_2$O$_3$ as a catalyst for skeletal isomerization of *n*-butylene to isobutylene at about 450 °C has been attributed to its medium–strong Brønsted acidity, which is sufficient to protonate *n*-butylenes at high temperature, thus producing carbenium ions, but too low to cause much cracking and coking *(303,304)*. The moderate acidity implies that aluminas become active only at high temperature *(305)*. The hydroxyl groups absorbing at wavenumbers below 3700 cm^{-1}, expected to be acidic, appear instead to not be attractive adsorption sites at all, neither for acidic nor for basic species *(285,296)*.

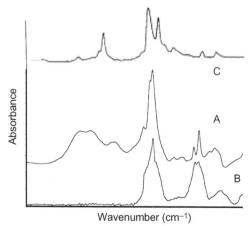

Figure 3.22 IR spectra of piperidine adsorbed on γ-Al$_2$O$_3$ (Akzo, 185 m^2 g^{-1}) after previous outgassing of the alumina at a temperature of 500 °C (a), of gaseous piperidine (b), and of piperidinium chloride (c). *From Ref.* (300).

The progressive decrease in intensity of the bands of the surface hydroxyl groups of γ-Al$_2$O$_3$ upon heating under vacuum is accompanied by the appearance of Lewis acid sites *(286)*. As shown by Wischert *et al.* *(192,306)*, the heat treatment of γ-Al$_2$O$_3$ in vacuum at 300–600 °C leads to substantial dehydroxylation with little or no surface area loss. Concomitantly, sites active for methane conversion appear. Outgassing at higher temperature causes little further dehydroxylation while the surface area decreases significantly and the transformation of γ-Al$_2$O$_3$ into δ-Al$_2$O$_3$ and θ-Al$_2$O$_3$ progresses. Because of the loss of surface area, the density of Lewis acid sites decreases.

Investigations of the conversion of "Puralox" γ-Al$_2$O$_3$ into δ-Al$_2$O$_3$ *(204)* and θ-Al$_2$O$_3$ *(257,285)* show that the band at about 3770 cm^{-1} disappears partially or completely during this process. Consistent with this observation, IR spectra of other δ-Al$_2$O$_3$ samples exhibit no such band or only a weak band *(98)*. Métivier *et al.* *(307)* covalently grafted PPTEOS (a fluorescent probe containing a pyrene moiety linked to a triethoxysilane group) onto the hydroxyl groups of γ-Al$_2$O$_3$ and δ-Al$_2$O$_3$ and analyzed the fluorescence spectra and determined the surface concentration. On the basis of the results, the authors concluded that the γ-Al$_2$O$_3$ surface is characterized by a higher density of reactive hydroxyl groups (absorbing at 3770 cm^{-1}) than that of δ-Al$_2$O$_3$, and that these groups were not evenly distributed on the surfaces. The fraction of hydroxyl groups that form areas of higher hydroxyl density appeared to be dependent on the type of alumina. The hydroxyl groups were indeed found to be more clustered on δ-Al$_2$O$_3$ than on γ-Al$_2$O$_3$, which can be explained by the disappearance of zones with a low density of hydroxyls during the γ-Al$_2$O$_3$ to δ-Al$_2$O$_3$ conversion process. It seems likely that dehydroxylation of intact planes occurs before that of defects, edges, and corners.

The IR spectra reported by Lodziana *et al.* *(257)* and by Busca *(285)* for θ-Al$_2$O$_3$ samples show the band at approximately 3725 cm^{-1} to be very intense and sharp, whereas lower frequency bands and also the band at about 3770 cm^{-1} are almost absent (Figure 3.23). However, other θ-Al$_2$O$_3$ samples are characterized by IR spectra very similar to those of γ-Al$_2$O$_3$ in terms of band intensities, and also show a pronounced band at about 3770 cm^{-1} *(285)*. Our interpretation of these seeming discrepancies is that the cation and vacancy ordering that accompanies the sequential conversion γ-Al$_2$O$_3 \rightarrow \delta$-Al$_2$O$_3 \rightarrow \theta$-Al$_2$O$_3$ actually starts in the bulk and progresses only later (i.e., at higher temperatures or longer heating times) to the surface. Thus, it may be possible to have a θ-Al$_2$O$_3$ with a γ-Al$_2$O$_3$-like surface, or

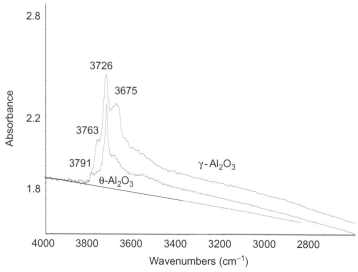

Figure 3.23 FTIR spectra of γ-Al₂O₃ (Puralox 200 from Sasol) and γ-Al₂O₃ (Puralox 90 from Sasol) after activation by outgassing at 500 °C.

one with a more "ordered" θ-Al₂O₃-like surface. Indeed, data show that the crystal structure of θ-Al₂O₃ samples may be quite complex. HRTEM images showed that θ-Al₂O₃ samples obtained by boehmite dehydration may contain multiply twinned lamellae arranged along the [001] direction *(308)*. The co-existence of γ-Al₂O₃ with θ-Al₂O₃ domains has also been reported *(309)*. A conclusion opposite to ours was made by Kwak *et al.* *(310)*, whose results also showed the co-existence of the two phases. These authors suggested that a core–shell structure can be formed during the γ-to-θ phase transformation process, with a θ-Al₂O₃ shell and a γ-Al₂O₃ core.

Although detailed investigations of the IR spectra of the hydroxyl groups of η-Al₂O₃ are scarce *(311–314)*, it appears that the spectra are very similar to those of γ-Al₂O₃. Nédez *et al.* *(313)* recorded the spectra of a sample dehydroxylated at 500 °C and observed four very intense bands at 3790 (sh), 3775, 3730, and 3682 cm⁻¹, with relative intensities similar to those usually reported for γ-Al₂O₃. Lundie *et al.* *(314)*, presenting only difference spectra, inferred that the band at 3770 cm⁻¹ is the most strongly affected band during pyridine adsorption experiments on η-Al₂O₃.

The spectra of the hydroxyl groups of α-Al₂O₃ that can be found in the literature vary substantially. In general, the spectra show a main band, some-times split, at 3745–3690 cm⁻¹. Any well-defined components above

3750 cm^{-1} may be entirely absent *(246–248)*; alternatively, there may be a pronounced component at 3770–3780 cm^{-1} *(98,249)*.

To our knowledge, the surface hydroxyl groups of χ-Al$_2$O$_3$ and κ-Al$_2$O$_3$ have never been reported.

A general interpretation of the spectra of the hydroxyl groups of aluminas has been attempted. A first interpretation was proposed by Peri *(315)*, who reported the presence of five infrared bands between 3700 and 3800 cm^{-1} and assigned them to isolated terminal hydroxyl ions differing in the number of oxide nearest neighbors. Knözinger and Ratnasamy *(294)* created a model for the interpretation of the observed spectra by refining the earlier model by Tsyganenko and Filimonov *(316)*. The latter authors pointed out that surface hydroxyl groups can be terminal (type I), bridging (type II), or triply bridging (type III) and that the stretching frequency should decrease in the order type I > type II > type III. Knözinger and Ratnasamy *(294)*, whose model became very influential, recognized that OH groups can be directly bonded either to tetrahedrally coordinated or to octahedrally coordinated aluminum cations (Al$_{IV}$ and Al$_{VI}$ respectively). They correlated the OH stretching frequency to the net charge of the hydroxyl groups, which they calculated as the sum of the negative charge of the OH$^-$ ion (-1) and the cumulative strengths of the electrostatic bonds between the anion and the adjacent cations. On the basis of this approach, they concluded that five possible scenarios exist for free (i.e., not H bonded) OH groups. In the following, the net charges and the proposed assignments are reported. Terminal OH groups on Al$_{VI}$ sites (type Ib) have a charge of -0.5 and were proposed to be responsible for the band with the highest frequency, located at 3800–3785 cm^{-1}. Terminal OH groups on Al$_{IV}$ sites (type Ia) have a charge of -0.25 and were associated with the band at 3760–3780 cm^{-1}. Hydroxyl groups bridging between two Al$_{VI}$ sites (type IIb) have no net charge and should produce a band at 3740–3745 cm^{-1}, whereas OH groups bridging one Al$_{IV}$ and one Al$_{VI}$ (type IIa) have a charge of $+0.25$ and may explain the band at 3730–3735 cm^{-1}. Triply bridging OH groups (type III) have a charge of $+0.5$, seen as consistent with a band position of 3700–3710 cm^{-1}. The lower frequency broader bands were assigned to H-bonded species. According to this model, terminal OH groups, characterized by a negative net charge, should have basic properties, whereas triply bridging OH groups, characterized by a positive charge should be acidic. Knözinger and Ratnasamy *(294)* also considered which terminations of spinel-type structures should be most abundant and what variations are introduced to the above described types of OH groups through these terminations, in the absence of restructuring. The most abundant (110) plane can have two terminations. If the termination is (110)C, type Ia and IIb OH groups are expected, whereas with a (110)D termination, only type Ib is expected. The (100) plane should only have type Ib OH groups. The (111)A termination should lead to type Ia and IIa hydroxyl groups, whereas the (111)B termination should produce IIb and III hydroxyl groups.

The considerations of Chen and Zhang *(317)* and the computational work of Alvarez *et al. (318)*, both relating to the same spinel terminations, led to the conclusion that the number of families of structurally different surface OH groups may be higher than that reported by Knözinger and Ratnasamy. Indeed, experimental IR spectra are more complex than anticipated by their model *(319,320)*.

The model by Knözinger and Ratnasamy has been modified by Busca *et al. (247,321,322)*. These authors suggested that the width of the OH stretching band is also of diagnostic value: they assumed that the natural "stoichiometric" structure for an OH group is a terminal configuration, as is typical of covalently bonded OH groups like those in organic molecules or the silanol groups on silica surfaces. Bridging and triply bridging configurations are "perturbations," implying they are unstable structures and produce broader bands. Additionally, these authors found that the band at 3770 cm^{-1} observed for spinels of the series $Ni_xAl_2O_{3+x}$ gradually disappeared as spinel stoichiometry (i.e., x equal to 1) was approached. On the basis of these considerations and observations, Busca *et al. (247)* concluded that all of the bands above 3710 cm^{-1} arise from terminal species with the following assignments: the highest frequency band (3795–3780 cm^{-1}) is associated with OH groups attached to tetrahedrally coordinated cations, the intermediate frequency band (about 3770 cm^{-1}) is attributed to similar species located near a cation vacancy, and the band at about 3730 cm^{-1} is assigned to terminal OH groups on octahedrally coordinated cations. Thus, in agreement with a previous assignment by Morterra *et al. (246)*, terminal OH groups on Al_{IV} sites should absorb at a higher frequency than terminal OH groups on Al_{VI} sites. This interpretation switches the assignments originally made by Knözinger and Ratnasamy *(294)*. The interpretation is corroborated by the observation that most α-Al_2O_3 samples, which contain aluminum ions only in sixfold coordination in their bulk structure, have the most intense IR band at 3730 cm^{-1}. The assignments of Busca *et al. (247)* were also based on the comparison of the spectra of alumina with those of a number of other oxides, among them the polymorphs of iron oxide (α-Fe_2O_3 and γ-Fe_2O_3), gallia (α-Ga_2O_3 and β-Ga_2O_3 *(169)*), and chromia (α-Cr_2O_3) as well as a number of aluminate, ferrite, and chromite spinels *(247)*.

Additional investigations were published by Tsyganenko and Mardilovich *(323)*, who inferred that on a completely hydrated surface, there should be no more than six configurations of OH groups characterized by differences in number and coordination of the surrounding metal atoms; this statement is valid for defect-free low index planes, but also when crystal edges and corners or cation vacancies are considered. However, the number of configurations increases dramatically upon dehydroxylation, and the authors concluded that the actual situation may be much more complex than assumed in earlier reports. In any case, they assigned the bands in the range of 3750–3700 cm^{-1} to bridging species, thus supporting the interpretation by Knözinger and Ratnasamy. These investigations were reviewed by Morterra and Magnacca *(293)* and Lambert and Che *(324)*.

Fripiat *et al. (325)* conducted molecular dynamics simulations, and, on the basis of the results, attributed all of the bands to bridging hydroxyl groups differing only in the coordination of the bridged cations. Digne *et al. (159)* approached the assignment of the OH groups of aluminas by density functional theory calculations and also attempted to model the adsorption of probe molecules using the nonspinel model for the γ-Al_2O_3 bulk and nonrelaxed (110) and (100) surfaces. The authors assigned the bands at approximately 3790, 3770, and 3730 cm^{-1} to three different terminal hydroxyl groups that are bonded to Al_{IV} sites on the (110) face, to Al_{VI} sites on (100) face, and to Al_V sites on the (110) face, respectively. However, this interpretation neglects the differing behavior of the band at 3770 cm^{-1}, that is, the higher reactivity of the corresponding OH group (as described above).

Dyan *et al.* modeled spinel-type surfaces and allowed relaxation *(326)*. According to these authors, even on ideal (i.e., nonrelaxed) surfaces, the local environments of the OH groups are more complicated than those formulated in the empirical models, and strongly influence the vibrational stretching modes of the hydroxyls. Cation vacancies were found to introduce surface disorder and influence the properties of the surface OH groups. These authors confirmed that terminal OH groups on tetrahedral aluminum ions should absorb at higher frequency than OH groups on octahedral aluminum ions and that vacancies nearby should lower the OH stretching frequency of terminal groups, as proposed by Busca *et al. (247)*.

Recently, Ferreira *et al. (284)* performed *ab initio* calculations with a nonspinel model and attempted an assignment of IR bands and NMR signals to OH groups on (100) and (110) faces. The results make the picture more complicated, because the trend of decreasing frequencies in the order terminal > bridging > triply bridging OH groups, on which all other authors agree, was not reproduced.

The author of this review *(285,304)* suggested that most if not all residual hydroxyl groups on outgassed, high-surface-area aluminas (1–3 OH nm^{-2}, 200 $m^2 g^{-1}$), can actually be located on corners and edges; thus, few or none would be on flat surfaces. These OH groups on corner and edge sites are assumed to be essentially terminal, and responsible for the three sharp and temperature-resistant bands at about 3790, 3770, and 3730 cm^{-1}. The assignment of the bands at 3790 and 3730 cm^{-1} to terminal species on nearly tetrahedral and nearly octahedral sites is well established, and the possible role of nearby cation vacancies in creating the "unstable" OH structure responsible for the band at 3770 cm^{-1} is quite likely. The location of very stable, terminal hydroxyl groups on corners seems indeed very likely, because the sites characterized by a high degree of coordinative unsaturation are found primarily at corner and edge positions, and to some extent, near defects. The density of active sites on alumina seems to be consistent with this hypothesis. The OH groups located on intact flat surfaces are proposed to be predominantly bridged and triply bridged, and also H bonded to each other.

In any case, as remarked by Dyan *et al.* *(326)*, it is difficult to attribute individual bands to particular species located on one of the predominant faces of alumina. There is ambiguity because several structure types can exist in very similar form on different faces. This problem is illustrated by the fact that γ-Al_2O_3 samples with very different morphologies have very similar spectra in the OH stretching region. On the other hand, as stated in Section 2.2.2, the most frequently exposed faces of aluminas like γ-Al_2O_3, δ-Al_2O_3, or η-Al_2O_3 are still not ascertained. The sites that are the most exposed, that is, those with the highest degree of under-coordination, should be readily hydroxylated, and the formed OH groups should be resistant to dehydroxylation. Such sites are likely to be located on edges rather than on plain faces. In fact, edges are possibly more relevant than previously thought, because recent TEM data show that the "plain faces" are actually faceted and characterized by a very high density of edges *(183)*.

As discussed above, the OH groups that are most stable (toward thermal decomposition) do not show the expected basicity; on the contrary, they show weakly acidic behavior. This result is not really surprising if the covalency of the Al—(OH) bond predominates over its ionicity.

A weakly acidic behavior is typical of the silanol groups of silica as well as of the B–OH groups of boria-containing materials. Considering that aluminum is at the border between metal and semimetal character and that aluminum hydroxides are typically amphoteric, both bonding concepts may apply. In other words, the properties of bulk alumina are better explained in terms of ionicity of the Al^{3+}—O^{2-} bonds, whereas the Al–OH groups at the surface may have a more covalent character.

As stated earlier, the IR spectra of aluminas outgassed at ambient or slightly higher temperature usually also show a prominent broader band at about 3680 cm^{-1} and a shoulder (sh) toward lower frequencies. The species associated with these bands are clearly less resistant to heating and outgassing than those discussed above, and are responsible for most of the water desorption in the range of stability of the γ- and η-alumina phases. Assignment of these bands to bridging and triply bridging species appears reasonable from the point of view of both chemistry and spectroscopy. The very broad absorption that extends into the region of $3600–2800 \text{ cm}^{-1}$ (Figures 3.11, 3.20, and 3.22) is certainly also due to OH stretching vibrations. The species responsible for this absorption are without doubt H bonded. They are not perturbed by adsorption of probe molecules and likely located in a deeper subsurface level. The species mostly disappear during the phase transition into θ-Al_2O_3 (Figure 3.23), and may contribute to the disorder of the spinel-type structures and to the defect fluorescence effect seen in Raman spectra. To our knowledge, the actual location and the structure of these species have not been addressed so far.

3.2. Lewis acid sites on alumina surfaces

The catalytic activity of "transition" aluminas, in particular of γ-, η-, δ-, and θ-Al_2O_3, is in most cases (i.e., reactions) undoubtedly related to the Lewis acidity of a small number of low coordinated surface aluminum ions. The Lewis acidity is associated with the ionic nature of the surface Al—O bond (301). The variations among these polymorphs regarding the catalytic activity for acid-catalyzed reactions (207) mainly arise from variations in surface area and morphology, since the quality of the acid sites is very similar according to spectroscopic experiments. For example, γ-Al_2O_3 is usually slightly less active than η-Al_2O_3, but both are more active than δ-Al_2O_3 or θ-Al_2O_3, which usually have lower surface areas. The Lewis acid sites have been extensively characterized by adsorption of probe molecules combined with analysis by spectroscopy or calorimetry. The surface density of very strong adsorption sites, defined as sites with a heat of adsorption for ammonia of more than 200 kJ mol^{-1}, is reported to be around 0.1 sites nm^{-2} *(229,327,328)* for aluminas outgassed at typical conditions (400–550 °C). Taking into account the bulk density of $\gamma\text{-Al}_2\text{O}_3$ ($\approx 3.65 \text{ g cm}^{-3}$), it is easy to calculate that, at most, one out of 50–100 surface cation sites acts as a strong Lewis site on γ-alumina outgassed at 400–550 °C, while the majority is still hydroxylated or not highly exposed (i.e., not highly undercoordinated). The total surface density of Lewis acid sites is on the order of 0.5 sites nm^{-2} and thus significantly higher than the surface density of strongly acidic Lewis sites. The number of strong Lewis acid sites is lower than the number of the surface hydroxyl groups eliminated by dehydroxylation, showing that only a fraction of Al^{3+} ions freed by water desorption become strong Lewis sites. This result suggests that part of the surface may reconstruct after dehydroxylation to lower the surface free energy.

The surface density of the strongest Lewis acid sites tends to slightly decrease with increasing annealing time and temperature; that is, the maximum temperature used during calcination of the alumina may play a role. The sequence of phase transformations $\gamma\text{-Al}_2\text{O}_3 \rightarrow \delta\text{-Al}_2\text{O}_3 \rightarrow \theta\text{-Al}_2\text{O}_3$ is accompanied by a decreasing surface area. The number of strongest acid sites per gram decreases in this sequence more than justified by the decrease of surface area, while catalyst stability increases. This trend can be ascribed to a decrease of the concentration of defects.

Investigations of the surface Lewis acidity of aluminas have mainly been performed by adsorbing basic probes after previous dehydroxylation of the samples by outgassing. Based on spectroscopic results, most authors agree that at least three different types of Lewis acid sites (with weak, medium, and high acid strength) exist on transition aluminas *(293)*.

The spectra of pyridine adsorbed on $\gamma\text{-Al}_2\text{O}_3$ (Figure 3.24) show three components for the 8a band at 1623, 1616–1610, and 1595–1590 cm^{-1}, which are usually attributed to three different species adsorbed on three different Lewis sites. Some authors simulated the adsorption of pyridine on aluminum oxide clusters and found that the calculated shifts of the vibrational modes of pyridine adsorbed on tri-coordinated Al^{3+} ions (giving a tetrahedral complex) agree with those measured experimentally for pyridine adsorbed on the strongest Lewis sites *(329,330)*; the agreement was seen for the 8a mode at 1623 cm^{-1} and the 19b mode at 1455 cm^{-1}. Liu and Truitt *(296)* emphasized the close proximity of Lewis acid sites to surface OH groups on $\gamma\text{-Al}_2\text{O}_3$. Lundie et al. *(314)* identified four different Lewis

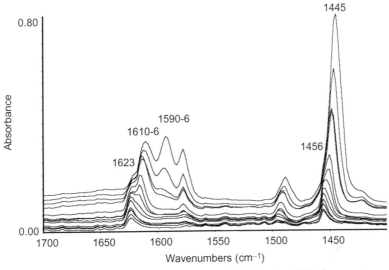

Figure 3.24 IR spectra of pyridine adsorbed on γ-Al$_2$O$_3$ (Puralox 200) after previous out-gassing of the alumina at 500 °C. Outgassing after pyridine adsorption was performed at increasing temperature from 30 to 250 °C (from top to bottom).

acid sites on η-Al$_2$O$_3$, with coordinatively unsaturated octahedral aluminum cations constituting the weakest acid sites and coordinatively unsaturated tetrahedral aluminum cations constituting the three stronger acid sites. Three of these sites were considered to be associated with three different types of hydroxyl groups, and the pyridine band at 1613 cm^{-1} was believed to be associated with two different sites.

Another commonly used probe molecule is CO, which is adsorbed at low temperature. CO adsorbed on the Lewis acid sites of γ-Al$_2$O$_3$ dehydroxylated at a moderate temperature (i.e., at less than 550 °C) is characterized by a band that continuously shifts with decreasing coverage from about 2185 cm^{-1} to a little above 2200 cm^{-1}. In addition, there are two weak components at about 2220 and 2230 cm^{-1} (Figure 3.25), which become more prominent if the preceding outgassing procedure is performed at higher temperatures. According to Gribov *et al.* *(186)*, the band at 2170–2205 cm^{-1} can be ascribed to CO adsorbed on Lewis sites located on the plain surfaces of the predominantly exposed faces, whereas the components at about 2230 and 2215 cm^{-1} may indicate CO interacting with Al^{3+} ions exposed on corners and edges or steps, respectively. The more strongly acidic sites are also able to attract molecular hydrogen (H$_2$) at 20 K, giving spectroscopically distinguishable species. Using molecular

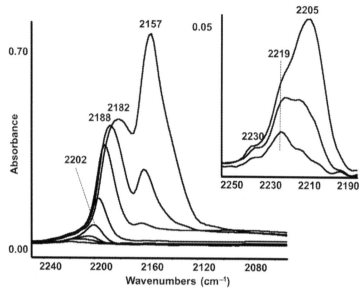

Figure 3.25 IR spectra of CO adsorbed on γ-Al$_2$O$_3$ (Puralox 200) at a temperature of 130 K after previous outgassing of the sample at 500 °C. After adsorption, the sample cell was evacuated at increasing temperatures from −143 to −93 °C (sequence of spectra from top to bottom). In the inset to the right, the spectra recorded at the higher outgassing temperatures (−123 to −93 °C) are expanded.

nitrogen (N$_2$) as a probe, Wischert *et al.* *(331)* demonstrated the presence of four different Lewis sites on a fumed alumina sample and proposed that the site with the highest acid strength involved tri–coordinated Al^{3+}.

In several recent publications, Peden and coworkers inferred that pentacoordinated aluminum ions, which are clearly visible in ^{27}Al NMR spectra of γ-Al$_2$O$_3$, determine most of the surface and bulk chemistry of this material. The authors proposed pentacoordinated aluminum ions to act as structural promoters for phase transitions *(150)*, for anchoring of platinum oxide *(332)*, as well as for the sintering of supported platinum metal particles *(333)*. The authors supported their conclusion with calculations based on the nonspinel structural model of γ-Al$_2$O$_3$ by Digne *et al.* *(158)*; the results indicated that pentacoordinated aluminum adsorption sites may be located on the (100) face. The same group used the (100) and (110) faces of this nonspinel-type structure to model the adsorption of NO$_x$ *(334)*.

It is evident that the Lewis acid sites are coordinatively unsaturated Al^{3+} ions at the surface, which, in principle, may be pentacoordinated (i.e., octahedral with one missing ligand and hence one free coordination site), tetracoordinated (normal tetrahedral sites that can expand their coordination or octahedral with two free coordination sites) and trigonal or tricoordinated (octahedral with three free coordination sites or tetrahedral with one free coordination site). Based on spectroscopic experiments, most authors *(186,314,331)* suggest that singly unsaturated, tetrahedral aluminum cations, that is, tricoordinated Al^{3+} ions, should constitute the strongest Lewis sites on aluminas. Al^{3+} ions bound to four or five oxygen ions should constitute the other sites, although some of these sites may involve nearest OH groups as proposed by Lundie *(314)* and also by Knözinger and Ratnasamy *(294)*. In their review *(294)*, the authors illustrated the possible structures of different Lewis sites on the most abundant terminations of spinel-type aluminas assuming no restructuring and full dehydroxylation. For a termination designated (110)C, tetrahedral cations with one free coordination and octahedral cations with one free coordination (pentacoordinated cations) are expected with fractions of 50% each. For the termination (110)D, only octahedral cations with two free coordination sites are expected. The (100) plane contains pentacoordinated cations, but fully saturated tetrahedral cations are located slightly recessed and could enlarge their coordination by adsorbing molecules. More recent theory-based calculations of the surfaces of spinel-type structures essentially agree with these conclusions *(335–337)*, although much disorder and a more complex situation was found by Dyan *et al.* *(179,326)*. A different view was presented by Vijay *et al.* *(338)*, who claimed that vacancies should be more stable in the bulk than at the surface, and massive surface reconstruction should occur.

Using their own nonspinel-type structure, Digne *et al.* *(159)* sought to corroborate the existence of different Lewis acid sites on γ-Al_2O_3 by modeling CO and pyridine adsorption on tricoordinated Al^{3+} ions on the (110) face, tetracoordinated Al^{3+} ion on the most abundant (110) face, and pentacoordinated Al^{3+} on the (110) and (100) faces. In a more recent paper *(192)*, based on calculations using the same nonspinel model, this research group concluded that tricoordinated Al^{3+} Lewis acid sites, which are present as metastable species exclusively on the major (110) termination of γ- and δ-Al_2O_3 particles, correspond to the "defect" sites, which are held responsible for the unique properties of "activated" (i.e., thermally pretreated) alumina. The authors hypothesized that these sites would form only by dehydroxylation at high temperature (700 °C), when surface reconstruction would become negligible. Moreover, they presumed tricoordinated Al^{3+} to be so acidic and the (110) surface so reactive that full dehydroxylation would be impossible under mild conditions. This behavior was cited as the reason for the usually abundant hydroxylation of the alumina surface. An open question, however, is why tricoordinated aluminum species on outgassed alumina surfaces have not been revealed by ^{27}Al NMR.

In this context, it is worth mentioning that a slightly different situation is found when pyridine is used, a much stronger and much less volatile basic probe than CO. When adsorbing pyridine, the strongest Lewis sites appear to be present after mild pretreatments or even without previous dehydroxylation. Additionally, alumina is used in a number of commercial processes and presumably acts as a Lewis acid catalyst, while the reaction conditions are such that the surface is certainly largely hydroxylated. This situation applies for the Claus reaction (performed at 200–300 °C with a wet feed), the synthesis of DME (performed at 250–280 °C and 0.04–0.05 MPa, with water being a main reaction product), the synthesis of monochloromethane from methanol and hydrogen chloride (carried out on alumina at 280–350 °C, 3–6 bar, producing also water) and alkylations of phenol to o-cresol and 2,6-xylenol (performed using methanol at 300–400 °C, again with water as a product of the reaction). Thus, Lewis acid catalysis occurs in wet conditions, provided the basicity and volatility of the reactants is comparable with or higher than that of water. Molecules with such properties compete with water for adsorption sites, and the adsorption–desorption equilibria of all involved species must be considered. Molecules such as light alcohols can adsorb on Lewis sites by displacing water and hydroxyl groups. In contrast, other reactions, such as double bond isomerization of olefins (performed at very low temperature but after activating the alumina catalyst *(339)*), olefin skeletal isomerization (performed at about 480 °C in dry conditions *(304)*), as well as H/D isotopic exchange in methane *(306)* need either high temperature or previous dehydroxylation and dry conditions. Since the basicity of these reactants (light olefins, methane) is much lower than that of water, dehydroxylation is needed for the appearance of Lewis acidity and catalytic activity. Remarkably, alumina exhibits Lewis acidity also in contact with liquid water *(340)*. These results imply that dehydroxylation can also occur by displacement, not only by water desorption.

Although, to our knowledge, a detailed investigation of the surface properties of χ-Al$_2$O$_3$ has never been reported, there are indications that it should have a surface acidity comparable to that of γ-Al$_2$O$_3$ *(208,215,224)*, consistent with the idea that these two polymorphs have similar structures *(178)*. As for κ-Al$_2$O$_3$, the presence of reasonably strong Lewis acid sites, comparable with the medium-strength sites found on γ-Al$_2$O$_3$ *(205)*, has been reported on the basis of an IR spectroscopic analysis of adsorbed pyridine. However, the surface concentration of these sites is reported to be low *(206)*, which may relate to the low ratio of tetrahedral to octahedral Al^{3+} ions in the bulk (ratio of 0.25).

Notwithstanding the high free surface energy of corundum, which is believed to "destabilize" fine powders of this polymorph *(114,160)*, the Lewis acid strength is usually reported to be weak *(186,207,247,248,250,251)*. The weak acidity has frequently been attributed to the exclusively octahedral coordination of aluminum ion in the bulk, and the presumed resulting similar configuration of unsaturated cations at the surface.

3.3. Basic sites and acid–base pairs on alumina surfaces

The origins of the peculiar properties of aluminas certainly are the ionicity of the Al^{3+}—O^{2-}—Al^{3+} bond and the small size of the Al^{3+} cation and its medium charge, which together make this ion very polarizing *(301)*. As a counterpart to these hard cations, oxide anions also exist on alumina surfaces, thus providing surface basicity *(341)*. Thus, the truly exceptional sites that aluminas can provide for catalytic reactions are very likely anion–cation pairs, which have a very high activity and work synergistically, as discussed by Knözinger and Ratnasamy *(294)*. The basic sites may be oxide anions or hydroxyl species. The weak adsorption of carbon dioxide on the surface of aluminas *(287)* is indicative of low basicity *(342)*. In fact, adsorption and mechanistic investigations suggest that the acidity of the Al^{3+} ions plays a larger role than the basicity of the anions. However, both together give cation–anion pairs that are very active for dissociating molecules. These pairs promote the dissociative adsorption of alcohols *(343,344)*, which is the first step in their catalytic dehydration to ethers and olefins *(187,345–347)* and in the hydrochlorination of methanol to methylchloride. Similarly, they serve as sites for alkylchloride adsorption and dehydrochlorination *(348)*. The very strongly polar acid–base pairs allow the formation of the anion of acetonitrile, $^-CH_2$–CN, *(349)* and of the anion of acetone, $^-CH_2COCH_3$ *(350)*, as well as the extraction of hydrogen from allyl positions of olefins, thus producing allyl-aluminum species *(351,352)*, which are likely intermediates in double bond isomerization. Moreover, such pairs can accommodate the intermediates formed through methane dissociative adsorption in isotopic exchange reactions *(192)*. According to the experimental and theoretical work of Wischert *et al.* *(192)*, outgassing at a temperature of 700 °C is needed to create active sites for methane activation.

Acid–base pairs are also likely to be involved in the adsorption of SO_2 *(353)*, which results in sulfite and bisulfite *(354–356)*. It has been reported that the adsorption of SO_2 on η-Al_2O_3, γ-Al_2O_3, and χ-Al_2O_3 is strongly influenced by sodium impurities that enhance the basicity (225). This phenomenon is closely related to the surface chemistry occurring on the alumina catalyst during Claus reaction (see Section 4.1).

3.4. Surface impurities and selective poisoning

The effect of surface impurities on the acidic properties and on the related spectroscopic data is frequently neglected. The content of sodium as well as of other alkali metals and alkaline-earth metals, such as potassium and calcium, in alumina strongly depends on the preparation method. Aluminas derived from aluminum metal via the alkoxide route have a low sodium content (≤ 40 ppm Na_2O), and fumed aluminas also have low sodium content (65 ppm (98)). In contrast, aluminas obtained by decomposition of boehmite previously precipitated from sodium aluminate solution, may have a significantly higher sodium content (3000 ppm) (176). Sodium (or other alkali metal or alkaline-earth metal) impurities decrease the number of the active sites and also possibly decrease the acid strength, as a result of induction effects (357). Thus, these cationic impurities can decrease the activity of alumina for acid-catalyzed reactions. Even when their total concentration is small, sodium cations have a significant poisoning effect: they concentrate at the surface because of their large size that does not allow them to penetrate into the cubic close packing of oxide anions.

For example, the activity of alumina for n-butylene isomerization shows a positive trend with increasing total integrated intensity of the OH stretching band of the surface hydroxyl groups and a negative trend with increasing sodium content (as derived by chemical analysis) (169). As already stated, the sodium content is critical for aluminas applied for the Claus reaction, because sodium enhances the basicity and thus promotes SO_2 adsorption, but it also decreases the reactivity of the adsorbed sulfite species (225).

Intentional doping with alkali metal ions is widely used to reduce the activity of alumina in supported catalysts. It has also been shown that doping with cations of alkali metals, alkaline earth metals, and rare-earth metals increases the stability of alumina with respect to phase transitions and loss of surface area. To reduce the surface activity and to increase the stability of alumina for use as a support of metal catalysts in commercial applications, phases that have already gone through most of the possible structural evolution (θ-Al_2O_3 or α-Al_2O_3) with moderate to low surface area (100 to 5 $m^2 g^{-1}$) are needed; doping with alkali metal, alkaline-earth metal, or rare earth metal cations may also be useful. Other impurities may be present in aluminas and may also affect the properties, for example chlorine in fumed aluminas *(97,98)* and titanium and iron in Ziegler-type aluminas *(73)*. However, hyperpure materials are now available, with a total impurity content of only a few ppm; these materials are prepared by a modified Ziegler-type procedure *(73)*.

4. INDUSTRIAL APPLICATIONS OF ALUMINAS AS CATALYSTS

There are not many industrial applications of alumina itself as a catalyst, but some of them are very important. It is remarkable that most applications are assumed to require the Lewis acidity or the acid–base pairs of alumina, but are performed in the presence of water at moderate temperatures, implying that the surface is still largely hydroxylated.

For many reactions, in particular alcohol dehydrations, olefin isomerization, methylchloride synthesis, and phenol alkylations, the proton forms of zeolites are reported to be more active than alumina. However, alumina is preferably applied because of fewer byproducts and slower coking. Thus, the strong Lewis acidity and perhaps also the medium–weak Brønsted acidity of alumina provide an optimal balance between activity and stability; these sites have sufficient activity while they deactivate only slowly.

4.1. The Claus reaction

The most important application of alumina as catalyst is in the catalytic steps of the Claus process (Equation 3.1), which is used to produce sulfur from H_2S in refineries *(358)*.

$$H_2S + 2SO_2 \rightarrow 3S_8 + H_2O \qquad (3.1)$$

Aluminas for this application are characterized by a high specific surface area of $300–400 \text{ m}^2 \text{ g}^{-1}$, pore volumes of 0.5 ml g^{-1}, of which 0.1 ml g^{-1} are due to macroporosity (>750 Å), and a loss on ignition of 5.5–6.5% wt/wt *(359,360)*. The alumina catalysts may be promoted by iron to reduce deactivation by sulfation. In the Claus process, several reactors are placed in series to progressively complete the reaction, with intermediate cooling, condensation of the sulfur product, and reheating. In the first reactor, titania is mostly used because it has a higher activity for COS and CS_2 hydrolysis than alumina, which is used in the other reactors at operating temperatures of about 200–250 °C. The reactors are horizontally oriented, to limit pressure drop as a result of deposition of sulfur on the catalyst.

The mechanism of the Claus reaction is complicated, and the sequence of the surface reaction steps is not fully elucidated *(361)*. The primary step is certainly the strong adsorption of SO_2 on acid–base pairs with formation of sulfite and bisulfite species *(354–356)*, which later react with gaseous or weakly adsorbed H_2S. The strength of the SO_2 adsorption depends on the surface basicity, which is influenced by the amount of sodium present *(225)*; however, it must be taken into account that the reactivity of the formed sulfites is lower on more basic surface sites. Thus, the acid–base properties of the catalyst play a key role in this reaction. According to Clark *et al. (362)*, sulfate and thiosulfate species apparently form in addition to the sulfites. It was proposed that sulfate and thiosulfate react to form $[HS_4O_4]^-$ ions, which then react with H_2S to form the S_3 sulfur polymorph, which is subsequently converted into cyclic S_6 and S_8 molecules.

4.2. Dehydration of alcohols

Aluminas are active and very selective catalysts for the dehydration of alcohols to olefins and to ethers *(346,347)*. In particular, alumina-based catalysts are reportedly used in industrial processes for the production of DME from methanol, as shown in Equation (3.2).

$$2CH_3OH \rightarrow CH_3OCH_3 + H_2O \qquad (3.2)$$

According to the literature, γ-alumina with controlled porosity and impurities is the catalyst in the Toyo process, performed at 300 °C, 1.1 MPag, and a gas hourly space velocity (GHSV) of $1700\ h^{-1}$ *(359)*. Alumina-based catalysts are reported to be used also in the Topsøe DME process *(364)* which can be the first step of the Tigas Methanol-to-Gasoline process *(365)*. The methanol dehydration using alumina-based catalysts is reported to be performed at 250–280 °C and 0.04–0.05 MPa, as a first step in the methanol-to-olefin process from Mobil *(366)*, and also in the methanol-to-propylene process *(367)*. To increase the rate of this reaction, it is necessary to increase the density of Lewis acid sites *(368)*, and a bimodal pore size distribution can be useful *(369)*. It has also been reported that for this reaction η-Al_2O_3 is more active than γ-Al_2O_3 *(207,215)*.

IR spectra show that methanol adsorbs on partially dehydrated Al_2O_3 in different forms, which include terminal and bridging methoxy groups, O-bonded undissociated chemisorbed methanol, and H-bonded undissociated methanol *(343,344)*. The DME synthesis reaction on aluminas has been the subject of several investigations, and two principal mechanisms have been proposed: the coupling of two methoxy species, or the reaction of methoxy species with adsorbed undissociated methanol species *(370–372)*.

In the 1960s, aluminas were used for the production of ethylene by dehydration of ethanol *(373,374)*.

$$CH_3CH_2OH \rightarrow H_2C = CH_2 + H_2O \quad\quad\quad (3.3)$$

This reaction (Equation 3.3), which occurs at 250–300 °C with almost total yield *(187)*, might find renewed interest in the future to convert bio-ethanol produced by fermentation into "bioethylene" *(375,376)* in the frame of a new industrial organic chemistry based on renewables. Ethanol dehydration has also been used recently as a test reaction for the investigation of the surface properties of aluminas *(187,377–379)*. At low conversions, ethanol can be converted into diethylether with high selectivitiy. IR spectra show that ethanol adsorbs in the form of ethoxy groups, which are formed either by dissociation on Lewis acid–base pairs or by substitution of hydroxyl groups *(187)*.

As reported by de Klerk *(380)*, aluminas are applied in the refining of High Temperature Fischer Tropsch (HTFT) syncrude. They catalyze several reactions, among them the dehydration of long-chain alcohols. For example, mixed alcohols from the HTFT aqueous product were dehydrated on an η-Al$_2$O$_3$ catalyst to produce mixed alkenes with high selectivity toward internal olefins when the reaction was performed at 350 °C *(381)*. η-Al$_2$O$_3$ catalyzes the conversion of terminal linear C5–C12 alcohols to linear ethers and to linear terminal olefins and some cis-2-olefins at reaction temperatures of 250–350 °C *(382)*, whereby the ratio of terminal to internal olefins increases with increasing temperature and time on stream. γ-Al$_2$O$_3$ was found to be active in catalyzing the conversion of the terminal alcohols 1-hexanol and 1-octanol to the corresponding terminal olefins at 300 °C *(383)*.

4.3. Skeletal and double bond isomerization of olefins

Alumina has been reported to be an active catalyst for olefin skeletal isomerization *(384)*, for example, for the production of isobutene from *n*-butenes or the production of isopentene from *n*-pentenes. Axens commercialized a process (called ISO5™) and an alumina catalyst (IS463 *(385)*) for this application. The reverse reaction may also be of interest, that is, the formation of linear from branched isomers. The skeletal isomerization occurs at relatively high temperature on alumina (450–550 °C), but with high selectivity and relatively little coking *(303,386)*. The alumina catalyst is tolerant of water and oxygenates and can be applied for hydrocarbon mixtures such as the

HTFT syncrude (380). It has been hypothesized that, in these conditions, the reaction may occur through a carbenium ion mechanism, thus demonstrating the significant Brønsted acidity of the alumina OH groups (303). Slight silication of the alumina improves the performance, likely by enhancing Brønsted acidity without creating strongly acidic sites (262). Treatment of alumina with HCl and H_2SO_4 also improves the catalytic activity but results in significant deactivation by coking (387).

The double bond isomerization of olefins can be performed at very low temperature, for example at 80 °C (339,388), but the alumina needs to be outgassed at temperatures above 300 °C prior to the catalytic reaction. This pretreatment implies that the surface is partly dehydrated and contains active cation–anion pairs. Infrared spectra clearly showed the formation of allyl species at room temperature on partially dehydrated γ-Al_2O_3 (351), whereas recent ^{13}C NMR spectroscopic investigations suggest that both allylic and alkoxy species are formed from propene and butenes on γ-Al_2O_3 (389).

As reported by de Klerk (380), aluminas are and have been widely used in the refining of Fischer Tropsch syncrude, catalyzing several reactions including double bond and skeletal isomerization of olefins. It has been shown that olefins present in the HTFT syncrude (which also contains oxygenates such as ketones and alcohols) tend to isomerize from terminal to internal olefins at 350 °C on η-Al_2O_3 (381).

4.4. Other reactions

Aluminas are reportedly used in the production of chloromethane from methanol and hydrogen chloride (390–392).

$$CH_3OH + HCl \rightarrow CH_3Cl + H_2O \qquad (3.4)$$

The reaction is performed with alumina as a catalyst at a temperature of 280–350 °C and a pressure of 3–6 bar. Both γ-Al_2O_3 (393) and η-Al_2O_3 (394) are active for this reaction. It was proposed that the reaction occurs mainly on Lewis acid sites, and the activity could be improved by the presence of $ZnCl_2$ (395).

Like the reactions discussed above, this reaction involves an alcohol as the reactant. Probably, these reactions proceed through a common first step, namely the activation of the alcohol through chemisorption on the Lewis acid sites of alumina. This interaction takes place via one of the oxygen lone pairs of the alcohol, and is the first step of alcohol dissociation on Lewis acid–base pairs.

Aluminas may be used for the dehydrofluorination of alkylfluorides, which are byproducts of the HF-catalyzed isobutane–butylene alkylation process. Fluoroalkanes are converted to olefins on alumina at temperatures of 170–220 °C. HF is adsorbed on the alumina and aluminum fluoride is formed; as a consequence, regeneration is needed every 6 months *(396)*.

Alkylations of phenol by alcohols such as methanol or isopropanol to produce cresols and xylenols and isopropyl phenols is performed in the vapor phase at temperatures of 200–400 °C using alumina catalysts *(397–399)*.

Interesting new applications of aluminas are in the field of catalysis for deoxygenation and upgrading of bio-oils *(400)*. This chemistry has parallels to the use of alumina in upgrading of HTFT syncrude, which is also rich in oxygenates *(380)*. The Lewis acidity of alumina enables the catalytic decomposition of esters such as triglycerides *(401,402)* without inducing much coking. Considering in addition its strong activity for the decomposition of alcohols, alumina may be a suitable catalytic material for the formation of hydrocarbons from oxygenates. A minor drawback of the use of alumina for this chemistry is its high catalytic activity for the ketonization of acids.

5. ALUMINAS AS SUPPORTS FOR CATALYSTS

5.1. Introduction into supported catalysts

Supported catalysts are produced for several reasons. One of them is that by supporting precious metals it is possible to reduce the amount needed to obtain the same catalytic effect, simply because the metal is maintained in a dispersed state. Often, there are additional benefits of supporting a catalytically active phase. The catalytic activity and/or selectivity may be improved as a result of the interaction with the support. Additionally, supported catalysts may be more stable than bulk catalysts; they may sinter more slowly (as a result of altered solid-state chemistry) and may be less prone to poisoning.

Aluminas are very good catalyst supports because of their ability to disperse the supported phase, their high thermal stability (depending on the crystal phase), and their moderate price. The ability to disperse is associated with the acid–basic character of Al_2O_3 surfaces. The impregnation or deposition of the active phase, or of its precursor, onto the support surface is a true chemical reaction. Alumina, with its very stable surface OH groups, its Lewis acidity, and the very high polarity of the surface acid–base pairs, provides specific sites for anchoring cationic, anionic, and metallic species. There are indications that supported species are located at specific sites on the alumina surface. Surface defects, corners, and edges are likely playing an important role for the dispersion of active species during the various preparation steps, including impregnation and deposition, subsequent drying and calcination, and reduction in the case of metal catalysts, or sulfidation in the case of supported sulfide catalysts. The presence of additional species during the contact of the precursor of the active phase with the support may also be a relevant factor. This question is currently researched, in particular in the field of hydrotreating catalysts (403), because it was found that the presence of organic species during the deposition of the active phase changed the distribution or location of the supported species and resulted in a significantly improved performance of the final catalyst (404). While a substantial amount of empirical information is available in the literature, the science of supporting catalysts is still in its early stages, and a detailed knowledge of the chemistry occurring during the preparation of supported catalysts has yet to be established.

5.2. Alumina-based catalyst supports

Aluminas find wide application as supports of catalysts. For applications requiring relatively low reaction temperatures (of less than 500 °C), high-surface-area γ-, δ-, or η-Al_2O_3 can be used. In many materials, some alkali metal or alkaline-earth metal species are present, that stabilize the support against sintering and phase transition, and decrease the surface acidity.

An important example are alumina-supported Co–Mo and Ni–Mo sulfides, which constitute the active phases in catalysts for hydrotreating of middle distillates (403). It appears that in such catalysts, mostly pseudoboehmite-derived γ-Al_2O_3 is used as the support. According to literature data, catalysts for fluid catalytic cracking (FCC) gasoline desulfurization, which is performed at 260–340 °C and 5–30 atm, may contain 5–11 wt% molybdenum and 2–3 wt% cobalt supported on Al_2O_3 with a surface area of 220–240 $m^2\,g^{-1}$ (405). Catalysts for diesel fuel desulfurization to low-sulfur diesel (performed at 300–400 °C, 30–100 atm) may contain 11–20 wt% MoO_3, 1.6–6.0 wt% CoO or NiO or both, and 0.05–2.0 wt% Na_2O on alumina, with the surface area being in the range of 230–250 $m^2\,g^{-1}$ (406).

Several hydrogenations of olefins and aromatics are performed at relatively low temperature using supported metal catalysts, mostly alumina-supported nickel, cobalt, palladium, or platinum. For example, Pt/Al_2O_3 catalysts are applied in various vapor-phase benzene hydrogenation technologies, to produce cyclohexane or to reduce benzene content in gasoline. The benzene hydrogenation reaction is carried out at temperatures of

200–350 °C and at a pressure of about 30 bar. The catalyst used in the BenSat™ benzene saturation technology from UOP is based on an alumina support characterized by spherical particles, a surface area of about 160–200 m^2 g^{-1}, and an apparent bulk density of about 0.45–0.6. The concentration of platinum metal present on the catalyst may vary from 0.375 to 0.75 wt% *(407, 408)*.

Cobalt-based low temperature Fischer–Tropsch catalysts, applied at approximately 220 °C and 30 atm, are usually supported on high-surface-area γ-Al$_2$O$_3$ (150–200 m^2 g^{-1}) and typically contain 15–30% weight of cobalt. To stabilize them and decrease selectivity to methane, these catalysts may contain small amounts of noble metal promoters (typically 0.05–0.1 wt% of ruthenium, rhodium, platinum, or palladium) or an oxide promoter (e.g., zirconia, lanthana, cerium oxide, in concentrations of 1–10 wt%) *(409)*.

For reactions occurring at high temperature, medium-to-low surface area aluminas are used, which sometimes contain alkali metal cations or silica as support stabilizers, as well as metal activators.

Chlorinated alumina-supported metal catalysts are the typical catalysts used today for catalytic naphtha reforming, which is performed at temperatures of 480–550 °C *(410)*. Modern versions of this type of catalyst are multimetallic; the catalytic properties of platinum are improved by the addition of another metal, often rhenium. Further elements that may be added are tin, silicon, germanium, lead, gallium, indium, iridium, thorium, lanthanum, cerium, cobalt, and nickel. All these components are supported on chlorinated γ-alumina (with a surface area of 150–300 m^2 g^{-1}), which provides the acid function *(411)*.

Chlorination of alumina may be performed by feeding chlorinated hydrocarbons together with the gasoline feed. Some patents state that at least 10 ppm (by weight) of one or several alkaline-earth metals selected from calcium, magnesium, barium, and strontium can be present, whereby the total amount of the modifier does not exceed about 5000 ppm *(412)*. These additives modify the activity, selectivity, and stability of the catalyst. The typical composition of the R-98 catalyst from UOP, an industrial catalyst designed for fixed-bed, semiregenerative reforming units, is claimed to be 0.25 wt% platinum, 0.25 wt% rhenium, with proprietary promoters *(413)*. For application in continuous regenerative reforming, catalysts R-232 (with 0.375% platinum) and R-234 (0.290% platinum) are produced by UOP *(414)*.

Other examples of the application of alumina as a support are light paraffin dehydrogenation catalysts, which are used to produce propene and isobutene from propane and isobutane. The reactors are operated at temperatures above 550 °C. An alkali metal-doped Pt/Al_2O_3 catalyst with 0.3–0.5% platinum (such as the UOP Oleflex™ DeH-14 catalyst) and tin, zinc, or copper promoters is part of the UOP Oleflex technology *(415)*. The metals are supported on a medium-surface-area alumina (i.e., $S_{BET} = 100$–200 m^2 g^{-1}) *(416)*. Paraffin dehydrogenation can also be performed on potassium-containing alumina-supported chromia. For this application, silica-stabilized alumina supports are used *(261)*, and the optimal support for a fluidized bed process is reported to contain some silica (1.5% wt/wt), be a mixture of δ-Al_2O_3 and θ-Al_2O_3, and have a surface area of 100 m^2 g^{-1} *(81)*.

High temperature methanation is conducted to produce substitute natural gas from syngas, and the Topsøe MCR catalyst is reported to have stable activity up to a temperature of 700 °C for this reaction *(417)*. The literature indicates that this catalyst contains 22 wt% nickel on a stabilized alumina support, has a surface area of 50 m^2 g^{-1} in its fresh state, and a surface area of 30 m^2 g^{-1} in its used state *(418,419)*.

When very high reaction temperatures are required, refractory support phases are used such as α-Al_2O_3 with very low surface area (1 m^2 g^{-1}), magnesium aluminate spinel ($MgAl_2O_4$), calcium aluminate $CaAl_{12}O_{19}$, or calcium potassium aluminate $CaK_2Al_{22}O_{34}$ *(420)*. An example for such a reaction is the steam reforming of natural gas to produce syngas and H_2, which is performed at 750–900 °C and 30–50 bar with nickel as the active phase.

Inert supports, usually with low surface area, are used for metallic catalysts even for low temperature applications. The supports are warranted in the case of very reactive feeds, to avoid unwanted side reactions. An example is the hydrogenation of acetylene in C2 cuts from steam cracking. Ideally, acetylene is completely converted to ethylene, with low conversion of acetylene or ethylene to high-molecular-weight hydrocarbons (known as "green oil"), and with high stability of the catalyst toward poisoning by traces of CO and sulfur in the feed. To accomplish these goals, the metal should be very active and the support should be inactive. For these reasons, palladium is the preferred metal. The loading is small, for example 0.04% wt, and the support is a low-surface-area alumina, for example α-Al_2O_3 with a BET surface area of about 20 m^2 g^{-1} *(421)*.

The total hydrogenation of butadiene in C4 cuts presents a similar challenge; in this case oligomerization of the diene on the support has to be avoided. According to the literature, suitable catalysts are composed of 0.15 wt% palladium supported on α-Al_2O_3 characterized by a surface area of $8 m^2 g^{-1}$ (422). This type of reaction occurs at temperatures of 50–200 °C under H_2 pressure.

Also for oxidation reactions, the choice of the alumina support mainly depends on two criteria: the stability of the phase at the reaction temperature and the reactivity (or better the lack thereof) toward feed components and products. For example, ethylene oxychlorination to ethylenedichloride is performed at approximately 220–250°C and 5–6 atm in the presence of a γ-Al_2O_3-supported catalyst, which has a surface area of from 100 to 200 $m^2 g^{-1}$ and contains 10 wt% $CuCl_2$ and 3 wt% KCl, (423,424). Another example is a process called ammonia selective oxidation (ASO, or also selective catalytic oxidation, SCO), which converts small amounts of NH_3 from waste gases to N_2 at reaction temperatures of 150–300 °C. The process is used to abate the ammonia slip after a selective catalytic reduction process with ammonia or urea in diesel-engine-exhaust after-treatment (425). The patented catalyst consists of γ-Al_2O_3 (60–300 $m^2 g^{-1}$) loaded with 0.5–4 wt% platinum and 0.5–4 wt% vanadia and is coated onto the surface of a ceramic or metallic monolithic structure (426).

The selective oxidation of ethylene to ethylene oxide (EO) is performed on supported silver catalysts at temperatures of 250–280 °C, and a pressure of roughly 20 bar. In this process, it is necessary to avoid secondary reactions of EO. Typical industrial catalysts may contain 8–15 wt% silver dispersed on low surface area α-Al_2O_3 (0.5–1.3 $m^2 g^{-1}$) with a porosity of about 0.2–0.7 $cm^3 g^{-1}$. In addition, the catalyst may contain several promoters in varying amounts (ppm by weight): 500–1200 ppm alkali metal (mostly cesium), 5–300 ppm of sulfur as cesium or ammonium sulfate, 10–300 ppm of fluorine as ammonium fluoride, or alkali metal fluoride (427).

Catalysts for total hydrocarbon and volatile organic compounds (VOC) combustion in waste gases contain noble metals supported on alumina. The noble metals are platinum, palladium, combinations of platinum and palladium, or rhodium; and the typical content is 0.3–0.5 wt%. The BASF RO-25 catalyst, specified for VOC combustion, is reported to contain 0.5% palladium on θ-Al_2O_3 characterized by a surface area of 109 $m^2 g^{-1}$ (428).

6. OTHER APPLICATIONS OF ALUMINAS IN CATALYSIS

Aluminas also find applications in adsorption processes (429) and as noncatalytic material in catalytic processes, namely as "binder" or an "active matrix." This usage of alumina as a "separated" component is realized in the catalyst mixture that constitutes the moving bed of the FCC process. In this process, the catalytically active component is a faujasites-type zeolite, usually a rare earth-Y or MgHY zeolite. However, the mixture of powders used in FCC reactors contains several other components as cocatalysts, binders, and active matrices. High-surface-area aluminas, or their precursors such as boehmites *(430, 431)*, are added to the catalysts to act as nickel scavengers. The feed for this process usually contains nickel porphyrine compounds that are deposited somewhere; subsequently nickel ions are reduced to metal, thus creating a catalyst with unwanted dehydrogenation activity. Nickel porphyrines tend to react preferentially with the alumina surface *(432)*, where they are partially stabilized in the bivalent state, thus reducing the amount of metallic nickel that is produced on the catalyst.

Aluminas also play an important part in the preparation of structured catalysts. They can be present on or constitute the *structured support* of catalysts. Common shapes are monolithic honeycombs or open-cell foams. Aluminas can be used as *washcoat*, a high-surface-area layer that gives the geometrical framework a suitable morphology to support and disperse the catalytically active phase, or they can be used as *primer*, an intermediate layer between the geometrical support and the washcoat that acts as "glue" between the two layers *(433)*. Alumina plays a central role for the preparation of efficient automotive three-way-catalysts *(434)*.

Aluminas are also the precursors of fluorided and chlorided aluminas, which may be produced *in situ* (i.e., in the catalytic reactor) upon halogenation, as well as for silicated aluminas, borated aluminas, and other "modified aluminas," which are produced *ex situ* by chemical treatments.

7. FUTURE WORK

The data discussed in this review give only a partial idea of the relevance of aluminas for the field of heterogeneous catalysis. In fact, most of the applications of aluminas have existed for several decades. Notwithstanding their wide-spread use, many details concerning the structure and the behavior of aluminas are still far from established. In recent years, a large number of new preparation procedures have been proposed, many focused on tuning particle size and shape as well as the porosity of aluminas. This kind of work was possible because of the availability of highly developed electron microscopy techniques. For many of these nanostructured materials, no obvious practical applications have emerged. The application of modern experimental techniques has not significantly improved the knowledge of the crystal structure and behavior of alumina-based materials. Also the development of new theoretical concepts and computational methods has, in the opinion of the present author, not led to major successes in regard to alumina materials, since reliable, well established, and unanimously accepted information is still lacking. In our opinion, the understanding of the properties of alumina and of its behavior as a catalyst has made relatively little progress in the last 20–30 years.

We propose that future work should have the following objectives:

1. Expanding the knowledge of the structure of the catalytically most relevant materials, γ-Al_2O_3 and η-Al_2O_3, with particular emphasis on the nature and origin of the disorder present in these structures.

2. Expanding the knowledge of χ-Al_2O_3, another potentially very interesting polymorph, in terms of phase structure, formation, morphology, thermal stability, and surface and catalytic properties.

3. Completing the assessment of the surface energies of the different phases in various conditions (dry and wet) and of their role in determining the existence and the stability of the different polymorphs.

4. Expanding the knowledge of the role of phase transition kinetics in determining the existence of the different polymorphs.

5. Improving the knowledge of the structure and the role of surface defects, including edges and corners, in determining the adsorption and catalytic properties of aluminas. Investigations should involve the synergistic use of both experimental techniques and modeling to approach this problem.

6. Revealing the exact structure of the Lewis sites, that is, determining the state of Al^{3+} surface species by improved techniques such as ^{27}Al NMR. A consistent interpretation of the ^1H NMR spectra of the surface hydroxyl groups should also be formulated.

7. Investigating the role of the surface structures (hydroxyl groups, Lewis acid sites, basic sites, and combinations thereof) and their location (faces, edges, corners, defects) in determining the dispersion and the structure of the supported species in alumina-supported catalysts. Also the nature of the anchoring bonds between surface species (e.g., metal particles) and the support should be clarified.

REFERENCES

1. Fierro, J. L. G. Ed.; *Metal Oxides: Chemistry and Applications*; Taylor & Francis: Boca Raton, FL, 2005.
2. Rouquerol, J.; Sing, K. S. W.; Llewellyn, P. *Adsorption by Powders and Porous Solids*, 2nd ed.; Elsevier: Amsterdam, 2014; pp 393–465.
3. Jackson, S. D., Hargreaves, J. S. J. Eds.; *Metal Oxide Catalysis*, Wiley-VCH: Weinheim, 2008.

4. Hsu, P. H. *Mineral Soil Environment*; In: Dixon, J. B., Weed, S. B. Eds.; Soil Science Society of America: Madison, WI, 1977; pp 99–143.

5. Digne, M.; Sautet, P.; Raybaud, P.; Toulhoat, H.; Artacho, E. *J. Phys. Chem. B* **2002**, *106*, 5155–5162.

6. Demichelis, R.; Noël, Y.; Ugliengo, P.; Zicovich-Wilson, C. M.; Dovesi, R. *J. Phys. Chem. C* **2011**, *115*, 13107–13134.

7. Zigan, F.; Joswig, W.; Burger, N. *Z. Kristallogr.* **1978**, *148*, 255–273.

8. Demichelis, R.; Civalleri, B.; Noel, Y.; Meyer, A.; Dovesi, R. *Chem. Phys. Lett.* **2008**, *465*, 220–225.

9. Saalfeld, H.; Wedde, M. *Z. Kristallogr.* **1974**, *139*, 129–135.

10. Clark, G. R.; Rodgers, K. A.; Henderson, G. S. *Z. Kristallogr.* **1998**, *213*, 96–100.

11. Demichelis, R.; Catti, M.; Dovesi, R. *J. Phys. Chem. C* **2009**, *113*, 6785–6791.

12. Saalfeld, H.; Jarchow, O. *Neues. Jahrb. Mineral* **1968**, *109*, 185–191.

13. Casassa, S.; Demichelis, R. *J. Phys. Chem. C* **2012**, *116*, 13313–13321.

14. Peintinger, M. F.; Kratz, M. J.; Bredow, T. *J. Mater. Chem. A* **2014**, *2*, 13143–13158.

15. Isobe, T.; Watanabe, T.; D'Espinose De La Caillerie, J. B.; Legrand, A. P.; Massiot, D. *J. Colloid Interface Sci.* **2003**, *261*, 320–324.

16. Vyalikh, A.; Zesewitz, K.; Scheler, U. *Magn. Reson. Chem.* **2010**, *48*, 877–881.

17. Damodaran, K.; Rajamohanan, P. R.; Chakrabarty, D.; Racherla, U. S.; Manohar, V.; Fernandez, C.; Amoureux, J.-P.; Ganapathy, S. *J. Am. Chem. Soc.* **2002**, *124*, 3200–3201.

18. Rodgers, K. A. *Clay Min.* **1993**, *28*, 85–99.

19. Ruan, H. D.; Frost, R. L.; Kloprogge, J. T. *J. Raman Spectrosc.* **2001**, *32*, 745–750.

20. Jodin-Caumon, M.-C.; Humbert, B.; Phambu, N.; Gaboriaud, F. *Spectrochim. Acta Mol. Biomol. Spectros.* **2009**, *72*, 959–964.

21. Ogorodova, L. P.; Kiseleva, I. A.; Sokolova, E. L.; Vigasina, M. F.; Kabalov, Y. K. *Geochem. Int.* **2012**, *50*, 90–94.

22. Balan, E.; Blanchard, M.; Hochepied, J.-F.; Lazzeri, M. *Phys. Chem. Miner.* **2008**, *35*, 279–285.

23. Noel, Y.; Demichelis, R.; Pascale, F.; Ugliengo, P.; Orlando, R.; Dovesi, R. *Phys. Chem. Miner.* **2009**, *36*, 47–59.

24. Tunega, D.; Păsalíc, H.; Gerzabek, M. H.; Lischka, H. *J. Phys. Condens. Matter* **2011**, *23*, 404201.

25. Bokhimi, X.; Toledo-Antonio, J. A.; Guzmán-Castillo, M. L.; Mar-Mar, B.; Hernández-Beltrán, F.; Navarrete, J. *J. Solid State Chem.* **2001**, *161*, 319–326.

26. Yang, H.; Lu, R.; Downs, R. T.; Costin, G. *Acta Crystallogr.* **2006**, *E62*, i250–i252.

27. Li, M.; Snoussi, K.; Li, L.; Wang, H.; Yang, W.; Gao, C. *Appl. Phys. Lett.* **2010**, *96*, 261902.

28. Hu, Y.; Liu, X.; Xu, Z. *Miner. Eng.* **2003**, *16*, 219–227.

29. Kloprogge, J. T.; Duong, L. V.; Wood, B. J.; Frost, R. L. *J. Colloid Interface Sci.* **2006**, *296*, 572–576.

30. Thomas, S.; Sherwood, P. M. A. *Anal. Chem.* **1992**, *64*, 2480–2495.

31. Demicheli, R.; Noel, Y.; Zicovich-Wilson, C. M.; Roetti, C.; Valenzano, L.; Dovesi, R. *J. Phys. Conf. Ser.* **2008**, *117*, 012013.

32. Sonthalia, R.; Behara, P.; Kumaresan, T.; Thakre, S. *Int. J. Miner. Process.* **2013**, *125*, 137–148.

33. Habashi, F. *Essential Readings in Light Metals*; In: Donaldson, D., Raahauge, B. Eds.; Alumina and Bauxite; Vol. 1, Wiley: Hoboken, USA, 2013; pp 85–93.

34. Helmboldt, O.; Hudson, L. K.; Misra, C.; Wefers, K.; Heck, W.; Stark, H.; Danner, M.; Rösch, N. In: *Ullmann's Encyclopedia of Industrial Chemistry*, Wiley-VCH: Weinheim, 2007. http://dx.doi.org/10.1002/14356007.a01_527.pub2.

35. Buchel, K. H.; Moretto, H. H.; Werner, D. *Industrial Inorganic Chemistry*, 2nd ed.; Wiley: New York, NY, 2000; pp. 247–256.

36. Floris, F.; Persico, M.; Tani, A.; Tomasi, J. *Chem. Phys.* **1995**, *195*, 207–220.
37. Wasserman, E.; Rustad, J. R.; Xantheas, S. S. *J. Chem. Phys.* **1997**, *106*, 9769–9780.
38. Casey, W. H. *Chem. Rev.* **2006**, *106*, 1–16.
39. Shafran, K. L.; Perry, C. C. *Dalton Trans.* **2005**, 2098–2105.
40. Al'myasheva, O. V.; Korytkova, E. N.; Maslov, A. V.; Gusarov, V. V. *Inorg. Mater.* **2005**, *41*, 460–467.
41. Wefers, K.; Mitra, C. *Oxides and Hydroxides of Aluminum*, Alcoa Laboratories: Pittsburgh, PA, 1987.
42. Violante, A.; Huang, P. M. *Clays Clay Miner.* **1985**, *33*, 181–192.
43. Pereira Antunes, M. L.; de Souza Santos, H.; de Souza Santos, P. *Mater. Chem. Phys.* **2002**, *76*, 243–249.
44. Cesteros, Y.; Salagre, P.; Medina, F.; Sueiras, J. E. *Chem. Mater.* **1999**, *11*, 123–129.
45. Liu, Ye; Ma, D.; Blackley, R. A.; Zhou, W.; Han, X.; Bao, X. *J. Phys. Chem. C* **2008**, *112*, 4124–4128.
46. Kumara, C. K.; Ng, W. J.; Bandara, A.; Weerasooriya, R. *J. Colloid Interface Sci.* **2010**, *352*, 252–258.
47. Kul'ko, E. V.; Ivanova, A. S.; Litvak, G. S.; Kryukova, G. N.; Tsybulya, S. V. *Kinet. Catal.* **2004**, *45*, 714–721.
48. Koga, N.; Fukagawa, T.; Tanaka, H. *J. Therm. Anal. Calorim.* **2001**, *64*, 965972.
49. Lefèvre, G.; Fédoroff, M. *Mater. Lett.* **2002**, *56*, 978–983.
50. You, S.; Li, Y.; Zhang, Y.; Yang, C.; Zhang, Yi. *Ind. Eng. Chem. Res.* **2013**, *52*, 12710–12716.
51. Lipin, V. A. *Russ. J. Appl. Chem.* **2001**, *74*, 184–187.
52. Peskleway, C. D.; Henderson, G. S.; Wicks, F. J. *J. Cryst. Growth* **2005**, *273*, 614–623.
53. Seo, C. W.; Jung, K. D.; Lee, K. Y.; Yoo, K. S. *J. Ind. Eng. Chem.* **2009**, *15*, 649–652.
54. Lee, H.; Jeon, Y.; Lee, S. U.; Sohn, D. *Chem. Lett.* **2013**, *42*, 1463–1465.
55. Oberlander, R. K., Leach, B. E. Eds.; Applied Industrial Catalysis; Vol. 3. Academic Press: Orlando, FL, 1984.
56. Shutz, A. A.; Cullo, L. A. Patent WO 1993/05010 to Aristech Chem. Co.
57. http://www.uop.com/wp-content/uploads/2012/12/UOP-Versal-Alumina-Brochure2.pdf (accessed 20 April 2013).
58. Peng, L.; Xu, X.; Zhi Lv, J.; Song, M.; He, Q.; Wang, L.; Yan, Y.; Li, Z.; Li, J. *J. Therm. Anal. Calorim.* **2012**, *110*, 749–754.
59. Kawano, M.; Tomita, K. *Clays Clay Miner.* **1997**, *45*, 365–377.
60. de Souza Santos, P.; Vieira Coelho, A. C.; de Souza Santos, H.; Kiyohara, P. K. *Mater. Res.* **2009**, *12*, 437–445.
61. Li, G.; Liu, Y.; Liu, C. *Micropor. Mesopor. Mater.* **2013**, *167*, 137–145.
62. Zhang, L.; Jiao, X.; Chen, D.; Jiao, M. *Eur. J. Inorg. Chem.* **2011**, 5258–5264.
63. Cho, S.; Jang, J.-W.; Park, J.; Jung, S.; Jeong, S.; Kwag, J.; Lee, J. S.; Kim, S. J. *J. Mater. Chem. C* **2013**, *1*, 4497–4504.
64. Chen, X. Y.; Zhang, Z. J.; Li, X. L.; Lee, S. W. *Solid State Commun.* **2008**, *145*, 368–373.
65. Zhao, Y.; Frost, R. L.; Martens, W. N. *J. Phys. Chem. C* **2007**, *111*, 5313–5324.
66. Cai, W.; Yu, J.; Gu, S.; Jaroniec, M. *Cryst. Growth Des.* **2010**, *10*, 3977–3982.
67. Carim, A. H.; Rohrer, G. S.; Dando, N. R.; Tzeng, S.-Y.; Rohrer, C. L.; Perrotta, A. J. *J. Am. Ceram. Soc.* **1997**, *80*, 2677–2680.
68. Perrotta, A. J.; Minnick, R. U.S. Patent 5,334,366 to Aluminum Co. of America, **1994**.
69. Pan, C.; Shen, P.; Huang, W. L.; Hwang, S. L.; Yui, T. F.; Chu, H. T. *J. Eur. Ceram. Soc.* **2006**, *26*, 2707–2717.
70. Kaiser, A.; Sporn, D.; Bertagnolli, H. *J. Eur. Ceram. Soc.* **1994**, *14*, 77–83.

71. Lazarev, V.; Panasyuk, G.; Voroshilov, I.; Danchevskaya, M.; Torbin, S.; Ivakin, Yu. *Ind. Eng. Chem. Res.* **1996**, *35*, 3721–3725.
72. Meyer, A.; Noweck, K.; Reichenauer, A. U.S. Patent No. 5,055,019 to Condea, **1989**.
73. Sasol, personal communications.
74. Pajonk, G. M. *Appl. Catal.* **1991**, *72*, 217–266.
75. Trombetta, M.; Busca, G.; Willey, R. J. *J. Colloid Interface Sci.* **1997**, *190*, 416–426.
76. Khaleel, A. A.; Klabunde, K. J. *Chem. Eur. J.* **2002**, *8*, 3991–3998.
77. Zu, G.; Shen, J.; Wei, X.; Ni, X.; Zhang, Z.; Wang, J.; Liu, G. *J. Non-Cryst. Solids* **2011**, *357*, 2903–2906.
78. Levin, I.; Brandon, D. *J. Am. Ceram. Soc.* **1998**, *81*, 1995–2012.
79. Koga, N.; Yamada, S. *Solid State Ionics* **2004**, *172*, 253–256.
80. Liu, C.; Liu, Y.; Ma, Q.; Ma, J.; He, H. *Ind. Eng. Chem. Res.* **2013**, *52*, 13377–13383.
81. Sanfilippo, D. *Catal. Today* **2011**, *178*, 142–150.
82. Chagas, L. H.; De Carvalho, G. S. G.; San Gil, R. A. S.; Chiaro, S. S. X.; Leitaoa, A. A.; Diniz, R. *Mater. Res. Bull.* **2014**, *49*, 216–222.
83. Brindley, F. G. W.; Choe, J. O. *J. Miner. Soc. Am.* **1961**, *46*, 771–785.
84. Chaitree, W.; Jiemsirilers, S.; Mekasuwandumrong, O.; Praserthdam, P.; Charinpanitkul, T.; Panpranotw, J. *J. Am. Ceram. Soc.* **2010**, *93*, 3377–3383.
85. Li, W.; Numako, C.; Koto, K.; Moriga, T.; Nakabayashi, I. *Int. J. Mod. Phys. B* **2003**, *17*, 1464–1469.
86. Louaer, S.; Wang, Y.; Lin Guo, L. *ACS Appl. Mater. Interfaces* **2013**, *5*, 9648–9655.
87. Tsuchida, T.; Ichikawa, N. *React. Solid* **1989**, *7*, 207–217.
88. Kano, J.; Saeki, S.; Saito, F.; Tanjo, M.; Yamazaki, S. *Int. J. Miner. Process.* **2000**, *60*, 91–100.
89. Du, X.; Su, X.; Wang, Y.; Li, J. *Mater. Res. Bull.* **2009**, *44*, 660–665.
90. Du, X.; Wang, Y.; Su, X.; Li, J. *Nanosci. Nanotechnol. Lett.* **2011**, *3*, 146–150.
91. MacKenzie, K. J. D.; Temuujin, J.; Okada, K. *Thermochim. Acta* **1999**, *327*, 103–108.
92. Pérez, L. L.; Zarubina, V.; Heeres, H. J.; Melián-Cabrera, I. *Chem. Mater.* **2013**, *25*, 3971–3978.
93. Sato, T. *J. Therm. Anal.* **1987**, *32*, 61–70.
94. Okumiya, M.; Yamaguchi, G.; Yamada, O.; Ono, S. *Bull. Chem. Soc. Jpn.* **1971**, *44*, 418–423.
95. Okumiya, M.; Yamaguchi, G. *Bull. Chem. Soc. Jpn.* **1971**, *44*, 1567–1570.
96. Sharanda, L. F.; Simansky, A. P.; Kulin, T. V.; Chiuko, A. A. *Colloids Surf. A Physicochem. Eng. Asp.* **1995**, *105*, 167–172.
97. Lavalley, J.-C.; Benaissa, M.; Busca, G.; Lorenzelli, V. *Appl. Catal.* **1986**, *24*, 249–255.
98. Srivanasan, S.; Narayanan, C. R.; Datye, A. K. *Appl. Catal. A Gen.* **1995**, *132*, 289–308.
99. Schumacher, K.; Golchert, R.; Schilling, R.; Batz-Sohn, R.; Moerters, C. US Patent 7749322 to Evonik Degussa GmbH, **2010**.
100. Pacewska, B.; Keshr, M. *Thermochim. Acta* **2002**, *385*, 73–80.
101. Hu, X.; Liu, Y.; Tang, Z.; Li, G.; Zhao, R.; Liu, C. *Mater. Res. Bull.* **2012**, *47*, 4271–4277.
102. Kiyohara, P. K.; Santos Souza, H.; Coelho Vieira, A. C.; Santos de Souza, P. *An. Acad. Bras. Cienc.* **2000**, *72*, 471–495.
103. Brand, P.; Troschke, R. *Cryst. Res. Technol.* **1989**, *24*, 671–675.
104. ÇIlgI, G. K.; Cetişli, H. *J. Therm. Anal. Calorim.* **2009**, *98*, 855–861.
105. Zecchina, A.; Escalona Platero, E.; Otero Arean, C. *J. Catal.* **1987**, *107*, 244–247.
106. Park, H. C.; Park, Y. J.; Stevens, R. *Mater. Sci. Eng. A* **2004**, *367*, 166–170.
107. Bhattacharya, I. N.; Gochhayat, P. K.; Mukherjee, P. S.; Paul, S.; Mitra, P. K. *Mater. Chem. Phys.* **2004**, *88*, 32–40.
108. Liu, G.; Li, J.; Chen, K. *Int. J. Refract. Met. Hard Mater.* **2013**, *39*, 90–102.

109. Toniolo, J. C.; Lima, M. D.; Takimi, A. S.; Bergmann, C. P. *Mater. Res. Bull.* **2005**, *40*, 561–571.

110. Abbattista, F.; Delmastro, S.; Gozzelino, G.; Mazza, D.; Vallino, M.; Busca, G.; Lorenzelli, V.; Ramis, G. *J. Catal.* **1989**, *117*, 42–51.

111. Lee, G. W. *Int. J. Chem. Mater. Sci. Eng.* **2013**, *7*, 210–213.

112. Lee, S. K.; Park, S. Y.; Yi, Y. S.; Moon, J. *J. Phys. Chem. C* **2010**, *114*, 13890–13894.

113. Sarou-Kanian, V.; Gleizes, A. N.; Florian, P.; Samélor, D.; Massiot, D.; Vahlas, C. *J. Phys. Chem. C* **2013**, *117*, 21965–21971.

114. Tavakoli, A. H.; Maram, P. S.; Widgeon, S. J.; Rufner, J.; van Benthem, K.; Ushakov, S.; Sen, S.; Navrotsky, A. *J. Phys. Chem. C* **2013**, *117*, 17123–17130.

115. Sakashita, S.; Araki, Y.; Shiomada, H. *Appl. Catal. A Gen.* **2001**, *215*, 101–110.

116. Sakashita, Y.; Yoneda, T. *J. Catal.* **1999**, *185*, 487–495.

117. Zhou, R.-S.; Snyder, R. L. *Acta Crystallogr. B* **1991**, *47*, 617–630.

118. Paglia, G.; Rohl, A. L.; Buckley, C. E.; Gale, J. D. *Phys. Rev. B* **2005**, *71*, 224115.

119. Krokidis, X.; Raybaud, P.; Gobichon, A.-E.; Rebours, B.; Euzen, P.; Toulhoat, H. *J. Phys. Chem. B* **2001**, *105*, 5121–5130.

120. Paglia, G.; Buckley, C. E.; Rohl, A. L.; Hart, R. D.; Winter, K.; Studer, A. J.; Hunter, B. A.; Hanna, J. V. *Chem. Mater.* **2004**, *16*, 220–236.

121. Yourdshahyan, Y.; Ruberto, C.; Bengtsson, L.; Lundqvist, B. I. *Phys. Rev. B* **1997**, *56*, 8553–8558.

122. Liu, P.; Skogsmo, J. *Acta Crystallogr.* **1991**, *B47*, 425–433.

123. Ollivier, B.; Retoux, R.; Lacorre, P.; Massiot, D.; Ferey, G. *J. Mater. Chem.* **1997**, *7*, 1049–1056.

124. Yourdshahyan, Y.; Ruberto, C.; Halvarsson, M.; Bengtsson, L.; Langer, V.; Lundqvist, B. I.; Ruppi, S.; Rolander, U. *J. Am. Ceram. Soc.* **1999**, *82*, 1365–1380.

125. Langer, V.; Smrčok, L.; Halvarsson, M.; Yourdshahyan, Y.; Ruberto, C.; Ruppi, S. *Mater. Sci. Forum* **2001**, *378–381*, 600–605.

126. Smrčok, L.; Langer, V.; Halvarsson, M.; Ruppi, S. *Z. Kristallogr.* **2001**, *216*, 409–412.

127. Ruberto, C.; Yourdshahyan, Y.; Lundqvist, B. I. *Phys. Rev.* **2003**, *67*, 195412.

128. Kogure, T. *J. Am. Ceram. Soc.* **1999**, *82*, 716–720.

129. Stumpf, H. C.; Russel, A. S.; Newsome, J. W.; Tucker, C. M. *Ind. Eng. Chem.* **1950**, *42*, 13981403.

130. Tournier, G.; Lacroix-Repellin, M.; Pajonk, G. M.; Teichner, S. J. *Preparation of Catalysts IV*; In: Delmon, B., Grange, P., Jacobs, P. A., Poncelet, G. Eds.; Elsevier: Amsterdam, 1987; pp 333–340.

131. Lee, S. K.; Ahn, C. W. *Nature Scientific Reports* **2014**, *4*, 4200.

132. Kim, H. J.; Kim, T. G.; Kim, J. J.; Park, S. S.; Hong, S. S.; Lee, G. D. *J. Phys. Chem. Solids* **2008**, *69*, 1521–1524.

133. Tsuchida, T. *Solid State Ionics* **1993**, *63–65*, 464–470.

134. Pajonk, G.; Repwellin, M.; Teichner, S. J. *Bull. Soc. Chim. Fr.* **1976**, 1333–1340.

135. Xiong, G.; Elam, J. W.; Feng, H.; Han, C. Y.; Wang, H.-H.; Iton, L. E.; Curtiss, L. A.; Pellin, M. J.; Kung, M.; Kung, H.; Stair, P. C. *J. Phys. Chem. B* **2005**, *109*, 14059–14063.

136. Nortier, P.; Fourre, P.; Mohammed Saad, A. B.; Saur, O.; Lavalley, J. C. *Appl. Catal.* **1990**, *61*, 141–160.

137. Hoang-Van, C.; Ghorbel, A.; Bandiera, J.; Teichner, S. J. *Bull. Soc. Chim. Fr.* **1973**, 841–845.

138. Abbattista, F.; Delmastro, A.; Gozzelino, G.; Mazza, D.; Vallino, M.; Busca, G.; Lorenzelli, V. *J. Chem. Soc. Faraday Trans.* **1990**, *86*, 3653–3658.

139. Lee, S. K.; Lee, S. B.; Park, S. Y.; Yi, S. Y.; Ahn, C. W. *Phys. Rev. Lett.* **2009**, *103*, 095501.

140. Lizárraga, R.; Holmström, E.; Parker, S. C.; Arrouvel, C. *Phys. Rev. B* **2011**, *83*, 094201.

141. Bloch, L.; Kauffmann, Y.; Pokroy, B. *Cryst. Growth Des.* **2014**, *14*, 3983–3989.
142. Dressler, M.; Nofz, M.; Malz, F.; Pauli, J.; Jäger, C.; Reinsch, S.; Scholz, G. *J. Solid State Chem.* **2007**, *180*, 2409–2419.
143. Hoang-Van, C.; Ghorbel, A.; Teichner, S. J. *Bull. Soc. Chim. Fr.* **1972**, 437–441.
144. Márquez-Alvarez, C.; Žilková, N.; Pérez-Pariente, J.; Čejka, J. *Catal. Rev. Sci. Eng.* **2008**, *50*, 222–286.
145. Verwey, E. J. W. *Z. Kristallogr.* **1935**, *91*, 65–69.
146. Lippens, B. C.; deBoer, J. H. *Acta Crystallogr.* **1964**, *17*, 1312–1321.
147. Gutiérrez, G.; Taga, A.; Johansson, B. *Phys. Rev. B* **2002**, *65*, 012101.
148. Wolverton, C.; Hass, K. C. *Phys. Rev. B* **2000**, *63*, 024102.
149. Kim, H. J.; Lee, H. C.; Lee, J. S. *J. Phys. Chem. C* **2007**, *111*, 1579–1583.
150. Kwak, J. H.; Hu, J.; Lukaski, A.; Kim, D. H.; Szanyi, J.; Peden, C. H. F. *J. Phys. Chem. C* **2008**, *112*, 9486–9492.
151. Düvel, A.; Romanova, E.; Sharifi, M.; Freude, D.; Wark, M.; Heitjans, P.; Wilkening, M. *J. Phys. Chem. C* **2011**, *115*, 22770–22780.
152. Soled, S. *J. Catal.* **1983**, *81*, 252–257.
153. Tsyganenko, A. A.; Smirnov, K. S.; Rzhevskij, A. M.; Mardilovich, P. P. *Mater. Chem. Phys.* **1990**, *26*, 35–46.
154. Sohlberg, K.; Pennycook, S. J.; Pantelides, S. T. *J. Am. Chem. Soc.* **1999**, *121*, 7493–7499.
155. Sohlberg, K.; Pennycook, S. J.; Pantelides, S. T. *Chem. Eng. Commun.* **2000**, *181*, 107–135.
156. Smrcok, L.; Langer, V.; Krestan, J. *Acta Crystallogr. C* **2006**, *62*, i83–i84.
156b. Rozita, Y.; Brydson, R.; Comyn, T. P.; Scott, A. J.; Hammond, C.; Brown, A.; Chauruka, S.; Hassanpour, A.; Young, N. P.; Kirkland, A. I.; Sawada, H.; Smith, R. I. *ChemCatChem* **2013**, *5*, 1–13.
157. Samain, L.; Jaworski, A.; Edén, M.; Ladd, D. M.; Seo, D.-K.; Javier Garcia-Garcia, F. J.; Häussermann, U. *J. Solid State Chem.* **2014**, *217*, 1–8.
158. Digne, M.; Sautet, P.; Raybaud, P.; Euzen, P.; Toulhoat, H. *J. Catal.* **2004**, *226*, 54–68.
159. Digne, M.; Sautet, P.; Raybaud, P.; Euzen, P.; Toulhoat, H. *J. Catal.* **2002**, *211*, 1–5.
160. McHale, J. M.; Auroux, A.; Perrotta, A. J.; Navrotsky, A. *Science* **1997**, *277*, 788791.
161. Ferreira, A. R.; Martins, M. J. F.; Konstantinova, E.; Capaz, R. B.; Souza, W. F.; Chiaro, S. S. X.; Laitãao, A. A. *J. Solid State Chem.* **2011**, *184*, 1105–1111.
162. Ferreira, A. R.; Küçükbenli, E.; Leitão, A. A. De; Gironcoli, S. *Phys. Rev. B* **2011**, *84*, 235119.
163. Streitz, F. H.; Mintmire, J. W. *Phys. Rev. B* **1999**, *60*, 773–777.
164. Okuyama, Y.; Kurita, N.; Fukatsu, N. *Solid State Ionics* **2006**, *177*, 59–64.
165. Miller, M. E.; Misture, S. T. *J. Phys. Chem. C* **2010**, *114*, 13039–13046.
166. Paglia, G.; Bozin, E. S.; Billinge, S. J. L. *Chem. Mater.* **2006**, *18*, 3242–3248.
167. Ishii, M.; Nakahira, M.; Yamanaka, T. *Solid State Commun.* **1972**, *11*, 209–212.
168. Saniger, J. M. *Mater. Lett.* **1995**, *22*, 109–113.
169. Busca, G. *Metal Oxide Catalysis*; In: Jackson, S. D., Hargreaves, J. S. J. Eds.; Vol. 1, Wiley-VCH: Weinheim, 2008; pp. 95–175.
170. Boumaza, A.; Favaro, L.; Lédion, J.; Sattonnay, G.; Brubach, J. B.; Berthet, P.; Huntz, A. M.; Royc, P.; Tétot, R. *J. Solid State Chem.* **2009**, *182*, 1171–1176.
171. Kawashita, M.; Kamitani, A.; Miyazaki, T.; Matsui, N.; Li, Z.; Kanetaka, H.; Hashimoto, M. *Mater. Sci. Eng. C* **2012**, *32*, 2617–2622.
172. Loyola, C.; Menéndez-Proupin, E.; Gutiérrez, G. *J. Mater. Sci.* **2010**, *45*, 5094–5100.
173. Kim, H.; Kosuda, K. M.; van Duyne, R. P.; Stair, P. C. *Chem. Soc. Rev.* **2010**, *39*, 4820–4844.

174. Chen, Y.; Hyldtoft, J.; Jacobsen, C. J. H.; Nielsen, O. F. *Spectrochim. Acta A* **1995**, *51*, 2161–2169.

175. Kim, H.-S.; Stair, P. C. *J. Phys. Chem. A* **2009**, *113*, 4346–4355.

176. Cava, S.; Tebcherani, S. M.; Souza, I. A.; Pianaro, S. A.; Paskocimas, C. A.; Longo, E.; Varela, J. A. *Mater. Chem. Phys.* **2007**, *103*, 394–399.

177. Hansen, M. R.; Jakobsen, H. J.; Skibsted, J. *J. Phys. Chem. C* **2008**, *112*, 7210–7222.

178. Tsybulya, S. V.; Kryukova, G. N. *Phys. Rev. B* **2008**, *77*, 024112.

179. Dyan, A.; Azevedo, C.; Cenedese, P.; Dubot, P. *Appl. Surf. Sci.* **2008**, *254*, 3819–3828.

180. Digne, M.; Revel, R.; Boualleg, M.; Chice, D.; Rebours, B.; Moreaud, M.; Celse, B.; Chaneac, C.; Jolivet, J. P. *Scientific Bases for the Preparation of Heterogeneous Catalysts*; In: Gaigneaux, E. M., Devillers, M., Hermans, S., Jacobs, P., Mertnes, J., Ruiz, P. Eds.; Elsevier: Amsterdam, 2010; pp. 127–134.

181. Klimov, O. V.; Leonova, K. A.; Koryakina, G. I.; Gerasimov, E. Yu.; Prosvirin, I. P.; Cherepanova, S. V.; Budukva, S. V.; Pereyma, V. Yu.; Dik, P. P.; Parakhin, O. A.; Noskov, A. S. *Catal. Today* **2014**, *220–222*, 66–77.

182. Bokhimi, X.; Sanchez-Valente, J.; Pedraza, F. *J. Solid State Chem.* **2002**, *166*, 182–190.

183. Kovarik, L.; Genc, A.; Wang, C.; Qiu, A.; Peden, C. H. F.; Szanyi, J.; Kwak, Ja-H. *J. Phys. Chem. C* **2013**, *117*, 179–186.

184. Hong, T. L.; Liu, H.-T.; Yeh, C. T.; Chen, S. H.; Sheu, F.-C.; Leu, L.-J.; Wang, C. I. *Appl. Catal. A Gen.* **1997**, *158*, 257–271.

185. http://www.sasoltechdata.com/tds/PURALOX_CATALOX.pdf (accessed 20 April 2013).

186. Gribov, E. N.; Zavorotynska, O.; Agostini, G.; Vitillo, J. G.; Ricchiardi, G.; Spoto, G.; Zecchina, A. *Phys. Chem. Chem. Phys.* **2010**, *12*, 6474–6482.

187. Phung, T. K.; Lagazzo, A.; Rivero Crespo, M. A.; Sánchez Escribano, V.; Busca, G. *J. Catal.* **2014**, *311*, 102–113.

188. Vigue, H.; Quintard, P.; Merle-Mejean, T.; Lorenzelli, V. *J. Eur. Ceram. Soc.* **1998**, *18*, 305–309.

189. Delgado, M.; Delbecq, F.; Santini, C. C.; Lefebvre, F.; Norsic, S.; Putaj, P.; Sautet, P.; Basset, J. M. *J. Phys. Chem. C* **2012**, *116*, 834–843.

190. Li, W.; Kenneth, J. T.; Livi, K. J. T.; Xu, W.; Siebecker, M. G.; Wang, Y.; Phillips, B. L.; Sparks, D. L. *Environ. Sci. Technol.* **2012**, *46*, 11670–11677.

191. Marchese, L.; Bordiga, S.; Coluccia, S.; Martra, G.; Zecchina, A. *J. Chem. Soc. Faraday Trans.* **1993**, *89*, 3483–3489.

192. Wischert, R.; Laurent, P.; Copéret, C.; Delbecq, F.; Sautet, P. *J. Am. Chem. Soc.* **2012**, *134*, 14430–14449.

193. Reller, A.; Cocke, D. L. *Catal. Lett.* **1989**, *2*, 91–95.

194. Rozita, Y.; Brydson, R.; Scott, A. J. *J. Phys. Conf. Ser.* **2010**, *241*, 012096.

195. Palmero, P.; Bonelli, B.; Lomello, F.; Garrone, E.; Montanaro, L. *J. Therm. Anal. Calorim.* **2009**, *97*, 223–229.

196. Liu, C.; Liu, Y.; Ma, Q.; He, H. *Chem. Eng. J.* **2010**, *163*, 133–142.

197. Wilson, S. J.; Mc Connel, J. D. C. *J. Solid State Chem.* **1980**, *34*, 315–322.

198. Wang, Y. G.; Bronsveld, P. M.; DeHosson, J. Th. M.; Djuričić, B.; McGarry, D.; Pickering, S. *J. Am. Ceram. Soc.* **1998**, *81*, 1655–1660.

199. Ghanti, E.; Tomar, N.; Gautam, P.; Nagarajan, R. *J. Sol-Gel Sci. Technol.* **2011**, *57*, 12–15.

200. Rooksby, H. P.; Rooymans, C. J. M. *Clay Miner. Bull.* **1961**, *4*, 234–238.

201. Kurumada, M.; Koike, C.; Kaito, C. *Mon. Not. R. Astron. Soc.* **2005**, *359*, 643–647.

202. Roy, A.; Sood Pramana, J. K. *J. Phys.* **1995**, *44*, 201–209.

203. Kovarik, L.; Bowden, M.; Genc, A.; Szanyi, J.; Peden, C. H. F.; Kwak, J. H. *J. Phys. Chem. C* **2014**, *118*, 18051–18058.

204. Métivier, R.; Leray, I.; Roy-Auberger, M.; Zanier-Szydlowskic, N.; Valeur, B. *New J. Chem.* **2002**, *26*, 411–415.

205. Healy, M. H.; Wieserman, L. F.; Arnett, E. M.; Wefers, K. *Langmuir* **1989**, *5*, 114–123.
206. Carre, S.; Gnep, N. S.; Revel, R.; Magnoux, P. *Appl. Catal. A Gen.* **2008**, *348*, 71–78.
207. Sung, D. M.; Kim, Y. H.; Park, E. D.; Yie, J. E. *Res. Chem. Intermed.* **2010**, *36*, 653–660.
208. Zhang, X.; Honkanen, M.; Levänen, E.; Mäntylä, T. *J. Cryst. Growth* **2008**, *310*, 3674–3679.
209. Pecharromán, C.; Sobrados, I.; Iglesias, J. E.; Gonzalez-Carreño, T.; Sanz, J. *J. Phys. Chem. B* **1999**, *103*, 6160–6170.
210. Pecharromán, C.; Gonzalez-Carreño, T.; Iglesias, J. E. *J. Mater. Res.* **1996**, *11*, 127–133.
211. John, C. S.; Alma, N. C. M.; Hays, G. R. *Appl. Catal.* **1983**, *6*, 341–346.
212. MacIver, D. S.; Tobin, H. H.; Barth, R. T. *J. Catal.* **1963**, *2*, 485–497.
213. Stone, F. S.; Whalley, L. *J. Catal.* **1967**, *8*, 173–182.
214. Della Gatta, G.; Fubini, B.; Ghiotti, G.; Morterra, C. *J. Catal.* **1976**, *43*, 90–98.
215. Maciver, D. S.; Wilmot, W. H.; Bridges, J. M. *J. Catal.* **1964**, *3*, 502–511.
216. Seo, C. W.; Jung, K. D.; Lee, K. Y.; Yoo, K. S. *Ind. Eng. Chem. Res.* **2008**, *47*, 6573–6578.
217. Sohlberg, K.; Pennycook, S. J.; Pantelides, S. T. *J. Am. Chem. Soc.* **2001**, *123*, 26–29.
218. Vieira Coelho, A. C.; de Souza Santos, H.; Kiyohara, P. K.; Pinto Marcos, K. N.; de Souza Santos, P. *Mater. Res.* **2007**, *10*, 183–189.
219. Malki, A.; Mekhalif, Z.; Detriche, S.; Fonder, G.; Boumaza, A.; Djelloul, A. *J. Solid State Chem.* **2014**, *215*, 8–15.
220. Chang, P.-L.; Wu, Y.-C.; Lai, S.-J.; Yen, F.-S. *J. Eur. Ceram. Soc.* **2009**, *29*, 3341–3348.
221. Mekasuwandumrong, O.; Pavarajarn, V.; Inoue, M.; Praserthdam, P. *Mater. Chem. Phys.* **2006**, *100*, 445–450.
222. Slade, R. C. T.; Southern, J. C.; Thompson, I. M. *J. Mater. Chem.* **1991**, *1*, 563–563.
223. Favaro, L.; Boumaza, A.; Roy, P.; Lédion, J.; Sattonnay, G.; Brubach, J. B.; Huntz, A. M.; Tétot, R. *J. Solid State Chem.* **2010**, *183*, 901–908.
224. Khom-ina, J.; Praserthdama, P.; Panpranot, J.; Mekasuwandumrong, O. *Catal. Commun.* **2008**, *9*, 1955–1958.
225. Zotin, J. L.; Faro, A. C., Jr. *Catal. Today* **1989**, *5*, 423–431.
226. Kim, H.-S.; Zygmunt, S. A.; Stair, P. C.; Zapol, P.; Curtiss, L. A. *J. Phys. Chem. C* **2009**, *113*, 8836–8843.
227. Narula, C. K.; Stocks, G. M. *J. Phys. Chem. C* **2012**, *116*, 5628–5636.
228. Morterra, C.; Bolis, V.; Magnacca, G. *Langmuir* **1994**, *10*, 1812–1824.
229. Bolis, V.; Cerrato, G.; Magnacca, G.; Morterra, C. *Thermochim. Acta* **1998**, *312*, 63–77.
230. Kim, H. N.; Lee, S. K. *Am. Mineral.* **2013**, *98*, 1198–1210.
231. Mo, S. D.; Ching, W. Y. *Phys. Rev. B* **1998**, *57*, 15219–15228.
232. Hyde, B. G.; Andersson, S. *Inorganic Crystal Structures*, Wiley: New York, NY, 1989; p 156.
233. Borosy, A. P.; Silvi, B.; Allavena, M.; Nortier, P. *J. Phys. Chem.* **1994**, *98*, 13189–13194.
234. O'Dell, L. A.; Savin, S. L. P.; Chadwick, A. V.; Smith, M. E. *Solid State Nucl. Magn. Reson.* **2007**, *31*, 169–173.
235. Levin, I.; Bendersky, L. A.; Brandon, D. G.; Rühle, M. *Acta Mater.* **1997**, *45*, 3659–3669.
236. Cai, S. H.; Rashkeev, S. N.; Pantelides, S. T.; Sohlberg, K. *Phys. Rev. B* **2003**, *67*, 224104.
237. Wen, H.-L.; Chen, Y. Y.; Yen, F.-S.; Huang, C. Y. *Nanostruct. Mater.* **1999**, *11*, 89–101.

238. Garcia-Guinea, J.; Correcher, V.; Rubio, J.; Valle-Fuentes, F. *J. Phys. Chem. Solids* **2005**, *66*, 1220.
239. Löffler, L.; Mader, W. *J. Am. Ceram. Soc.* **2003**, *86*, 534–540.
240. Perrotta, A. J. *Mater. Res. Innov.* **1998**, *2*, 33–38.
241. Adair, J. D.; Cho, S. B.; Bell, N. S.; Perrotta, A. J. *J. Dispers. Sci. Technol.* **2001**, *22*, 143–165.
242. Kim, A. Y.; Kim, H. S.; Park, N. K.; Lee, T. J.; Lee, W. G.; Kim, H. D.; Park, J. W.; Kang, M. *J. Nanomater.* **2012**, 907503.
243. Kelber, J. A. *Surf. Sci. Rep.* **2007**, *62*, 271–303.
244. Ma, G.; Liu, D.; Allen, H. C. *Langmuir* **2004**, *20*, 11620–11629.
245. Al-Abadleh, H. A.; Grassian, V. H. *Langmuir* **2003**, *19*, 341–347.
246. Morterra, C.; Ghiotti, G.; Garrone, E.; Boccuzzi, F. *J. Chem. Soc. Faraday Trans. 1* **1976**, *72*, 2722–2734.
247. Busca, G.; Lorenzelli, V.; Ramis, G.; Willey, R. J. *Langmuir* **1993**, *9*, 1492–1499.
248. Morterra, C.; Magnacca, G.; DeMaestri, P. P. *J. Catal.* **1995**, *152*, 384–395.
249. Shirai, T.; Ishizaki, C.; Ishizaki, K. *J. Ceram. Soc. Jpn.* **2006**, *114*, 415–417.
250. Morterra, C.; Magnacca, G.; Del Favero, N. *Langmuir* **1993**, *9*, 642–645.
251. Morterra, C.; Cerrato, G.; Meligrana, G. *Langmuir* **2001**, *17*, 7053–7060.
252. Morterra, C.; Coluccia, S.; Chiorino, A.; Boccuzzi, F. *J. Catal.* **1978**, *54*, 348–364.
253. Hicks, R. W.; Pinnavaia, T. J. *Chem. Mater.* **2003**, *15*, 78–82.
254. Zhang, Z.; Pinnavaia, T. J. *Angew. Chem. Int. Ed.* **2008**, *47*, 7501–7504.
255. El-Nadjar, W.; Bonne, M.; Trela, E.; Rouleau, L.; Mino, A.; Hocine, S.; Payen, E.; Lancelot, C.; Lamonier, C.; Blanchard, P.; Courtois, X.; Can, F.; Duprez, D.; Royer, S. *Micropor. Mesopor. Mater.* **2012**, *158*, 88–98.
256. Yang, R.-J.; Yu, P.-C.; Chen, C. C.; Yen, F. S. *J. Eur. Ceram. Soc.* **2012**, *32*, 2153–2162.
257. Lodziana, Z.; Topsoe, N. Y.; Nørskov, J. K. *Nat. Mater.* **2004**, *3*, 289–293.
258. Yang, R. J.; Yu, P. C.; Chih-Cheng Chen, C.-C.; Yen, F.-S. *Cryst. Growth Des.* **2009**, *9*, 1692–1697.
259. Bagwell, R. B.; Messing, G. L. *J. Am. Ceram. Soc.* **1999**, *82*, 825–832.
260. Zhao, X.; Cong, Y.; Huang, Y.; Liu, S.; Wang, X.; Zhang, T. *Catal. Lett.* **2011**, *141*, 128–135.
261. Ruettinger, W. Patent application number: 20130072739, **2013**.
262. Finocchio, E.; Busca, G.; Rossini, S. A.; Cornaro, U.; Piccoli, V.; Miglio, R. *Catal. Today* **1997**, *33*, 335–352.
263. Arnone, S.; Busca, G.; Lisi, L.; Milella, F.; Russo, G.; Turco, M. *Symp. Combust.* **1998**, *27*, 2293–2299.
264. Castro, R. H. R.; Ushakov, S. V.; Gengembre, L.; Gouvea, D.; Navrotsky, A. *Chem. Mater.* **2006**, *18*, 1867–1872.
265. Rose, V.; Podgursky, V.; Costina, I.; Franchy, R. *Surf. Sci.* **2003**, *541*, 128–136.
266. Becker, C.; Kandler, J.; Raaf, H.; Linke, R.; Pelster, T.; Dräger, M.; Tanemura, M.; Wandelt, K. *J. Vac. Sci. Technol. A Vac. Surf. Films* **1998**, *16*, 1000–1005.
267. Nemsak, S.; Skala, T.; Yoshitake, M.; Tsud, N.; Kim, T.; Yagyua, S.; Matolin, V. *Surf. Interface Anal.* **2010**, *42*, 1581–1584.
268. Podgursky, V.; Rose, V.; Costina, J.; Franchy, R. *Surf. Sci.* **2007**, *601*, 3315–3323.
269. Graupner, H.; Hammer, L.; Heinz, K.; Zehner, D. M. *Surf. Sci.* **1997**, *380*, 335–351.
270. Napetschnig, E.; Schmid, M.; Varga, P. *Surf. Sci.* **2008**, *602*, 1750–1756.
271. Simon, G. H.; Konig, T.; Heinke, L.; Lichtenstein, L.; Heyde, M.; Freund, H.-J. *New J. Phys.* **2011**, *13*, 123028.
272. Klimenkov, M.; Nepijko, S.; Kuhlenbeck, H.; Freund, H.-J. *Surf. Sci.* **2011**, *385*, 66–76.

273. Heemeier, M.; Frank, M.; Libuda, J.; Wolter, K.; Kuhlenbeck, H.; Bäumer, M.; Freund, H.-J. *Catal. Lett.* **2000**, *68*, 19–24.

274. Stierle, A.; Renner, F.; Streitel, R.; Dosch, H.; Drube, W.; Cowie, B. C. *Science* **2004**, *303*, 1652–1656.

275. Ivey, M. M.; Layman, K. A.; Avoyan, A.; Allen, H. C.; Hemminger, J. C. *J. Phys. Chem. B* **2003**, *107*, 6391–6400.

276. Cai, Na; Qin, H.; Tong, X.; Zhou, G. *Surf. Sci.* **2013**, *618*, 20–26.

277. Gassmann, P.; Franchy, R.; Ibach, H. *Surf. Sci.* **1994**, *319*, 95–109.

278. Haeberle, J.; Henkel, K.; Gargouri, H.; Naumann, F.; Gruska, B.; Arens, M.; Tallarida, M.; Schmeißer, D. *Beilstein J. Nanotechnol.* **2013**, *4*, 732–742.

279. Mastikhin, V. M.; Mudrakovsky, I. L.; Nosov, A. V. *Prog. Nucl. Magn. Reson. Spectrosc.* **1991**, *23*, 259–299.

280. DeCanio, E. C.; Edwards, J. C.; Bruno, J. W. *J. Catal.* **1994**, *148*, 76–83.

281. Deng, F.; Wang, G.; Du, Y.; Ye, C.; Kong, Y.; Li, X. *Solid State Nucl. Magn. Reson.* **1997**, *7*, 281–290.

282. Qu, L.; Zhang, W.; Kooyman, P. J.; Prins, R. *J. Catal.* **2003**, *215*, 7–13.

283. Huittinen, N.; Sarv, P.; Lehto, J. *J. Colloid Interface Sci.* **2011**, *361*, 252–258.

284. Ferreira, A. R.; Küçükbenli, E.; de Gironcoli, S.; Souza, W. F.; Chiaro, S. S. X.; Konstantinova, E.; Leitão, A. A. *Chem. Phys.* **2013**, *423*, 62–72.

285. Busca, G. *Catal. Today* **2014**, *226*, 2–13.

286. Ballinger, T. H.; Yates, J. T., Jr. *Langmuir* **1991**, *7*, 3041–3045.

287. Montanari, T.; Castoldi, L.; Lietti, L.; Busca, G. *Appl. Catal. A Gen.* **2011**, *400*, 61–69.

288. Digne, M.; Raybaud, P.; Sautet, P.; Guillaume, D.; Toulhoat, H. *Phys. Chem. Chem. Phys.* **2007**, *9*, 2577–2582.

289. Scokart, P. O.; Amin, A.; Defossa, C.; Rouxhet, P. G. *J. Phys. Chem.* **1981**, *85*, 1406–1412.

290. Phung, T.K.; Herrera, C., Larrubia, M.A., García-Diéguez, M.; Finocchio, E., Alemany, L.J., Busca, G., *Appl. Catal. A: Gen.* **2014**, *483*, 41–51.

291. Busca, G.; Ramis, G.; Lorenzelli, V.; Rossi, P. F.; La Ginestra, A.; Patrono, P. *Langmuir* **1989**, *5*, 911–916.

292. Sahibet-Dine, A.; Aboulayt, A.; Bensitel, M.; Mohammed-Saad, A. B.; Daturi, M.; Lavalley, J. C. *J. Mol. Catal. A Chem.* **2000**, *162*, 125–134.

293. Morterra, C.; Magnacca, G. *Catal. Today* **1996**, *27*, 497–532.

294. Knözinger, H.; Ratnasamy, P. *Catal. Rev. Sci. Eng.* **1978**, *17*, 31–70.

295. Malpartida, I.; Larrubia Vargas, M. A.; Alemany, L. J.; Finocchio, E.; Busca, G. *Appl. Catal. B* **2008**, *80*, 214–225.

296. Liu, X.; Truitt, R. E. *J. Am. Chem. Soc.* **1997**, *119*, 9856–9860.

297. Busca, G.; Montanari, T.; Bevilacqua, M.; Finocchio, E. *Colloids Surf. A Physicochem. Eng. Asp.* **2008**, *320*, 205–212.

298. Busca, G. *Catal. Today* **1998**, *41*, 191–206.

299. Ramis, G.; Busca, G. *J. Mol. Struct.* **1989**, *193*, 93–100.

300. Cabana, A.; Sandorfy, C. *Can. J. Chem.* **1962**, *40*, 615–621.

301. Busca, G. *Phys. Chem. Chem. Phys.* **1999**, *1*, 723–736.

302. Tretyakov, N. E.; Filimonov, V. N. *Kinet. Katal.* **1972**, *13*, 815.

303. Trombetta, M.; Busca, G.; Rossini, S.; Piccoli, V.; Cornaro, U. *J. Catal.* **1997**, *168*, 334–348.

304. Busca, G. *Metal Oxides Chemistry and Applications*; In: Fierro, J. L. G. Ed.; CRC Press: Boca Raton, FL, 2005; pp 247–318.

305. Knözinger, H.; Kaerlein, C.-P. *J. Catal.* **1972**, *25*, 436–438.

306. Wischert, R.; Copéret, C.; Delbecq, F.; Sautet, P. *Angew. Chem. Int. Ed.* **2011**, *50*, 3202–3205.

307. Métivier, R.; Leray, I.; Lefèvre, J. P.; Roy-Auberger, M.; Zanier-Szydlowskic, N.; Valeur, B. *Phys. Chem. Chem. Phys.* **2003**, *5*, 758–766.

308. Wang, Y. G.; Bronsveld, P. M.; De Hosson, J. Th. M.; Duričić, B.; McGarry, D.; Pickering, S. *J. Eur. Ceram. Soc.* **1998**, *18*, 299–304.

309. Podgursky, V.; Rose, V.; Costina, J.; Franchy, R. *Appl. Surf. Sci.* **2006**, *252*, 8394–8398.

310. Kwak, J. H.; Peden, C. H. F.; Szanyi, J. *J. Phys. Chem.* C **2011**, *115*, 12575–12579.

311. Borello, E.; Della Gatta, G.; Fubini, B.; Morterra, C.; Venturello, G. *J. Catal.* **1974**, *35*, 1–10.

312. Morterra, C.; Chiorino, A.; Ghiotti, G.; Garrone, E. *J. Chem. Soc. Faraday Trans. 1* **1979**, *75*, 271–288.

313. Nédez, C.; Lefebvre, M.; Choplin, A.; Niccolai, G. P.; Basset, J.-M.; Benazzi, E. *J. Am. Chem. Soc.* **1994**, *116*, 8638–8646.

314. Lundie, D. T.; McInroy, A. R.; Marshall, R.; Winfield, J. M.; Jones, P.; Dudman, C. C.; Parker, S. F.; Mitchell, C.; Lennon, D. *J. Phys. Chem.* B **2005**, *109*, 11592–11601.

315. Peri, J. B. *J. Phys. Chem.* **1965**, *69*, 220–230.

316. Tsyganenko, A.; Filimonov, V. N. *Spectrosc. Lett.* **1972**, *5*, 477–487.

317. Chen, Y.; Zhang, Li. *Catal. Lett.* **1992**, *12*, 51–62.

318. Alvarez, L. J.; Sanz, J. F.; Capitan, M. J.; Centeno, M. A.; Odriozola, J. A. *J. Chem. Soc. Faraday Trans.* **1993**, *89*, 3623–3628.

319. Millman, W. S.; Crespin, M.; Cirillo, A. C.; Abdo, S., Jr.; Hall, W. K. *J. Catal.* **1979**, *60*, 404–416.

320. Okamoto, Y.; Imanaka, T. *J. Phys. Chem.* **1988**, *92*, 7102–7112.

321. Busca, G.; Lorenzelli, V.; Sanchez Escribano, V.; Guidetti, R. *J. Catal.* **1991**, *131*, 167–177.

322. Busca, G.; Lorenzelli, V.; Sanchez Escribano, V. *Chem. Mater.* **1992**, *4*, 595–605.

323. Tsyganenko, A. A.; Mardilovich, P. P. *J. Chem. Soc. Faraday Trans. 1* **1996**, *92*, 4843–4852.

324. Lambert, J. F.; Che, M. *J. Mol. Catal. A Chem.* **2000**, *162*, 5–18.

325. Fripiat, J.; Alvarez, L.; Sanchez Sanchez, S.; Martinez Morades, E.; Saniger, J.; Sanchez, N. *Appl. Catal. A Gen.* **2001**, *215*, 91–100.

326. Dyan, A.; Cenedese, P.; Dubot, P. *J. Phys. Chem.* B **2006**, *110*, 10041–10050.

327. Auroux, A.; Gervasini, A. *J. Phys. Chem.* **1990**, *94*, 6371.

328. Guillaume, D.; Gautier, S.; Alario, F.; Devès, J.-M. *Oil Gas Sci. Technol.* **1999**, *54*, 537–545.

329. Nastova, I.; Skapin, T.; Pejov, L. *Surf. Sci.* **2011**, *605*, 1525–1536.

330. Kassab, E.; Castellà-Ventura, M. *J. Phys. Chem.* B **2005**, *109*, 13716–13728.

331. Wischert, R.; Copéret, C.; Delbecq, F.; Sautet, P. *Chem. Commun.* **2011**, *47*, 4890–4892.

332. Kwak, J. H.; Hu, J. Z.; Mei, D.; Yi, C. W.; Kim, D. H.; Peden, C. H. F.; Allard, L. F.; Szanyi, J. *Science* **2009**, *325*, 1670–1673.

333. Mei, D.; Kwak, J. H.; Hu, J. Z.; Cho, S. J.; Szanyi, J.; Allard, L. F.; Peden, C. H. F. *J. Phys. Chem. Lett.* **2010**, *1*, 2688–2691.

334. Mei, D.; Ge, Q.; Szanyi, J.; Peden, C. H. F. *J. Phys. Chem.* C **2009**, *113*, 7779–7789.

335. Ouyang, C. Y.; Sljivancanin, Z.; Baldereschi, A. *Phys. Rev.* B **2009**, *79*, 235410.

336. Loviat, F.; Czekaj, I.; Wambach, J.; Wokaun, A. *Surf. Sci.* **2009**, *603*, 2210–2217.

337. Dabbagh, H. A.; Taban, K.; Zamami, M. *J. Mol. Catal. A Chem.* **2010**, *326*, 55–68.

338. Vijay, A.; Mills, G.; Metiu, H. *J. Chem. Phys.* **2002**, *117*, 4509–4516.

339. Guisnet, M.; Lamberton, J. L.; Perot, G.; Maurel, R. *J. Catal.* **1977**, *48*, 166–176.

340. Kasprzyk-Hordern, B. *Adv. Colloid Interface Sci.* **2004**, *110*, 19–48.

341. Ono, Y.; Hattori, H. *Solid Base Catalysis*, Tokyo Institute of Technology Press: Tokyo, 2011; pp 118–127.

342. Busca, G. *Chem. Rev.* **2010**, *110*, 2217–2249.

343. Busca, G.; Rossi, P. F.; Lorenzelli, V.; Benaissa, M.; Travert, J.; Lavalley, J. C. *J. Phys. Chem.* **1985**, *89*, 5433–5439.
344. McInroy, A. R.; Lundie, D. T.; Winfield, J. M.; Dudman, C. C.; Jones, P.; Lennon, D. *Langmuir* **2005**, *21*, 11092–11098.
345. Busca, G. *Catal. Today* **1996**, *27*, 457–496.
346. Pines, H.; Manassen, J. *Advan. Catal. Relat. Subj.* **1966**, *16*, 49–93.
347. Knözinger, H.; Kohne, R. *J. Catal.* **1966**, *5*, 264–270.
348. Pistarino, C.; Finocchio, E.; Romezzano, G.; Brichese, F.; DiFelice, R.; Busca, G.; Baldi, M. *Ind. Eng. Chem. Res.* **2000**, *39*, 2752–2760.
349. Knözinger, H.; Krietenbrink, H. *J. Chem. Soc. Faraday Trans. 1* **1975**, *71*, 2421.
350. Zaki, M. I.; Hasan, M. A.; Pasupulety, L. *Langmuir* **2001**, *17*, 768–774.
351. Busca, G.; Finocchio, E.; Lorenzelli, V.; Trombetta, M.; Rossini, S. A. *J. Chem. Soc. Faraday Trans. 1* **1996**, *92*, 4687–4693.
352. Hong, Y.; Chen, F. R.; Fripiat, J. *J. Catal. Lett.* **1993**, *17*, 187–195.
353. Gervasini, A.; Auroux, A. *J. Phys. Chem.* **1993**, *97*, 2628–2639.
354. Mohammed Saad, A. B.; Saur, O.; Wang, Y.; Tripp, C. P.; Morrow, B. A.; Lavalley, J. C. *J. Phys. Chem.* **1995**, *99*, 4620–4625.
355. Piéplu, A.; Saur, O.; Lavalley, J. C.; Legendre, O.; Nédez, C. *Catal. Rev. Sci. Eng.* **1998**, *40*, 409–450.
356. Ziolek, M.; Kujawa, J.; Saur, O.; Aboulayt, A.; Lavalley, J. C. *J. Mol. Catal. A Chem.* **1996**, *112*, 125–132.
357. Mohammed Saad, A. B.; Ivanov, V. A.; Lavalley, J. C.; Nortier, P.; Luck, F. *Appl. Catal. A Gen.* **1993**, *94*, 71–83.
358. Zagoruiko, A. N.; Shinkarev, V. V.; Vanag, S. V.; Bukhtiyarova, G. A. *Catal. Ind.* **2010**, *2*, 343–352.
359. BASF Claus Catalysts, brochure. Availble from: www.basf.com.
360. Li, J.; Huang, L.; He, J. *Adv. Mater. Res.* **2014**, *884–885*, 182–185.
361. Khanmamedov, T. H.; Weiland, R. H. *Sulphur* **2013**, *345*, 62–68.
362. Clark, P. D.; Dowling, N. I.; Huang, M.; Okemona, O.; Butlin, G. D.; Hou, R.; Kijlstra, W. S. *Appl. Catal. A Gen.* **2002**, *235*, 61–69.
363. Muroy, T.; Naojiri, N.; Deguchi, T. *Appl. Catal. A Gen.* **2010**, *389*, 27–45.
364. Juul Dahl, P.; Stahl, O., WIPO Patent Application WO/2011/095270 to Topsoe, **2011**.
365. http://www.topsoe.com/business_areas/~/media/PDF%20files/tigas/10198_TIGAS_brochure_low%20rez.ashx.
366. Liu, Z.; Sun, C.; Wang, G.; Wang, Q.; Cai, G. *Fuel Process. Technol.* **2000**, *62*, 161.
367. Liebner, W. *Handbook of Petrochemicals Production Processes*; In: Meyers, R. A. Ed.; Mc Graw Hill: New York, NY, 2005; pp 10.3–10.14.i.
368. Kim, S.-M.; Lee, Y.-J.; Bae, J.-W.; Potdar, H. S.; Jun, K.-W. *Appl. Catal. A Gen.* **2008**, *348*, 113–120.
369. Vinogradov, V. V.; Vinogradov, A. V.; Kraev, A. S.; Agafonov, A. V.; Kessler, V. G. *J. Sol-Gel Sci. Technol.* **2013**, *68*, 155–161.
370. Schiffino, R. S.; Merrill, R. P. *J. Phys. Chem.* **1993**, *97*, 6425–6435.
371. Tamm, S.; Ingelsten, H. H.; Skoglundh, M.; Palmqvist, A. E. C. *J. Catal.* **2010**, *276*, 402–411.
372. Akarmazyan, S. S.; Panagiotopoulou, P.; Kambolis, A.; Papadopoulou, C.; Kondarides, D. I. *Appl. Catal. B Environ.* **2014**, *145*, 136–148.
373. Hu, Y. C. *Chemical Processing Handbook*; In: McKetta, J. J. Ed.; Dekker: New York, NY, 1993; pp 768–769.
374. http://www.chematur.se/sok/download/Ethylene_rev_0904.pdf (accessed 20 May 2013).
375. Zhang, M.; Yu, Y. *Ind. Eng. Chem. Res.* **2013**, *52*, 9505–9514.
376. Fan, D.; Dai, D.-J.; Wu, H.-S. *Materials* **2013**, *6*, 101–115.

377. Kwak, J. H.; Mei, D.; Peden, C. H. F.; Rousseau, R.; Szanyi, J. *Catal. Lett.* **2011**, *141*, 649–655.
378. Roy, S.; Mpourmpakis, G.; Hong, D.-Y.; Vlachos, D. G.; Bhan, A.; Gorte, R. J. *ACS Catal.* **2012**, *2*, 1846–1853.
379. DeWilde, J. F.; Chiang, H.; Hickman, D. A.; Ho, C. R.; Bhan, A. *ACS Catal.* **2013**, *3*, 798–807.
380. de Klerk, A. *Catalysis* **2011**, *23*, 1–49.
381. Bolder, F. H. A.; Mulder, H. *Appl. Catal. A Gen.* **2006**, *300*, 36–40.
382. Nel, R. J. J.; de Klerk, A. *Ind. Eng. Chem. Res.* **2009**, *48*, 5230–5238.
383. Makgoba, N. P.; Sakuneka, T. M.; Koortzen, J. G.; van Schalkwyk, C.; Botha, J. M.; Nicolaides, C. P. *Appl. Catal. A Gen.* **2006**, *297*, 145–150.
384. Butler, A. C.; Nicolaides, C. P. *Catal. Today* **1993**, *18*, 443–471.
385. http://www.axens.net/product/catalysts-a-adsorbents/105/is-463.html.
386. Lamberov, A. A.; Sitnikova, E. Yu.; Mukhambetov, I. N.; Zalyaliev, R. F.; Gil'mullin, R. R.; Gilmanov, Kh. Kh. *Catal. Ind.* **2012**, *4*, 141–149.
387. Macho, V.; Králik, M.; Jurecek, Lu.; Jurecekova, E.; Balazova, J. *Appl. Catal. A Gen.* **2000**, *203*, 5–14.
388. Corado, A.; Kiss, A.; Knözinger, H.; Mueller, H. D. *J. Catal.* **1975**, *37*, 68–80.
389. Gabrienko, A. A.; Arzumanov, S. S.; Toktarev, A. V.; Stepanov, A. G. *J. Phys. Chem. C* **2012**, *116*, 21430–21438.
390. Ivanov, S. I.; Makhlin, V. A. *Kinet. Catal.* **1996**, *37*, 812–818.
391. Schmidt, S. A.; Kumar, N.; Zhang, B.; Eränen, K.; Murzin, D. Yu.; Salmi, T. *Ind. Eng. Chem. Res.* **2012**, *51*, 4545–4555.
392. Schmidt, S. A.; Kumar, N.; Reinsdorf, A.; Eränen, K.; Wärnå, J.; Murzin, D. Yu.; Salmi, T. *Chem. Eng. Sci.* **2013**, *95*, 232–245.
393. Thyagarajan, M. S.; Kumar, R.; Kuloor, N. R. *Ind. Eng. Chem. Proc. Des. Dev.* **1966**, *5*, 209–213.
394. McInroy, A. R.; Lundie, D. T.; Winfield, J. M.; Dudman, C. C.; Jones, P.; Lennon, D. *Appl. Catal. B Environ.* **2007**, *70*, 606–610.
395. Schmidt, S. A.; Kumar, N.; Shchukarev, K.; Eränen, K.; Mikkola, J. P.; Murzin, D. Yu.; Salmi, T. *Appl. Catal. A Gen.* **2013**, *468*, 120–134.
396. http://www.basf.com/group/corporate/en/brand/HF_200XPHF (accessed 10 December 2012).
397. Frank, H. G.; Stadelhofer, J. W. *Industrial Aromatic Chemistry*, Springer Verlag: Berlin, Germany, 1988.
398. Tleimat-Manzalji, R.; Bianchi, D.; Pajonk, G. M. *Appl. Catal. A Gen.* **1993**, *101*, 339–350.
399. Grabowska, A.; Syper, L.; Zawadzki, M. *Appl. Catal. A Gen.* **2004**, *277*, 91–97.
400. Bulusheva, D. A.; Ross, J. R. H. *Catal. Today* **2011**, *171*, 1–13.
401. Vonghia, E.; Boocock, D. G. B.; Konar, Samir K.; Leung, Anna. *Energy Fuel* **1995**, *9*, 1090–1096.
402. Phung, T. K.; Casazza, A. A.; Aliakbarian, B.; Finocchio, E.; Perego, P.; Busca, G. *Chem. Eng. J.* **2013**, *215–216*, 838–848.
403. Toulhoat, H., Raybaud, P. Eds. *Catalysis by Transition Metal Sulphides*; Edition Technip: Paris, France, 2013.
404. Gutierrez Alejandre, A.; Laurrabaquio Rosas, G. U.; Ramirez, J.; Busca, G. *Appl. Catal. B: Envir.* in press.
405. Hancsók, J.; Marsi, G.; Kasza, T.; Kallo, D. *Top. Catal.* **2011**, *54*, 1102–1109.
406. Klimov, O. V.; Pashigreva, A. V.; Bukhtiyarova, G. A.; Kashkin, V. N.; Nioskov, A. S.; Polunkin, Ya. M. *Catal. Ind.* **2010**, *2*, 101–107.
407. Haizmann, R. S.; Rice, L. H.; Turowicz, M. S. US patent 5453552 to UOP, **1995**.
408. Schiavone, B. J. NIPRA 2007 O&A and Technology Forum, **2007**, available on the web.

409. Diehl, F.; Khodakov, A. Y. *Oil Gas Sci. Technol.* **2009**, *64*, 11–24.
410. Antos, G. J., Aitani, A. A. Eds. *Catalytic Naphtha Reforming*, 2nd ed.; M. Dekker: New York, NY, 2004, Revised and Expanded.
411. Modica, F. S.; Mcbride, T. K.; Galperin, L. B. Patent WO/2006/078240 to UOP LLC, **2006**.
412. Shoukri, E. I. Patent 5013704 to Exxon Res & Eng. Co, **1991**.
413. http://www.uop.com/pr/releases/Hunt.pdf.
414. http://www.uop.com/objects/R%20230%20Series%20Catalyst.pdf.
415. Bhasin, M. M.; McCain, J. H.; Vora, B. V.; Imai, T.; Pujadó, P. R. *Appl. Catal. A Gen.* **2001**, *221*, 397–419.
416. Dyroff, D. R. U.S. Patent No. 6,700,028 B2 to Huntsmann Petrochem. Co., **2004**.
417. http://www.topsoe.com/products/CatalystPortfolio.aspx.
418. Røstrup-Nielsen, J. R.; Pedersen, K.; Sehested, J. *Appl. Catal. A Gen.* **2007**, *330*, 134–138.
419. Nguyen, T. T. M.; Wissing, L.; Skjøth-Rasmussen, M. S. *Catal. Today* **2013**, *215*, 233–238.
420. Aasberg-Petersen, K.; Dybkjær, I.; Ovesen, C. V.; Schjødt, N. C.; Sehested, J.; Thomsen, S. G. *J. Nat. Gas Sci. Eng.* **2011**, *3*, 423–459.
421. Schbib, N. S.; Garcia, M. A.; Gigola, C. E.; Errazu, A. F. *Ind. Eng. Chem. Res.* **1996**, *35*, 1496–1505.
422. Berhault, G.; Bisson, L.; Thomazeau, C.; Verdon, C.; Uzio, D. *Appl. Catal. A Gen.* **2007**, *327*, 32–43.
423. Carmello, D.; Finocchio, E.; Marsella, A.; Cremaschi, B.; Leofanti, G.; Padovan, M.; Busca, G. *J. Catal.* **2000**, *191*, 354–363.
424. Gianolio, D.; Muddad, N. B.; Olsbye, U.; Lamberti, C. *Nucl. Instrum. Methods Phys. Res. B* **2012**, *284*, 53–57.
425. Jabłońska, M.; Chmielarz, L.; Węgrzyn, A. *Chemik* **2013**, *67*, 701–710.
426. Harrison Tran, P.; Chen, J. M. H.; Lapadula, G. D.; Blute, T. US patent 7410626 B2 to BASF Catalysts LLC, **2008**.
427. Lauritzen, A. M. US patent 4,833,261 to Shell, 1989.
428. Hurtado, P.; Ordóñez, S.; Vega, A.; Diez, F. *Chemosphere* **2004**, *55*, 681–689.
429. http://www.catalysts.basf.com/p02/USWeb-Internet/catalysts/en/content/microsites/catalysts/prods-inds/adsorbents/act-alum-adsorb (accessed 10 December 2012).
430. Madon, R.; Harris, D. H.; Xu, M.; Stockwell, D.; Lerner, B.; Dodwell, G. W. US patent 6716338 to Engelhard Corporation.
431. Harris, D. H.; Xu, M.; Stockwell, D.; Madon, R. J. US patent 6673235 to Engelhard Corporation.
432. Busca, G.; Riani, P.; Garbarino, G.; Ziemacki, G.; Gambino, L.; Montanari, E.; Millini, R. *Appl. Catal. A Gen.* **2014**, *486*, 176–186.
433. Visconti, C. G. *Trans. Ind. Ceram. Soc* **2012**, *71*, 123–136.
434. Twigg, M. V. *Catal. Today* **2011**, *163*, 33–41.

INDEX

Note: Page numbers followed by "*f*" indicate figures and "*t*" indicate tables.

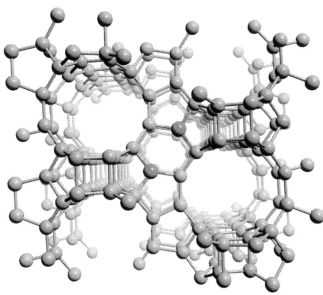

Chapter 1, Figure 1.3 (See page 37 of this volume.)

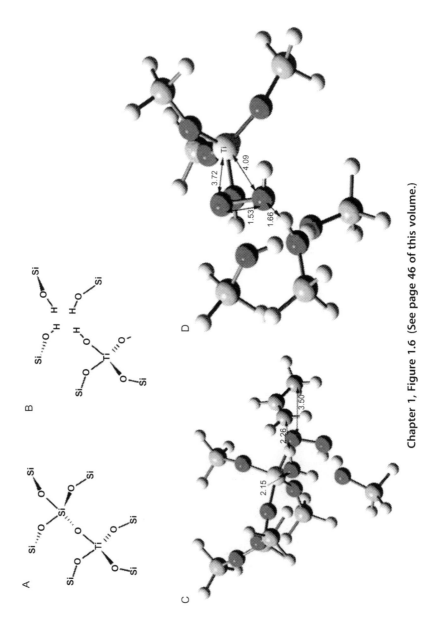

Chapter 1, Figure 1.6 (See page 46 of this volume.)

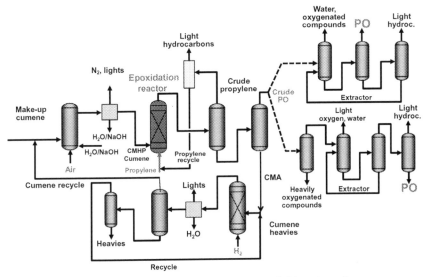

Chapter 1, Figure 1.15 (See page 71 of this volume.)

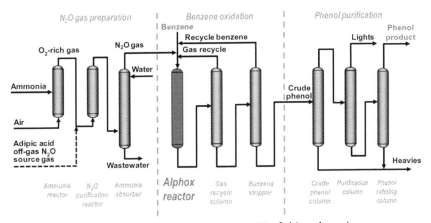

Chapter 1, Figure 1.18 (See page 76 of this volume.)

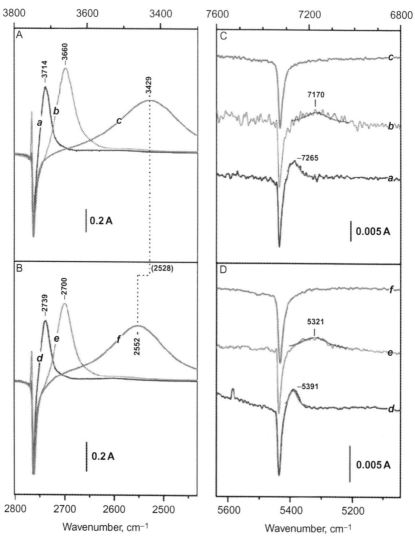

Chapter 2, Figure 2.5 (See page 118 of this volume.)

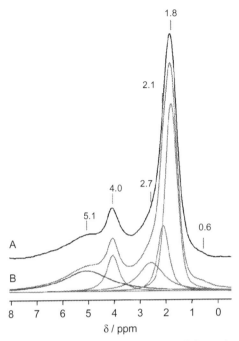

Chapter 2, Figure 2.13 (See page 128 of this volume.)

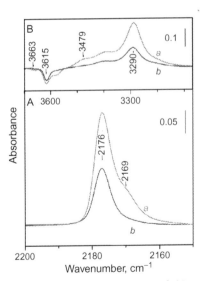

Chapter 2, Figure 2.20 (See page 151 of this volume.)

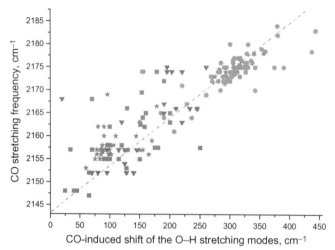

Chapter 2, Figure 2.24 (See page 173 of this volume.)

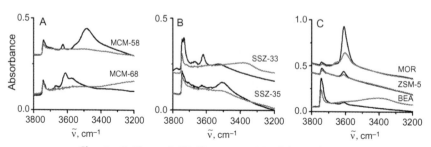

Chapter 2, Figure 2.45 (See page 238 of this volume.)

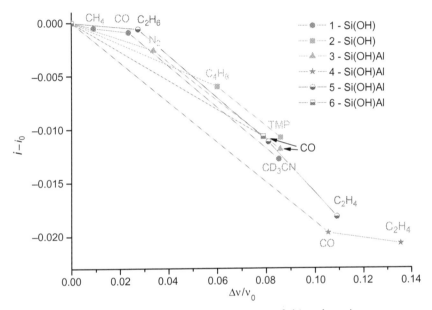

Chapter 2, Figure 2.49 (See page 257 of this volume.)

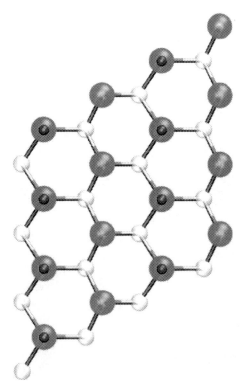

Chapter 2, Figure 2.53 (See page 276 of this volume.)

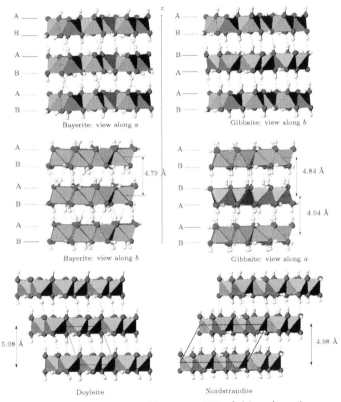

Chapter 3, Figure 3.2 (See page 323 of this volume.)

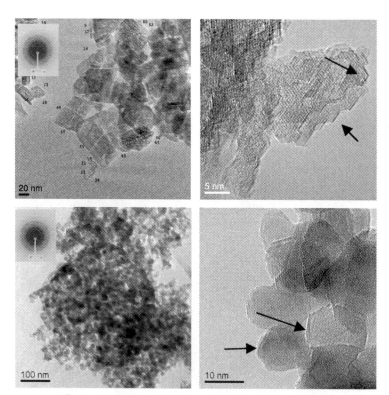

Chapter 3, Figure 3.14 (See page 346 of this volume.)